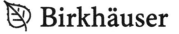

Pseudo-Differential Operators
Theory and Applications
Vol. 11

Managing Editor

M.W. Wong (York University, Canada)

Editorial Board

Pseudo-Differential Operators: Theory and Applications is a series of moderately priced graduate-level textbooks and monographs appealing to students and experts alike. Pseudo-differential operators are understood in a very broad sense and include such topics as harmonic analysis, PDE, geometry, mathematical physics, microlocal analysis, time-frequency analysis, imaging and computations. Modern trends and novel applications in mathematics, natural sciences, medicine, scientific computing, and engineering are highlighted.

André Unterberger

Pseudodifferential Operators with Automorphic Symbols

 Birkhäuser

André Unterberger
Université de Reims
Laboratoire de Mathématiques
Reims, France

ISSN 2297-0355   ISSN 2297-0363   (electronic)
Pseudo-Differential Operators
ISBN 978-3-319-18656-6   ISBN 978-3-319-18657-3   (eBook)
DOI 10.1007/978-3-319-18657-3

Library of Congress Control Number: 2015943692

Mathematics Subject Classification (2010): 11F03, 11F37, 47G30

Springer Cham Heidelberg New York Dordrecht London

Printed on acid-free paper

Springer International Publishing AG Switzerland is part of Springer Science+Business Media (www.birkhauser-science.com)

Preface

The present book, just as some of its forerunners, grew out of the author's desire to find a different approach to analytic number theory, certainly one of the more aesthetically satisfying branches of mathematics: this is especially true so far as modular form and L-function theory are concerned. That we relied on pseudodifferential analysis to try such an approach was initially due to the fact that we had practiced the latter since its beginnings, half a century ago. But we discovered, on the way, that it has deep connections with many areas: quantization theory and mathematical Physics, the theory of symmetric spaces, representation theory and, our main interest here, modular form and L-function theory. Our aim in this preface is to convey to interested readers, not necessarily familiar with pseudodifferential analysis, the realization that some of their interests are closer to it than what they probably believe.

Pseudodifferential operators were first devised as a help in partial differential equations: the analysis of such problems required a constant use of auxiliary (non differential) operators, and pseudodifferential operators, of a more and more general type, soon provided the quite adaptable box of tools desired. At the same time, the structure of pseudodifferential analysis became a subject of interest of its own. It starts with a specific way (the Weyl calculus) of representing linear operators on functions of n variables by functions of $(2n)$ variables, called their symbols. Such a correspondence is a linear isomorphism, but it cannot be an algebra isomorphism if, on the symbol side, the (commutative) pointwise product is considered. This is why much emphasis has always been put on the "sharp composition formula" expressing the symbol $h_1 \# h_2$ of the composition of two operators with given symbols h_1 and h_2. Such a formula is in general based on that valid for differential operators, which led many people to believe that asymptotic, no to say formal, expansions are a fixture of pseudodifferential analysis.

This is not the case and, in situations of interest (staying within the Weyl calculus or not), what is required instead is to combine the sharp product with the decomposition of symbols into irreducible parts provided by some relevant representation-theoretic notions. These come into play through the two properties of covariance of the Weyl calculus, which express that the symbol of an operator transforms by a translation or by a symplectic linear change of coordinates if the operator undergoes a conjugation under an element of the Heisenberg representation or of the metaplectic representation. It is only the latter representation that will concern us here: actually, we shall specialize in this book in the one-dimensional case, so that the group of interest is simply $SL(2, \mathbb{R})$. As another major simplification, the only arithmetic subgroup of $SL(2, \mathbb{R})$ considered here will be $SL(2, \mathbb{Z})$.

By definition, automorphic symbols will be symbols invariant under linear changes of coordinates associated to matrices in $SL(2, \mathbb{Z})$: they must be distributions, since no non-constant continuous function can qualify. Thanks to the covariance property, automorphic symbols can be expected to make up an algebra under the sharp product: there are difficulties, linked to the fact that two operators with automorphic symbols cannot be composed in the quite usual sense, but these are fully solved. Automorphic distributions which are at the same time homogeneous of some degree are to be called modular distributions: they are introduced, together with their associated L-functions, in an independent way. They can then be shown to be a notion only slightly more precise than that of non-holomorphic modular form (in the hyperbolic half-plane), and a complete two-to-one dictionary is provided. The main question solved in the book can then be formulated as follows: given two modular distributions $\mathfrak{N}_1$ and $\mathfrak{N}_2$, decompose their sharp product $\mathfrak{N}_1 \# \mathfrak{N}_2$ as a linear superposition (both integrals and series are needed) of modular distributions (Eisenstein distributions and Hecke distributions); then, express the coefficients of the decomposition in terms of L-functions, Rankin-Selberg product L-functions and related notions.

Developing this program necessitated some machinery, in particular a reinterpretation in terms of pseudodifferential analysis of the Radon transform from the homogeneous space G/MN of $G = SL(2, \mathbb{R})$ (functions on this space can be identified with even functions on $\mathbb{R}^2 \backslash \{0\}$) to the homogeneous space G/K (the hyperbolic half-plane). Of course, under this transformation, the non-commutative sharp product does not transfer to the pointwise product but if the two entries, as well as the output, of the sharp operation are reduced to homogeneous components, it becomes almost true: analyzing the terms of the sharp composition reduces ultimately to the analysis, in the hyperbolic half-plane, of the pair made up of the pointwise product and Poisson bracket, after functions in the hyperbolic half-plane have been decomposed into generalized eigenfunctions of the hyperbolic Laplacian. This is proved first in a non-automorphic environment, next in the automorphic case.

We hope, and believe, that several categories of readers may find useful developments in this volume. New methods had to be developed, of course, in pseudodifferential analysis, while new perspectives in classical (i.e., non-adelic) non-holomorphic modular form theory might be gained from the automorphic distribution point of view. So-called invariant triple kernels have been considered lately by several authors, with or without number-theoretic applications in mind:

the end of Section 3.3 will explain to what extent some of this this appears as a byproduct of pseudodifferential analysis. Connections with quantization (not only geometric quantization) theory, a subject which has been this author's central interest for a few years, are hinted at in a short last chapter. In view of the current interest in Rankin-Cohen brackets, we have also included a short review showing that series of such constitute the right-hand side of the sharp product formula in a genuine symbolic calculus, a counterpart of the Weyl calculus in which non-holomorphic modular forms are traded for modular forms of the holomorphic type: much remains to be done in this direction. One of our other projects would be to understand more about some easier cases of Langland's L-function theory: to this end, some generalization of the present theory to the n-dimensional case might prove manageable and helpful.

Contents

Introduction

Automorphic distributions are distributions in the plane invariant under the action of the group $SL(2,\mathbb{Z})$ by linear changes of coordinates, and automorphic pseudodifferential operators are the operators from Schwartz' space $\mathcal{S}(\mathbb{R})$ to $\mathcal{S}'(\mathbb{R})$ with automorphic distributions for symbols. Our interest in considering these operators is twofold. First, it offers a new vantage point on non-holomorphic modular forms, from which these objects can be interpreted as operators, reinvigorating the subject in a novel direction. Let us immediately mention that the case of modular forms of holomorphic type requires drastic changes and will be approached in a short last chapter. Next, the species of symbols we have to consider here are by nature very singular distributions: the analytical difficulties one comes across when dealing with these distributions, clarified throughout the central chapters of the book, are important.

Automorphic function theory is commonly introduced as a study of functions in the hyperbolic half-plane $\Pi = \{z \in \mathbb{C} : \text{Im } z > 0\}$ invariant under the action of some arithmetic group Γ (say, $SL(2,\mathbb{Z})$, the only one considered here) by fractional-linear changes of the complex coordinate. As a major starting point, we here trade automorphic functions in Π for automorphic distributions in the plane $\mathbb{R}^2$, as just defined. Besides making it feasible in a natural way, thanks to the Weyl pseudodifferential calculus, to associate operators to such objects, this shift has several important advantages, of which we will try, in this introduction, to convince the reader: note that, in the automorphic situation, the word "distribution" (resp. "function") will always refer to the theory in $\mathbb{R}^2$ (resp. Π).

First, the two theories are closely related. Given any tempered distribution $\mathfrak{S}$ in the plane, define the function $\Theta_0\mathfrak{S}$ in Π by the equation

$$(\Theta_0\mathfrak{S})\,(z) = \left\langle \mathfrak{S},\, (x,\xi) \mapsto 2\,\exp\left(-2\pi\frac{|x - z\xi|^2}{\text{Im } z}\right) \right\rangle, \quad z \in \Pi. \tag{0.0.1}$$

Then, $\mathfrak{S}$ is characterized by its pair of transforms $(\Theta_0\mathfrak{S}, \Theta_1\mathfrak{S})$, where $\Theta_1\mathfrak{S}$ is the Θ_0-transform of the image of $\mathfrak{S}$ under the Euler operator $2i\pi\mathcal{E} = x\frac{\partial}{\partial x} + \xi\frac{\partial}{\partial \xi} + 1$. Moreover, both $\Theta_0\mathfrak{S}$ and $\Theta_1\mathfrak{S}$ are automorphic functions if $\mathfrak{S}$ is an automorphic distribution.

Recall that non-holomorphic modular forms (for a given group Γ) are classically defined as C^∞ automorphic functions which are for some complex number ν in the nullspace of the operator $\Delta - \frac{1-\nu^2}{4}$, where $\Delta = (z - \bar{z})^2 \frac{\partial^2}{\partial z \partial \bar{z}}$ is the hyperbolic Laplacian. A full list of non-holomorphic modular forms is made of the so-called non-holomorphic Eisenstein series $E_{\frac{1-\nu}{2}}$, together with the mysterious Hecke eigenforms, a notion slightly more precise than that of (Maass) cusp-forms: this will be briefly recalled in Chapter 2. Now, for every tempered distribution $\mathfrak{S}$, one has the identity

$$\Theta_0\left(\pi^2 \mathcal{E}^2 \mathfrak{S}\right) = \left(\Delta - \frac{1}{4}\right) \Theta_0 \mathfrak{S}. \tag{0.0.2}$$

It follows that if the distribution $\mathfrak{S}$ is homogeneous of degree $-1 - \nu$ for some $\nu \in \mathbb{C}$, its Θ_0-transform lies in the nullspace of $\Delta - \frac{1-\nu^2}{4}$. In particular, if $\mathfrak{S}$ is also an automorphic distribution (we say, then, that it is a modular distribution of degree $-1 - \nu$), its Θ_0-transform will be a non-holomorphic modular form. This does not imply that the theory of modular distributions (in the plane) and that of non-holomorphic modular forms (in Π) are fully equivalent: the first one is slightly more precise, since any tempered distribution has the same image, under Θ_0, as its symplectic Fourier transform (the symplectic Fourier transformation is the one, with the right combination of signs in the exponent, that commutes with the linear action of $SL(2, \mathbb{R})$ in $\mathbb{R}^2$, not only that of $SO(2)$). It is for this reason that a pair $(\Theta_0 \mathfrak{S}, \Theta_1 \mathfrak{S})$ of transforms of $\mathfrak{S}$ is required to characterize it. One advantage of the concept of automorphic distribution is immediately apparent: one can substitute, for the problem of decomposing any given automorphic function into non-holomorphic modular forms (integral superpositions of Eisenstein series and series of Eisenstein and Hecke eigenforms are generally needed), the often easier problem of decomposing any given automorphic distribution into homogeneous components. In other words, spectral decompositions relative to a first-order, rather than second-order, operator, are now required.

We shall honestly admit that Hilbert space facts, in particular the properties of self-adjoint operators, are used much more easily in Π than in $\mathbb{R}^2$. The reason is that, while there are nice fundamental domains (for instance, $\{z \colon |\mathrm{Re}\, z| < \frac{1}{2}, |z| > 1\}$) for the action of $\Gamma = SL(2, \mathbb{Z})$ in Π, no such thing can exist in $\mathbb{R}^2$, since most orbits are everywhere dense. There does exist an independently defined Hilbert space substitute $L^2(\Gamma \backslash \mathbb{R}^2)$ for $L^2(\Gamma \backslash \Pi)$, but it is by no means so simple to use. We shall actually rely on the spectral decomposition theorem for the automorphic Laplacian (the Roelcke-Selberg theorem) to prove something similar, in the plane, involving the decomposition of sufficiently general automorphic distributions into modular distributions. In the problems to be tackled here, it is actually better to stay within a rather general distribution environment, since terms (Eisenstein) outside what would be the spectral line in a Hilbert space frame will present themselves too.

We shall introduce the full collection of modular distributions (for the group $SL(2,\mathbb{Z})$), to wit Eisenstein distributions $\mathfrak{E}_\nu$ and Hecke distributions $\mathfrak{N}$, without appealing to any previous knowledge of automorphic function theory (in Π), in Chapter 1. It is only in Chapter 2 that, after some reminders about non-holomorphic modular form theory, we shall relate the two kinds of concepts under Θ_0, more precisely under the pair Θ_0, Θ_1. Needless to say, the vocabulary relative to automorphic distributions (Eisenstein distributions, Hecke distributions) has been chosen so that, under Θ_0, it should correspond to that in use in automorphic function theory. But, while $\mathfrak{E}_\nu$ and $\mathfrak{E}_{-\nu}$ are the images of each other under the symplectic Fourier transformation, their images under Θ_0 are proportional since

$$\Theta_0\left(2^{\frac{-1\mp\nu}{2}}\mathfrak{E}_{\pm\nu}\right) = \zeta^*(1-\nu)E_{\frac{1-\nu}{2}}, \text{ with } \zeta^*(s) = \pi^{-\frac{s}{2}}\Gamma(\tfrac{s}{2})\zeta(s):$$

now, the classical functional equation of non-holomorphic Eisenstein series says precisely that $\zeta^*(1-\nu)E_{\frac{1-\nu}{2}}$ is an even function of ν. Modular distributions, which are already globally homogeneous functions of (x,ξ), can be further decomposed into (non-automorphic) bihomogeneous functions: this introduces L-function theory at an early stage of the theory, in a spectral-theoretic role.

In Chapter 3, we present a short introduction to (one-dimensional) pseudodifferential analysis. In the most important pseudodifferential analysis, to wit the Weyl calculus, tempered distributions $\mathfrak{S}$ in the plane give rise to operators $\mathrm{Op}(\mathfrak{S})$ from $\mathcal{S}(\mathbb{R})$ to $\mathcal{S}'(\mathbb{R})$: the correspondence is linear and one-to-one, and $\mathfrak{S}$ is called the symbol of the operator $\mathrm{Op}(\mathfrak{S})$. Higher-dimensional pseudodifferential analysis has been for decades a major tool in partial differential equations, and innumerable papers, as well as important books [12, 29, 21] have been written on this subject and its applications. Nevertheless, Chapter 3 is not to be skipped by specialists of the field, since methods completely foreign to the well-known ones, in particular regarding the sharp composition of symbols (the operation defined when possible by the rule $\mathrm{Op}(\mathfrak{S}_1)\,\mathrm{Op}(\mathfrak{S}_2) = \mathrm{Op}(\mathfrak{S}_1 \,\#\, \mathfrak{S}_2)$), are necessary here. Another topic treated in this chapter concerns a relation between the sharp product of symbols on one side, the pointwise product and Poisson bracket of functions in Π on the other side.

We have arrived at a formulation of the main question addressed in this book: to develop pseudodifferential analysis when dealing with automorphic symbols exclusively. However, this program meets with considerable difficulties (which we regard as solved here), the main one being that two operators with automorphic symbols cannot, in fact, be composed at all in general. One cannot fail to regret this state of affairs since (a consequence of the covariance of the Weyl calculus under the metaplectic representation) operators with automorphic symbols are exactly those which commute on one hand with the (one-dimensional) Fourier transformation, on the other hand with the multiplication by the function $x \mapsto e^{i\pi x^2}$, a property which would be preserved under composition if this operation were possible. We would then obtain a general formula for the sharp composition $\mathfrak{S}_1 \,\#\, \mathfrak{S}_2$ of any two modular distributions, expressing it as a sum of explicit integrals and series of modular distributions.

In [35], under the pressure of our desire for such a formula, we were led to define a composition of operators with automorphic symbols in an extremely weak sense (dubbed the "minimal" sense there): a few words about it will be useful. Given $z \in \Pi$, define the pair of functions

$$\phi^0_{-z-1}(x) = 2^{\frac{1}{4}} (\operatorname{Im} z)^{\frac{1}{4}} e^{-i\pi \bar{z} x^2}, \quad \phi^1_{-z-1}(x) = 2^{\frac{5}{4}} \pi^{\frac{1}{2}} (\operatorname{Im} z)^{\frac{3}{4}} x \, e^{-i\pi \bar{z} x^2} : \quad (0.0.3)$$

when $z = i$, these functions are the first two normalized eigenstates of the so-called harmonic oscillator. For every distribution $\mathfrak{S} \in \mathcal{S}'(\mathbb{R}^2)$, one has

$$(\Theta_0 \mathfrak{S})(z) = (\phi^0_z \,|\, \operatorname{Op}(\mathfrak{S}) \phi^0_z), \quad (\Theta_1 \mathfrak{S})(z) = (\phi^1_z \,|\, \operatorname{Op}(\mathfrak{S}) \phi^1_z). \quad (0.0.4)$$

It is thus pseudodifferential analysis that led to the construction of the pair Θ_0, Θ_1, but these transformations relate also to the Radon transformation between two homogeneous spaces of $SL(2, \mathbb{R})$, the simplest case of an extensive theory [11]: more general transformations, still linked to the Radon transformation, will be indicated in Chapter 3, and will be useful when proving the decomposition theorem of automorphic distributions. It is true, as already said, that a distribution $\mathfrak{S}$ is characterized by its pair of Theta-transforms: but going in the reverse direction is an extremely discontinuous operation in any useful sense. In [35], we systematically analyzed automorphic distributions by the exclusive means of their Theta-transforms. This did not suffice to make the sharp product of any two modular distributions meaningful, and some other tricks were needed as well [35, p.133]. While we certainly obtained, up to some extent, an explicit sharp-composition formula for automorphic symbols, we did not reach any understanding of the associated operators.

In this book, we shall analyze exactly what makes it impossible to compose two operators with automorphic symbols and what, at best, can remain of this composition. Just like a non-holomorphic modular form, any modular distribution, homogeneous of degree $-1 - \nu$, admits a Fourier series decomposition: the terms of this series are distributions $h_{\nu,k}(x, \xi) = |\xi|^{-1-\nu} \exp\left(2i\pi \frac{kx}{\xi}\right)$ with $k \in \mathbb{Z}$. The question of analyzing the sharp product of any two modular distributions thus begins with that of analyzing a sharp product such as $h_{\nu_1,k_1} \# h_{\nu_2,k_2}$. This will be done in Chapter 4, the most technical of the book. The main difficulty originates from the fact that, in the case when $k_1 + k_2 = 0$, $k_1 \neq 0$, the composition $A_1 A_2$ of two operators A_1 and A_2 with symbols h_{ν_1,k_1} and h_{ν_2,k_2} does not act from $\mathcal{S}(\mathbb{R})$ to $\mathcal{S}'(\mathbb{R})$, hence does not admit any symbol in the Weyl calculus. However, denoting as $P = \frac{1}{2i\pi} \frac{d}{dx}$ and Q (the multiplication by x) the infinitesimal operators of the Heisenberg representation, one observes that both operators $A_1 A_2 P$ and $P A_1 A_2$ act from $\mathcal{S}(\mathbb{R})$ to $\mathcal{S}'(\mathbb{R})$, which makes it possible to define in particular, even when $k_1 + k_2 = 0$, the symbol $\mathcal{M}(h_{\nu_1,k_1}, h_{\nu_2,k_2})$ of the operator $2i\pi [P(A_1 A_2)Q - Q(A_1 A_2)P]$. Our interest in this special combination stems from the fact that, in the case when an operator A acts from $\mathcal{S}(\mathbb{R})$ to $\mathcal{S}'(\mathbb{R})$, hence admits a symbol $h \in \mathcal{S}'(\mathbb{R}^2)$, the symbol of the operator $2i\pi(PAQ - QAP)$ is $(2i\pi\mathcal{E})\,h$, which is of course especially useful in automorphic theory since the Euler operator commutes with the action of $SL(2,\mathbb{R})$.

Then, assuming that $|\mathrm{Re}\,(\nu_1 \pm \nu_2)| < 1$, we give an integral decomposition into functions $h_{i\lambda,k_1+k_2}$ of the sharp product $h_{\nu_1,k_1} \# h_{\nu_2,k_2}$ if $k_1 + k_2 \neq 0$, or of $\mathcal{M}\,(h_{\nu_1,k_1}, h_{\nu_2,k_2})$ if dispensing with this assumption. This is based on the general composition formula (quite different from the ones pseudodifferential practition- ers are familiar with) recalled in Chapter 3, and detailed computations involving hypergeometric functions. Dispensing with the assumption that $|\mathrm{Re}\,(\nu_1 \pm \nu_2)| < 1$ is possible, at the price of replacing the symbol $\mathcal{M}\,(h_{\nu_1,k_1}, h_{\nu_2,k_2})$ by its image under some polynomial, of Pochhammer's style, in the operator $2i\pi\mathcal{E}$. Different explanations of this necessity are given: ultimately, it comes down to the fact that, while the resolvent $(2i\pi\mathcal{E} - \mu)^{-1}$ is well-defined, in the space $L^2(\mathbb{R}^2)$, as soon as $\mathrm{Re}\,\mu \neq 0$, applying it to the tempered distributions we shall come across here will be possible only for values of μ avoiding a small set of integers.

In Chapter 5, we arrive at what was the major aim of the book (Theorems 5.3.4 and 5.4.1), obtaining in the case of two modular distributions $\mathfrak{S}_1$ and $\mathfrak{S}_2$ a full decomposition of $\mathcal{M}\,(\mathfrak{S}_1, \mathfrak{S}_2)$ into modular distributions: we assume in the case of two Eisenstein distributions $\mathfrak{E}_{\nu_1}$ and $\mathfrak{E}_{\nu_2}$ that $|\mathrm{Re}\,(\nu_1 \pm \nu_2)| < 1$. The various terms of the decomposition are an integral superposition of the Eisenstein distributions $\mathfrak{E}_{i\lambda}$, $\lambda \in \mathbb{R}$, a series of Hecke distributions, finally, in the case of two Eisenstein distributions only, a sum of 4 exceptional Eisenstein distributions. The coefficients of the decomposition are quite interesting: they involve repeatedly products of values of the zeta function and of L-functions, as well as "product L-functions". Let us emphasize that this is an exact formula, providing a sophisticated example of how far composition formulas, in pseudodifferential analysis, may differ from decompositions into asymptotic series, a rather popular if definitely too narrow concept: more will be said about this in Sections 3.3 and 7.1.

In Chapter 6, we show that, for any odd eigenstate $\phi^{(2n+1)}$ of the harmonic oscillator $\pi(P^2 + Q^2)$, the distribution $\mathrm{Op}\,(\mathfrak{E}_\nu)\,\phi^{(2n+1)}$ lies in $L^2(\mathbb{R})$ if $|\mathrm{Re}\,\nu| < \frac{1}{2}$: the corresponding fact is false if $|\mathrm{Re}\,\nu| > \frac{1}{2}$, or if we take in place of $\phi^{(2n+1)}$ an even eigenstate $\phi^{(2n)}$ of the harmonic oscillator. Then, in the odd case, we extend the formula for $\mathcal{M}\,(\mathfrak{E}_{\nu_1}, \mathfrak{E}_{\nu_2})$, at the price of having to insert the extra factor $(2i\pi\mathcal{E})^2 - 4$, beyond the line $\mathrm{Re}\,(\nu_1 + \nu_2) = 1$. A consequence of the for- mula obtained will consist in an interpretation of the function $|\zeta(\nu)|^2$, on the critical line $\mathrm{Re}\,\nu = \frac{1}{2}$, as the discontinuity of a function of interest, defined on both sides of the critical line. Finally, for $0 < \mathrm{Re}\,\nu < \frac{1}{2}$, we consider the series $\sum_{n \geq 0} e^{-(2n+\frac{3}{2})\alpha} \| \mathrm{Op}\,(\mathfrak{E}_\nu)\,\phi^{(2n+1)} \|^2$, a convergent one for $\alpha > 0$. With $\sigma = \mathrm{Re}\,\nu$, we show that it is the sum of a main term, the product of $\alpha^{-1-\sigma}$ by an explicit constant, and of an error term, in general a $\mathrm{O}(\alpha^{-1})$: the error term is a $\mathrm{O}(\alpha^{-\frac{1}{2}})$ if and only if ν (which lies on the left of the critical line) is a zero of zeta. Now, it is not uncommon to give a characterization of the Riemann hypothesis in terms of a bound for a certain error term: the best-known example is of course the expression of the number $\pi(N)$ of primes less than some large number N as the sum of the integral $\int_2^N \log t \, dt$ and of an error term, which is for every $\varepsilon > 0$ a $\mathrm{O}\left(N^{\frac{1}{2}+\varepsilon}\right)$ if and only if the Riemann hypothesis holds. But the present characterization of individual zeros of zeta on the left of the critical line is different: the Riemann hypothesis, on the contrary, is equivalent here to the fact that a certain error term is never too small.

After some thought, we decided against fully developing, in a way analogous to that expounded in the book in the non-holomorphic case, a similar theory in the holomorphic case. Such a theory exists, and is just as rich in symmetries as the Weyl calculus together with its satellites. But the metaplectic representation and Weyl pseudodifferential calculus have to be replaced by the anaplectic representation and alternative pseudodifferential analysis, as will be briefly recalled in Section 7.2. Decompositions of automorphic symbols into holomorphic modular forms then substitute for decompositions of the type considered in the greater part of this book. The anaplectic theory, introduced in [38], is very unusual: but, in a sense to be made precise, there is no room for any other choice, and the alternative pseudodifferential analysis is the sole possible competitor of the one-dimensional Weyl calculus, if one wishes to extend the two fundamental covariance properties of this calculus (under the metaplectic as well as under the Heisenberg representation). Also, the right-hand side of the alternative sharp composition formula, which is just a series (a convergent one, not a formal or asymptotic one) of Rankin-Cohen brackets, has met with some popularity recently.

The central point of the present book is that non-commutative algebras of non-holomorphic modular forms appear naturally in connection with the Weyl calculus: in the holomorphic case, such algebras appear in connection with alternative pseudodifferential analysis. One may thus find it highly unsatisfactory that arithmeticians should be familiar only with algebras of the second kind, and analysts only with the first-mentioned symbolic calculus. With the purpose, among others, to make our solution to this dilemma perfectly clear, a small first section in the last chapter was included for the benefit of people interested in combining pseudodifferential analysis, or quantization theory, with harmonic analysis and arithmetic.

Chapter 1

Basic modular distributions

The theory of L-functions is a subject for experts [10, 17], involving deep questions of algebraic number theory. We make here the simple observation that, defining in the usual way the L-function of a Maass form from its collection of Fourier coefficients, one can also give it in a direct way a natural spectral interpretation. This demands shifting from modular form theory in the hyperbolic half-plane to modular distribution theory in the plane and may be an introduction as good as any to this latter topic.

Consider the distribution $\mathfrak{d}_\chi$ on the line, defined as

$$\mathfrak{d}_\chi(x) = \sum_{m \in \mathbb{Z}} \chi(m)\, \delta\left(x - \frac{m}{\sqrt{N}}\right), \tag{1.0.1}$$

where χ is a primitive character with conductor N. It is an easy matter ([37, p. 18]: not needed in the sequel) to obtain a decomposition of $\mathfrak{d}_\chi$ into homogeneous components, to wit an integral expression, valid in the weak sense in the space $\mathcal{S}'(\mathbb{R})$ of tempered distributions,

$$\mathfrak{d}_\chi(x) = \int_{-\infty}^{\infty} c(\lambda)\, |x|_\varepsilon^{-\frac{1}{2} - i\lambda}\, d\lambda : \tag{1.0.2}$$

here, $\varepsilon = 0$ or 1 is defined so that $(-1)^\varepsilon = \chi(-1)$, and one sets $|x|_\varepsilon^{-\frac{1}{2}-i\lambda} = |x|^{-\frac{1}{2}-i\lambda} (\operatorname{sign} x)^\varepsilon$. The coefficient c of the decomposition is found to be

$$c(\lambda) = \frac{1}{2\pi} N^{\frac{1}{4} - \frac{i\lambda}{2}}\, L\left(\frac{1}{2} - i\lambda, \chi\right) \tag{1.0.3}$$

in terms of the Dirichlet L-function associated to the character χ. Defining the Fourier transformation $\mathcal{F}$ by means of the integral kernel $e^{-2i\pi xy}$, and using the equation

$$\mathcal{F}\left(|x|_\varepsilon^{-\frac{1}{2}-i\lambda}\right) = (-i)^\varepsilon \pi^{i\lambda}\, \frac{\Gamma(\frac{1}{4} + \frac{\varepsilon - i\lambda}{2})}{\Gamma(\frac{1}{4} + \frac{\varepsilon + i\lambda}{2})}\, |x|_\varepsilon^{-\frac{1}{2}+i\lambda}, \tag{1.0.4}$$

one obtains from the decomposition $(1.0.2), (1.0.3)$ that the classical functional equation of Dirichlet L-functions with primitive characters [4, p. 10] or [16, p. 84] is equivalent to the equation

$$\mathcal{F}^{-1}\mathfrak{d}_\chi = \frac{\tau(\chi)}{\sqrt{N}}\, \mathfrak{d}_{\bar\chi}, \tag{1.0.5}$$

where $\tau(\chi)$ is the Gauss sum $\sum_{m \bmod N} \chi(m)\, e^{\frac{2i\pi m}{N}}$: another expression of the same identity is the twisted Poisson formula, as to be found in [4, p. 8].

Automorphic distributions in the plane are tempered distributions invariant under the action of $SL(2, \mathbb{Z})$ by linear changes of coordinates. Modular distributions are distributions both automorphic and homogeneous of some degree: they make up the collection of Eisenstein distributions and Hecke distributions. On the other hand, homogeneous distributions in the plane can be further decomposed into bihomogeneous components, which are separately homogeneous with respect to each of the two coordinates of $(x, \xi) \in \mathbb{R}^2$. A canonical set of such functions $(1.1.22)$ is denoted as $\hom_{\rho,\nu}^{(\varepsilon)}$: the number $-1 + \nu$ stands for the global degree of homogeneity, and $\rho - 1$ stands for the difference between the degrees of homogeneity with respect to x and ξ separately. We show that the coefficient, with respect to the family of functions just referred to, of the decomposition into bihomogeneous components of an Eisenstein distribution or of a Hecke distribution, coincides as a function of ρ with the L-function naturally associated to it: in this sense, this decomposition may thus be regarded as analogous to $(1.0.2), (1.0.3)$, the L-function of Maass type taking the role previously taken by a Dirichlet L-function. It is only in Chapter 2 that these concepts will be linked to the more familiar ones, in which the analysis takes place in the hyperbolic half-plane instead.

1.1 Eisenstein distributions

We need to start this section with a shorthand, the reference to which will be permanent. We define, when $\varepsilon = 0$ or 1,

$$B_\varepsilon(\mu) = (-i)^\varepsilon\, \pi^{\mu - \frac{1}{2}}\, \frac{\Gamma(\frac{1-\mu+\varepsilon}{2})}{\Gamma(\frac{\mu+\varepsilon}{2})}, \quad \mu \neq \varepsilon + 1, \varepsilon + 3, \ldots. \tag{1.1.1}$$

One has

$$B_\varepsilon(\mu) B_\varepsilon(1 - \mu) = (-1)^\varepsilon \tag{1.1.2}$$

and one may note the relations

$$B_0(\mu) = \frac{\zeta(\mu)}{\zeta(1-\mu)} \quad \text{and} \quad B_1(\mu) = 2^{2\mu-1} i \frac{L(\mu, \chi)}{L(1-\mu, \chi)} \tag{1.1.3}$$

if χ is the non-trivial Dirichlet character mod 4: the first one (only) will be used on occasions.

We use, for power functions, the notation

$$|t|_\varepsilon^\alpha = \begin{cases} |t|^\alpha & \text{if } \varepsilon = 0, \\ |t|^\alpha \text{sign } t & \text{if } \varepsilon = 1. \end{cases} \tag{1.1.4}$$

The equation

$$|t|_\varepsilon^{-\nu-1} = -\frac{1}{\nu} \frac{d}{dt} |t|_{1-\varepsilon}^{-\nu}, \qquad \varepsilon = 0 \text{ or } 1, \tag{1.1.5}$$

makes it possible, by induction, to give the left-hand side a meaning as a tempered distribution on the line provided that $\nu \neq \varepsilon, \varepsilon + 2, \dots$. The function $B_\varepsilon(\mu)$ is a necessary factor when dealing with the Fourier transform of (signed or not) power functions, since one has

$$\int_{-\infty}^{\infty} |s|_\varepsilon^{-\mu} e^{-2i\pi s\sigma} \, ds = B_\varepsilon(\mu) \, |\sigma|_\varepsilon^{\mu-1}, \tag{1.1.6}$$

a semi-convergent integral if $0 < \text{Re } \mu < 1$: if this is not the case, the Fourier transformation in the space of tempered distributions makes it possible to extend the identity, provided that $\mu \neq \varepsilon + 1, \varepsilon + 3, \dots$ and $\mu \neq -\varepsilon, -\varepsilon - 2, \dots$. As functions of μ with values in $\mathcal{S}'(\mathbb{R}^2)$, both sides of this equation are holomorphic in the domain just indicated.

We shall use without further reference, throughout the book, the formula of complements of the Gamma function, as well as its duplication formula [22, p. 3]

$$\Gamma(z)\Gamma(z + \frac{1}{2}) = (2\pi)^{\frac{1}{2}} 2^{\frac{1}{2}-2z} \Gamma(2z). \tag{1.1.7}$$

We shall also make use, consistently, of the following asymptotics for the Gamma function on vertical lines [22, p. 13]

$$|\Gamma(x + iy)| \sim (2\pi)^{\frac{1}{2}} e^{-\frac{\pi}{2}|y|} |y|^{x-\frac{1}{2}}, \qquad |y| \to \infty. \tag{1.1.8}$$

The following lemma will be useful.

Lemma 1.1.1. *If* $-1 < a < -\frac{1}{2}$, *one has, for* $t \in \mathbb{R}^\times$,

$$e^{2i\pi t} - 1 = \frac{1}{4i\pi} \sum_{\varepsilon=0,1} (-1)^\varepsilon \int_{\text{Re } \nu=a} B_\varepsilon(1 - \nu) \, |t|_\varepsilon^{-\nu} \, d\nu. \tag{1.1.9}$$

Proof. Start from the right-hand side. As $|\text{Im } \nu| \to \infty$, it follows from (1.1.1) and (1.1.8) that the factor $B_\varepsilon(1 - \nu)$ is of the order of $|\text{Im } \nu|^{a-\frac{1}{2}}$, hence integrable if $a < -\frac{1}{2}$. We write

$$B_\varepsilon(1 - \nu) = \frac{(-i)^\varepsilon (2\pi)^{1-\nu}}{2 \sin \frac{\pi(\nu+\varepsilon)}{2}} \times \frac{1}{\Gamma(1 - \nu)} \tag{1.1.10}$$

so as to take advantage of the periodicity of the sine factor, and we move the line of integration to the line $\mathrm{Re}\,\nu = 1 - 4k$ with $k = 1, 2, \ldots$. The new line integral goes to zero as $k \to \infty$, and one obtains for the right-hand side of (1.1.9) the expression

$$\frac{1}{2} \sum_{\varepsilon=0,1} (-1)^\varepsilon \sum_{j=0}^{\infty} \mathrm{Res}_{\nu=-\varepsilon-2j} \left(B_\varepsilon(1-\nu)|t|_\varepsilon^{-\nu} \right) - \frac{1}{2} \mathrm{Res}_{\nu=0} \left(B_0(1-\nu)|t|^{-\nu} \right).$$

$$(1.1.11)$$

Now, one has

$$\frac{1}{2} \mathrm{Res}_{\nu=-\varepsilon-2j} \left(B_\varepsilon(1-\nu) \right) = (-i)^\varepsilon \frac{(-1)^j}{j!} \frac{\pi^{\frac{1}{2}+\varepsilon+2j}}{\Gamma(\frac{1}{2}+\varepsilon+j)}$$

$$= \frac{(-i)^\varepsilon(-1)^j \pi^{\frac{1}{2}+\varepsilon+2j}}{\Gamma(1+j)\Gamma(\frac{1}{2}+\varepsilon+j)} = (-i)^\varepsilon(-1)^j \frac{(2\pi)^{2j+\varepsilon}}{(2j+\varepsilon)!}.$$

$$(1.1.12)$$

Hence, the right-hand side of (1.1.9) is

$$\sum_{\varepsilon=0,1} i^\varepsilon \sum_{j\geq 0} (-1)^j \frac{|2\pi t|_\varepsilon^{\varepsilon+2j}}{(2j+\varepsilon)!} - 1 = e^{2i\pi t} - 1. \qquad (1.1.13)$$

$$\square$$

One of the nice things about power functions is the way they extend, when considered as tempered distributions, as meromorphic functions of the exponent in the full complex plane, as it follows from (1.1.5). This makes it possible to give the right-hand side of (1.1.9) a meaning under more general conditions regarding a, provided it is interpreted as an integral convergent in the weak sense in the space $\mathcal{S}'(\mathbb{R})$, *i.e*, if what is really meant is that the integrand is to be tested against an arbitrary function in $\mathcal{S}(\mathbb{R})$. The first thing to do, to make sure singularities of the functions $\nu \mapsto B_\varepsilon(1-\nu)$ and $\nu \mapsto |t|_\varepsilon^{-\nu}$ are avoided on the line of integration $\mathrm{Re}\,\nu = a$, is to assume that $a \notin \mathbb{Z}$. Then, the integral automatically converges in the sense indicated, in view of the identity

$$|t|_\varepsilon^{-\nu} = \frac{\Gamma(1-\nu)}{\Gamma(1-\nu+k)} \left(\frac{d}{dt}\right)^k |t|_{\varepsilon'}^{-\nu+k}, \quad \varepsilon' \equiv \varepsilon + k \bmod 2 : \qquad (1.1.14)$$

indeed, whatever the value of $a \notin \mathbb{Z}$, if $k > a + \frac{1}{2}$ and $\mathrm{Re}\,\nu = a$, the distribution $|t|_{\varepsilon'}^{-\nu+k}$ is a locally summable function while, at the same time, the integral

$$\int_{\mathrm{Re}\,\nu=a} \frac{\Gamma(1-\nu)}{\Gamma(1-\nu+k)} B_\varepsilon(1-\nu) |t|_\varepsilon^{-\nu} \, d\nu \qquad (1.1.15)$$

converges since, as already used in the proof of Lemma 1.1.1, $B_\varepsilon(1-\nu)$ is of the order of $|\mathrm{Im}\,\nu|^{a-\frac{1}{2}}$ as $|\mathrm{Im}\,\nu| \to \infty$.

Lemma 1.1.2. *One has, in the weak sense in* $S'(\mathbb{R})$,

$$\frac{1}{4i\pi} \sum_{\varepsilon=0,1} (-1)^\varepsilon \int_{\text{Re } \nu=a} B_\varepsilon(1-\nu) |t|_\varepsilon^{-\nu} \, d\nu = \begin{cases} e^{2i\pi t} - 1 & \text{if } -1 < a < 0, \\ e^{2i\pi t} & \text{if } a > 0. \end{cases}$$

(1.1.16)

Also,

$$\frac{1}{4i\pi} \int_{\text{Re } \nu=a} B_0(1-\nu) |t|^{-\nu} \, d\nu = \begin{cases} \cos(2\pi t) - 1 & \text{if } -2 < a < 0, \\ \cos(2\pi t) & \text{if } a > 0, \end{cases}$$

(1.1.17)

and

$$\frac{1}{4i\pi} \int_{\text{Re } \nu=a} B_1(1-\nu) |t|_1^{-\nu} \, d\nu = \begin{cases} -i \sin(2\pi t) + 2i\pi t & \text{if } -3 < a < -1, \\ -i \sin(2\pi t) & \text{if } a > -1. \end{cases}$$

(1.1.18)

Proof. The considerations which precede the lemma show that the only thing needed is to locate the poles and compute residues. Within the half-plane $\text{Re } \nu > -2$, the product $B_\varepsilon(1-\nu)|t|_\varepsilon^{-\nu} = \mathcal{F}(|t|_\varepsilon^{\nu-1})$ has a pole at $\nu = 0$, at $\nu = -1$ if $\varepsilon = 1$. According to (1.1.12), one has $\text{Res}_{\nu=0} B_0(1-\nu) = 2$ and $\text{Res}_{\nu=-1} B_1(1-\nu) = -4i\pi$. The first part of the lemma follows, and the rest of the lemma is immediate. $\square$

Definition 1.1.3. A distribution $\mathfrak{S} \in S'(\mathbb{R}^2)$ will be called an automorphic distribution if it is invariant under the action by linear changes of coordinates of the group $\Gamma = SL(2, \mathbb{Z})$. It will be called a modular distribution if, moreover, it is homogeneous of degree $-1 - \nu$ for some $\nu \in \mathbb{C}$.

A distribution $\mathfrak{S}$ is homogeneous of degree $-1 - \nu$ if and only if it satisfies, in the distribution sense, the equation $2i\pi\mathcal{E}\mathfrak{S} = -\nu\mathfrak{S}$, where the Euler operator is defined as

$$2i\pi\mathcal{E} = x \frac{\partial}{\partial x} + \xi \frac{\partial}{\partial \xi} + 1 \tag{1.1.19}$$

in terms of the coordinates x, ξ on $\mathbb{R}^2$. An equivalent, somewhat more useful, characterization is by means of the identity $t^{2i\pi\mathcal{E}}\mathfrak{S} = t^{-\nu}\mathfrak{S}$, where the operator $t^{2i\pi\mathcal{E}}$ on tempered distributions is defined for $t > 0$ by the equation

$$\langle t^{2i\pi\mathcal{E}}\mathfrak{S}, h \rangle = \langle \mathfrak{S}, t^{-2i\pi\mathcal{E}}h \rangle = \langle \mathfrak{S}, t \mapsto t^{-1}h(t^{-1}x, t^{-1}\xi) \rangle, \quad h \in S(\mathbb{R}^2). \tag{1.1.20}$$

Remarks 1.1.1. (i) That we limit ourselves to the case of the group $SL(2, \mathbb{Z})$ is of course a matter of convenience: definitions would be meaningful in relation to any arithmetic subgroup Γ of $SL(2, \mathbb{Z})$, while it should not be too difficult to extend *some* results to the case of congruence subgroups. The emphasis of this book is on analysis more than on arithmetic.

(ii) As we wish to give automorphic distribution theory (in the plane) a role of its own, it is only in the next chapter that we shall relate it to automorphic function theory (in the hyperbolic half-plane). The correspondence will not be absolutely one-to-one, since the first concept is more precise, only slightly so but in a way which will turn out to be very useful.

Besides the operator $2i\pi\mathcal{E}$, consider in the plane the operator

$$2i\pi\mathcal{E}^{\natural} = x\frac{\partial}{\partial x} - \xi\frac{\partial}{\partial\xi}. \tag{1.1.21}$$

The globally even bihomogeneous function

$$\hom_{\rho,\nu}^{(\varepsilon)}(x,\xi) = |x|_{\varepsilon}^{\frac{\rho+\nu-2}{2}}|\xi|_{\varepsilon}^{\frac{-\rho+\nu}{2}}, \qquad x \neq 0, \, \xi \neq 0, \tag{1.1.22}$$

a generalized eigenfunction of the pair $(2i\pi\mathcal{E}, 2i\pi\mathcal{E}^{\natural})$ for the eigenvalues $(\nu, \rho-1)$, makes sense as a distribution in the plane provided that $\rho+\nu \neq -2\varepsilon, -2\varepsilon - 4, \dots$ and $2-\rho+\nu \neq -2\varepsilon, -2\varepsilon-4, \dots$. The functions $\hom_{\rho,\nu}^{(\varepsilon)}$ with $\mathrm{Re}\,\nu = 0$ and $\mathrm{Re}\,\rho = 1$ constitute a set of joint generalized eigenfunctions of the pair of operators $\mathcal{E}, \mathcal{E}^{\natural}$ for the realizations of these as unbounded self-adjoint operators on $L_{\mathrm{even}}^2(\mathbb{R}^2)$. Observe that $\hom_{\rho,\nu}^{(\varepsilon)}$ is homogeneous of degree $-1+\nu$, while $\mathfrak{E}_\nu$, as defined below, is homogeneous of degree $-1-\nu$. The following easy facts will be used time and again.

Lemma 1.1.4. *The product $B_\varepsilon(\frac{\rho-\nu}{2})\hom_{\rho,-\nu}^{(\varepsilon)}$ is an analytic function of ρ outside the points $\rho = 2 \pm \nu + 2\varepsilon, 6 \pm \nu + 2\varepsilon, \dots$ (so that the poles of $\hom_{\rho,\pm\nu}^{(\varepsilon)}$ of the first species have disappeared from the product). One has the special values*

$$\hom_{-\nu,-\nu}^{(0)}(x,\xi) = |x|^{-\nu-1}, \qquad \hom_{2+\nu,-\nu}^{(0)}(x,\xi) = |\xi|^{-\nu-1}, \tag{1.1.23}$$

while special residues are

$$\left[\mathrm{Res}_{\rho=\nu}\hom_{\rho,-\nu}^{(0)}\right](x,\xi) = 4\,\delta(x)\,|\xi|^{-\nu}, \qquad \left[\mathrm{Res}_{\rho=2-\nu}\hom_{\rho,-\nu}^{(0)}\right](x,\xi) = -4\,|x|^{-\nu}\delta(\xi). \tag{1.1.24}$$

Proof. Most of this is proved by inspection, while the computation of the residues follows from the identities

$$\hom_{\rho,-\nu}^{(0)}(x,\xi) = \frac{2}{\rho-\nu}\frac{d}{dx}\left(|x|_1^{\frac{\rho-\nu}{2}}|\xi|^{\frac{-\rho-\nu}{2}}\right) = \frac{2}{2-\rho-\nu}\frac{d}{d\xi}\left(|x|^{\frac{\rho-\nu-2}{2}}|\xi|_1^{\frac{2-\rho-\nu}{2}}\right): \tag{1.1.25}$$

considered for ρ close to ν, the first one yields

$$\left[\mathrm{Res}_{\rho=\nu}\hom_{\rho,-\nu}^{(0)}\right](x,\xi) = 2\frac{d}{dx}\left(\mathrm{sign}\,x\,|\xi|^{-\nu}\right), \tag{1.1.26}$$

and the second equation furnishes in the same way the second residue.   □

That, with a few exceptions (including the definition of the Weyl calculus in Chapter 3), we interest ourselves in globally even distributions only is a simplification we could dispense with. However, as will be seen in the next chapter, no other distributions need be considered as long as, in the hyperbolic half-plane, we limit our considerations to Maass forms of weight zero (cf. [4, p. 129] for the definition of more general ones: Maass forms of weight one are linked to automorphic distributions of an odd type in [35, section 18]). Considering only even distributions will put a desirable limit on the number of parameters necessary in the composition formula to be developed in Section 3.3.

Definition 1.1.5. If $\nu \in \mathbb{C}$, Re $\nu < -1$, we define the Eisenstein distribution $\mathfrak{E}_\nu$ by the equation, valid for every $h \in \mathcal{S}(\mathbb{R}^2)$,

$$\langle \mathfrak{E}_\nu, h \rangle = \frac{1}{2} \sum_{|m|+|n| \neq 0} \int_{-\infty}^{\infty} |t|^{-\nu} h(mt, nt) \, dt. \tag{1.1.27}$$

It is immediate that the series of integrals converges if Re $\nu < -1$, in which case $\mathfrak{E}_\nu$ is well defined as a tempered distribution. Obviously, it is $SL(2, \mathbb{Z})$–invariant as a distribution, in particular an even distribution (i.e., it vanishes on odd functions), finally it is homogeneous of degree $-1 - \nu$. The relation between this notion and the classical one of non–holomorphic Eisenstein series will be recalled in Proposition 2.1.1.

Theorem 1.1.6. *As a tempered distribution, $\mathfrak{E}_\nu$ extends as a meromorphic function of $\nu \in \mathbb{C}$, whose only poles are at $\nu = \pm 1$: these poles are simple, and the residues of $\mathfrak{E}_\nu$ there are*

$$\mathrm{Res}_{\nu=-1} \, \mathfrak{E}_\nu = -1 \quad \text{and} \quad \mathrm{Res}_{\nu=1} \, \mathfrak{E}_\nu = \delta, \tag{1.1.28}$$

the unit mass at the origin of $\mathbb{R}^2$. Let $\mathcal{F}^{\mathrm{symp}}$ be the so-called symplectic Fourier transformation on $\mathcal{S}'(\mathbb{R}^2)$ (an advantage of which is that it commutes with the action of $SL(2, \mathbb{R})$ by linear changes of coordinates, not only that of $SO(2)$), defined by the equation

$$(\mathcal{F}^{\mathrm{symp}} h)(x, \xi) = \int_{\mathbb{R}^2} h(y, \eta) \, e^{2i\pi(x\eta - y\xi)} \, dy \, d\eta : \tag{1.1.29}$$

then, one has

$$\mathcal{F}^{\mathrm{symp}} \, \mathfrak{E}_\nu = \mathfrak{E}_{-\nu} \quad \text{for} \quad \nu \neq \pm 1. \tag{1.1.30}$$

Proof. For the sake of completeness, let us reproduce the proof given in [39, p. 93]. Denote as $(\mathfrak{E}_\nu)_{\mathrm{princ}}$ (*resp.* $(\mathfrak{E}_\nu)_{\mathrm{res}}$) the distribution defined in the same way as $\mathfrak{E}_\nu$, except for the fact that the integral on the line in (1.1.27) is replaced by the same integral taken from -1 to 1 (*resp.* on $] - \infty, -1] \cup [1, \infty[$), and observe that the distribution $(\mathfrak{E}_\nu)_{\mathrm{res}}$ extends as an entire function of ν. As a consequence of Poisson's formula, one has when Re $\nu < -1$ the identity

$$\int_1^\infty t^\nu \sum_{(n,m)\in\mathbb{Z}^2} (\mathcal{F}^{\mathrm{symp}} h)\,(tn,\, tm)\, dt = \int_1^\infty t^\nu \sum_{(n,m)\in\mathbb{Z}^2} t^{-2}\, h(t^{-1}n,\, t^{-1}m)\, dt$$

$$= \int_0^1 t^{-\nu} \sum_{(n,m)\in\mathbb{Z}^2} h(tn,\, tm)\, dt, \quad (1.1.31)$$

from which one obtains that

$$\langle \mathcal{F}^{\mathrm{symp}}\, (\mathfrak{E}_{-\nu})_{\mathrm{res}}\,,\, h \rangle = \langle (\mathfrak{E}_\nu)_{\mathrm{princ}}\,,\, h \rangle + \frac{h(0,0)}{1-\nu} + \frac{(\mathcal{F}^{\mathrm{symp}}h)(0,0)}{1+\nu}. \quad (1.1.32)$$

From this identity, one finds the meromorphic continuation of the function $\nu \mapsto \mathfrak{E}_\nu$, including the residues at the two poles, as well as the fact that $\mathfrak{E}_\nu$ and $\mathfrak{E}_{-\nu}$ are the images of each other under $\mathcal{F}^{\mathrm{symp}}$.   □

Eisenstein distributions alone already make it possible to decompose some automorphic distributions into homogeneous components: as a basic example, defining the Dirac comb $\mathfrak{D}$ as

$$\mathfrak{D}(x,\xi) = 2\pi \sum_{|m|+|n|\neq 0} \delta(x-n)\delta(\xi-m), \quad (1.1.33)$$

one has

$$\mathfrak{D} = 2\pi + \int_{-\infty}^\infty \mathfrak{E}_{i\lambda}\, d\lambda. \quad (1.1.34)$$

The short proof [39, p. 95] relies on the decomposition of test functions into homogeneous components, as will be systematically used later. Starting from (3.2.1) and using a change of contour, one may write, for any $a > 1$ and any function $h \in \mathcal{S}_{\mathrm{even}}(\mathbb{R}^2)$,

$$h = \frac{1}{i} \int_{\mathrm{Re}\ \nu=-a} h_{-\nu}\, d\nu \quad \text{with} \quad h_{-\nu}(x,\xi) = \frac{1}{2\pi} \int_0^\infty t^{-\nu} h(tx,\, t\xi)\, dt. \quad (1.1.35)$$

Then,

$$\langle \mathfrak{D}\,,\, h_{-\nu} \rangle = \sum_{|m|+|n|\neq 0} \int_0^\infty t^{-\nu}\, h(tn,\, tm)\, dt \quad (1.1.36)$$

and, using (1.1.27),

$$\langle \mathfrak{D}\,,\, h \rangle = \frac{1}{i} \int_{\mathrm{Re}\ \nu=-a} \langle \mathfrak{E}_\nu\,,\, h \rangle\, d\nu, \quad (1.1.37)$$

after which it suffices to move the contour of integration to the line Re $\nu = 0$, using the first equation (1.1.28).

We give now two distinct decompositions of Eisenstein distributions: into Fourier series, and as integral superpositions of bihomogeneous functions. Both kinds of developments will be available too in the case of Hecke distributions, in the next section. The following was already proved and used in [35, 39], but we give here a direct proof, staying within automorphic distribution theory (in the plane). The reason why we consider $\frac{1}{2}\mathfrak{E}_\nu$, rather than $\mathfrak{E}_\nu$, in the next formula, will be explained in Remarks 1.2.1(v) and 2.1.1(iii).

Theorem 1.1.7. *For Re $\nu < 0$, $\nu \neq -1$, and $h \in \mathcal{S}(\mathbb{R}^2)$, one has [39, p. 93]*

$$\frac{1}{2}\langle \mathfrak{E}_\nu, h \rangle = \frac{1}{2}\zeta(-\nu) \int_{-\infty}^{\infty} |t|^{-\nu-1} \left(\mathcal{F}_1^{-1}h\right)(0,t)\, dt$$

$$+ \frac{1}{2}\zeta(1-\nu) \int_{-\infty}^{\infty} |t|^{-\nu} h(t,0)\, dt$$

$$+ \frac{1}{2} \sum_{n \neq 0} \sigma_\nu(|n|) \int_{-\infty}^{\infty} |t|^{-\nu-1} \left(\mathcal{F}_1^{-1}h\right)\left(\frac{n}{t}, t\right) dt, \qquad (1.1.38)$$

where $\mathcal{F}_1^{-1}$ denotes the inverse Fourier transform with respect to the first variable and $\sigma_\nu(|n|) = \sum_{1 \leq d | n} d^\nu$. After the power function $t \mapsto |t|^\mu$ has been given a meaning, as a distribution on the line, for $\mu \neq -1, -3, \ldots$, this decomposition is actually valid for $\nu \neq \pm 1$, $\nu \neq 0$.

Proof. Isolating the terms with $n = 0$ in (1.1.27), we write, after a change of variable,

$$\frac{1}{2}\langle \mathfrak{E}_\nu, h \rangle = \frac{1}{2}\zeta(1-\nu) \int_{-\infty}^{\infty} |t|^{-\nu} h(t,0)\, dt + \frac{1}{4} \sum_{m \in \mathbb{Z},\, n \neq 0} \int_{-\infty}^{\infty} |t|^{-\nu} h(mt, nt)\, dt$$

$$= \frac{1}{2}\zeta(1-\nu) \int_{-\infty}^{\infty} |t|^{-\nu} h(t,0)\, dt$$

$$+ \frac{1}{4} \sum_{m \in \mathbb{Z},\, n \neq 0} \int_{-\infty}^{\infty} |t|^{-\nu-1} \left(\mathcal{F}_1^{-1}h\right)\left(\frac{m}{t}, nt\right) dt, \qquad (1.1.39)$$

where we have used Poisson's formula at the end. Isolating now the terms such that $m = 0$, we obtain

$$\frac{1}{4} \sum_{m \in \mathbb{Z},\, n \neq 0} \int_{-\infty}^{\infty} |t|^{-\nu-1} \left(\mathcal{F}_1^{-1}h\right)\left(\frac{m}{t}, nt\right) dt$$

$$= \frac{1}{2}\zeta(-\nu) \int_{-\infty}^{\infty} |t|^{-\nu-1} \left(\mathcal{F}_1^{-1}h\right)(0,t)\, dt$$

$$+ \frac{1}{4} \sum_{mn \neq 0} \int_{-\infty}^{\infty} |t|^{-\nu-1} \left(\mathcal{F}_1^{-1}h\right)\left(\frac{m}{t}, nt\right) dt, \qquad (1.1.40)$$

from which the main part of the theorem follows after we have made the change of variable $t \mapsto \frac{t}{n}$ in the main term. The last part of the statement uses also the fact that the product $\zeta(-\nu)|t|^{-\nu-1}$ is regular at $\nu = 2, 4, \ldots$ and the product $\zeta(1 - \nu)|t|^{-\nu}$ is regular at $\nu = 3, 5, \ldots$, thanks to the trivial zeros of zeta: for instance, the value of the second one at $\nu = 3$ is the distribution $\frac{1}{2}\zeta'(-2)\delta''$.

Let us remark also that, even though neither of the first two terms of (1.1.38) is the image on h of a meaningful distribution when $\nu = 0$, their sum is still an analytic function of ν near that point. Indeed, since $\zeta(0) = -\frac{1}{2}$ and the residue at $\nu = 0$ of the distribution $|\xi|^{-\nu-1} = -\frac{1}{\nu}\frac{d}{d\xi}\left(|\xi|_1^{-\nu}\right)$ is $-\frac{d}{d\xi}\operatorname{sign}\xi = -2\delta(\xi)$ (a fact to be used time and again), the sum $\zeta(-\nu)|\xi|^{-\nu-1} + \zeta(1 - \nu)|x|^{-\nu}\delta(\xi)$ is regular at $\nu = 0$. $\qquad\Box$

So as to obtain the decomposition of $\mathfrak{E}_\nu$ into bihomogeneous components, we first decompose the individual terms of its Fourier series expansion. Fixing ν such that $\operatorname{Re}\nu < 0$, consider for every $n \geq 1$ the distribution $\mathfrak{S}_n$ such that, for $h \in \mathcal{S}(\mathbb{R}^2)$,

$$\langle \mathfrak{S}_n, h \rangle = \frac{1}{2}\int_{-\infty}^{\infty}|t|^{-\nu-1}\left[\left(\mathcal{F}_1^{-1}h\right)\left(\frac{n}{t}, t\right) + \left(\mathcal{F}_1^{-1}h\right)\left(-\frac{n}{t}, t\right)\right]dt$$

$$= \int_{-\infty}^{\infty}|t|^{-\nu-1}\,dt\int_{-\infty}^{\infty}h(x,t)\cos\frac{2\pi nx}{t}\,dx, \qquad (1.1.41)$$

a superposition of integrals (starting from the right) which can be transformed into a genuine double integral by means of an integration by parts, to wit

$$\langle \mathfrak{S}_n, h \rangle = -\frac{1}{2\pi n}\int_{\mathbb{R}^2}|t|_1^{-\nu}\frac{\partial h}{\partial x}(x,t)\sin\frac{2\pi nx}{t}\,dx : \qquad (1.1.42)$$

in other words,

$$\mathfrak{S}_n(x, \xi) = |\xi|^{-\nu-1}\cos\frac{2\pi nx}{\xi} = \frac{1}{2\pi n}\frac{\partial h}{\partial x}\left(|\xi|_1^{-\nu}\sin\frac{2\pi nx}{\xi}\right) : \qquad (1.1.43)$$

iterating this integration by parts shows that, as a tempered distribution in (x, ξ), this is an entire function of ν (in contrast to the factor $|\xi|^{-\nu-1}$, only meaningful for $\nu \neq 0, 2, \ldots$). The equation $|\xi|^{-\nu-1}\left|\frac{x}{\xi}\right|^{-\mu} = |x|^{-\mu}|\xi|^{\mu-\nu-1}$, certainly true for $x \neq 0, \xi \neq 0$, involves locally summable functions of $x, \xi, \frac{x}{\xi}$ in the case when $\operatorname{Re}\nu < 0$, $\operatorname{Re}\mu < 1$ and $\operatorname{Re}(\mu - \nu) > 0$. If $\operatorname{Re}\nu < 0$ and $0 < a < 1$, one may then write in the weak sense in $\mathcal{S}'(\mathbb{R}^2)$, in view of (1.1.17), the decomposition

$$\mathfrak{S}_n(x, \xi) = \frac{1}{4i\pi}\int_{\operatorname{Re}\mu=a}|n|^{-\mu}B_0(1 - \mu)|x|^{-\mu}|\xi|^{\mu-\nu-1}\,d\mu. \qquad (1.1.44)$$

Since the left-hand side is an entire function of ν, this remains true, dropping the assumption that $\operatorname{Re}\nu < 0$, under the sole assumption, besides the condition $0 < a < 1$, that $a > \operatorname{Re}\nu$. Set $\mu = \frac{2-\rho+\nu}{2}$, so that the new line of integration is

$\mathrm{Re}\ \nu = c$ with $c = -2a + 2 + \mathrm{Re}\ \nu$, the constraints on c being thus $\mathrm{Re}\ \nu < c < 2 \pm \mathrm{Re}\ \nu$: noting that $|x|^{\frac{\rho-\nu-2}{2}}|\xi|^{-\rho-\nu} = \hom^{(0)}_{\rho,-\nu}(x,\xi)$, one obtains

$$\mathfrak{S}_n = \frac{1}{8i\pi}\int_{\mathrm{Re}\ \rho = c} n^{\frac{\rho-\nu-2}{2}}\, B_0(\frac{\rho-\nu}{2})\, \hom^{(0)}_{\rho,-\nu}\, d\rho, \quad \mathrm{Re}\ \nu < c < 2 \pm \mathrm{Re}\ \nu. \quad (1.1.45)$$

Theorem 1.1.8. *Assume that* $-1 < \mathrm{Re}\ \nu < 1$, $\nu \neq 0$. *Then, in the weak sense in* $\mathcal{S}'(\mathbb{R}^2)$,

$$\frac{1}{2}\,\mathfrak{E}_\nu(x,\xi) = \frac{1}{2}\zeta(-\nu)\left[\,|x|^{-\nu-1} + |\xi|^{-\nu-1}\right] + \frac{1}{2}\zeta(1-\nu)\left[\,|x|^{-\nu}\delta(\xi) + \delta(x)\,|\xi|^{-\nu}\,\right]$$

$$+ \frac{1}{8i\pi}\int_{\mathrm{Re}\ \rho = 1} \zeta(\frac{2-\rho-\nu}{2})\,\zeta(\frac{\rho-\nu}{2})\,\hom^{(0)}_{\rho,-\nu}(x,\xi)\, d\rho. \quad (1.1.46)$$

One can also, when desirable, replace the line $\mathrm{Re}\ \rho = 1$ *by any line* $\mathrm{Re}\ \rho = c$ *with* $c > 2 + |\mathrm{Re}\ \nu|$, *provided that one deletes from the right-hand side the terms* $\frac{1}{2}\zeta(-\nu)|\xi|^{-\nu-1}$ *and* $\frac{1}{2}\zeta(1-\nu)|x|^{-\nu}\delta(\xi)$.

Proof. We rewrite (1.1.38) as a decomposition

$$\frac{1}{2}\,\mathfrak{E}_\nu(x,\xi) = \frac{1}{2}\zeta(-\nu)\,|\xi|^{-\nu-1} + \frac{1}{2}\zeta(1-\nu)\,|x|^{-\nu}\delta(\xi) + \frac{1}{2}\,\mathfrak{E}^{\mathrm{main}}_\nu(x,\xi) \quad (1.1.47)$$

and from (1.1.45), assuming that $\mathrm{Re}\ \nu < c < 2 \pm \mathrm{Re}\ \nu$ (a condition certainly verified if $c = 1$),

$$\frac{1}{2}\,\mathfrak{E}^{\mathrm{main}}_\nu = \frac{1}{8i\pi}\sum_{n\geq 1}\sigma_\nu(n)\int_{\mathrm{Re}\ \rho = c} n^{\frac{\rho-\nu-2}{2}}\, B_0(\frac{\rho-\nu}{2})\,\hom^{(0)}_{\rho,-\nu}\, d\rho. \quad (1.1.48)$$

We have shown in (1.1.14) in which sense this type of integral is always convergent: but we must still arrange for summability with respect to n. The product

$$B_0(\frac{\rho-\nu}{2})\,\hom^{(0)}_{\rho,-\nu} = B_0(\frac{\rho-\nu}{2})\,|x|^{\frac{\rho-\nu-2}{2}}|\xi|^{\frac{-\rho-\nu}{2}}, \quad (1.1.49)$$

contrary to its second factor, is regular at $\rho = \nu$. This makes it possible, in (1.1.48), to move the line of integration to any line $\mathrm{Re}\ \rho = c$ with $c < \pm\mathrm{Re}\ \nu$. The right-hand side of (1.1.48) then becomes a convergent series of integrals, and one has for ρ on the new line of integration

$$\sum_{n\geq 1}\sigma_\nu(n)\, n^{\frac{\rho-\nu-2}{2}} = \sum_{d,k\geq 1} d^\nu (kd)^{\frac{\rho-\nu-2}{2}} = \zeta(\frac{2-\rho-\nu}{2})\,\zeta(\frac{2-\rho+\nu}{2}). \quad (1.1.50)$$

Using this identity together with the functional equation (1.1.3), one obtains the weak integral decomposition in $\mathcal{S}'(\mathbb{R}^2)$

$$\frac{1}{2}\,\mathfrak{E}^{\mathrm{main}}_\nu(x,\xi) = \frac{1}{8i\pi}\int_{\mathrm{Re}\ \rho = c} \zeta(\frac{2-\rho-\nu}{2})\,\zeta(\frac{\rho-\nu}{2})\,\hom^{(0)}_{\rho,-\nu}(x,\xi)\, d\rho, \quad (1.1.51)$$

provided that Re $\rho < -|\text{Re } \nu|$. What has to be done finally is to move back the line of integration to the line Re $\rho = 1$, paying attention to the poles ρ of the integrand such that Re $\rho < 1$. The first zeta factor contributes a simple pole at $\rho = -\nu$, and the factor $\text{hom}_{\rho,-\nu}^{(0)}$ is singular when $\rho = \nu, \nu - 4, \ldots$: but the simple poles $\nu - 4, \nu - 8, \ldots$ can be discarded since they are (trivial) zeros of the second zeta factor. We shall have to add to the new integral obtained the product of the sum of residues by $-2i\pi$. The residue at $\rho = -\nu$ is

$$(-2)\,\zeta(-\nu)\,\text{hom}_{-\nu,-\nu}^{(0)}(x,\xi) = -2\,\zeta(-\nu)\,|x|^{-\nu-1} \qquad (1.1.52)$$

while, as a consequence of (1.1.24) and of the equation $\zeta(0) = -\frac{1}{2}$, the residue of the integrand at $\rho = \nu$ is $-2\,\zeta(1-\nu)\,\delta(x)\,|\xi|^{-\nu}$. This leads to the decomposition (1.1.46).

   If, as will be helpful later, one wishes to replace the line Re $\rho = 1$ by a line Re $\rho = c$ with c large, one must take into consideration the poles of the product $\zeta(\frac{\rho-\nu}{2})\,\text{hom}_{\rho,-\nu}^{(0)}$ with Re $\rho > 1$. There is one at $\rho = 2 + \nu$ because of the first factor, while the second factor has poles at $\rho = 2 - \nu, 6 - \nu, \ldots$: but the poles $6 - \nu, 10 - \nu, \ldots$ are killed by zeros of the other zeta factor $\zeta(\frac{2-\rho-\nu}{2})$. Finally, the residues at the only two remaining poles are obtained from an application of Lemma 1.1.4, together with the fact that $\zeta(0) = -\frac{1}{2}$.   □

Remark 1.1.2. Since $\text{hom}_{\rho,-\nu}^{(0)}(-\xi,x) = \text{hom}_{2-\rho,-\nu}^{(0)}(x,\xi)$, the fact that $\mathfrak{E}_\nu$ is invariant under the action of the map $(x,\xi) \mapsto (-\xi,x)$ remains apparent under the decomposition: not so the invariance under the action of the matrix $\left(\begin{smallmatrix} 1 & 1 \\ 0 & 1 \end{smallmatrix}\right)$. It is the other way around when the Fourier expansion (1.1.38) is used instead.

Corollary 1.1.9. *Theorem 1.1.8 extends to the values of ν such that Re $\nu \neq -1$, Re $\nu \neq 1, 5, \ldots$ and $\nu \neq 0, 1, \ldots$.*

Proof. The left-hand side can be continued to values of ν distinct from ± 1. On the first line of the right-hand side, it suffices to assume that, moreover, $\nu \neq 0, 2, \ldots$ and $\nu \neq 0$, $\nu \neq 1, 3, \ldots$. So far as the integrand on the right-hand side is concerned, it suffices to manage so that one will always have $\rho + \nu \neq 0$, $\rho - \nu \neq 2$ (considering the zeta factors) and $\rho - \nu \neq 0, -4, \ldots$, $2 - \rho + \nu \neq 0, -4, \ldots$ so that $\text{hom}_{\rho,-\nu}^{(0)}$ should be well-defined: this will be the case for every ρ on the line Re $\rho = 1$ provided that Re $\nu \neq -1$ and Re $\nu \neq 1, 5, \ldots$.   □

Remarks 1.1.3. (i) If ν lies on one of the lines Re $\nu = -1$ or Re $\nu = 1, 5, \ldots$ but $\nu \neq \pm 1$, a decomposition of $\mathfrak{E}_\nu(x,\xi)$ is still possible, only turning slightly around the point ρ on the line Re $\rho = 1$ responsible for the singularity, and computing a residue: this is trivial when the singularity originates from a zeta factor, and can be obtained, when the singularity originates from the factor $\text{hom}_{\rho,-\nu}^{(0)}(x,\xi)$, from an application of (1.1.14).

(ii) It is natural, imitating a definition given for Hecke (rather than Eisenstein) distributions later, in Theorem 2.1.2, to set, with $b_k = k^{-\frac{\nu}{2}}\sigma_\nu(k)$,

$$L\left(s, \frac{1}{2}\mathfrak{E}_\nu\right) = \sum_{k \geq 1} b_k\, k^{-s} = \zeta\left(s - \frac{\nu}{2}\right)\zeta\left(s + \frac{\nu}{2}\right): \qquad (1.1.53)$$

then, rewriting (1.1.46) in terms of this function, to wit

$$\frac{1}{2}\mathfrak{E}_\nu(x,\xi) = \frac{1}{2}\zeta(-\nu)\left[|x|^{-\nu-1} + |\xi|^{-\nu-1}\right] + \frac{1}{2}\zeta(1-\nu)\left[|x|^{-\nu}\delta(\xi) + \delta(x)|\xi|^{-\nu}\right]$$
$$+ \frac{1}{8i\pi}\int_{\mathrm{Re}\,\rho=1} B_0\left(\frac{\rho-\nu}{2}\right) L\left(\frac{2-\rho}{2}, \frac{1}{2}\mathfrak{E}_\nu\right)\,\mathrm{hom}_{\rho,-\nu}^{(0)}(x,\xi)\,d\rho,$$

$$(1.1.54)$$

it will be seen in the next section that a fully analogous formula holds for Hecke distributions, except that there are no longer "cuspidal" terms (the analogues, in modular distribution theory, of the two terms, in the Fourier expansion of non-holomorphic Eisenstein series, not rapidly decreasing at the cusp of $\Gamma\backslash\Pi$). Note the functional equation

$$B_0\left(\frac{\rho-\nu}{2}\right) L\left(\frac{2-\rho}{2}, \frac{1}{2}\mathfrak{E}_\nu\right) = B_0\left(\frac{2-\rho-\nu}{2}\right) L\left(\frac{\rho}{2}, \frac{1}{2}\mathfrak{E}_\nu\right), \qquad (1.1.55)$$

a consequence of (1.1.3).

1.2 Hecke distributions

We introduce here Hecke operators acting on automorphic distributions: these will be related later to the more traditional notion of Hecke operator acting on automorphic functions.

Definition 1.2.1. Given an automorphic distribution $\mathfrak{S}$, we set, for $N \geq 1$,

$$\left\langle T_N^{\mathrm{dist}}\mathfrak{S}, h\right\rangle = N^{-\frac{1}{2}} \sum_{\substack{ad=N,\,d>0 \\ b\,\mathrm{mod}\,d}} \left\langle \mathfrak{S}, (x,\xi) \mapsto h\left(\frac{dx - b\xi}{\sqrt{N}}, \frac{a\xi}{\sqrt{N}}\right)\right\rangle \qquad (1.2.1)$$

and

$$\left\langle T_{-1}^{\mathrm{dist}}\mathfrak{S}, h\right\rangle = \left\langle \mathfrak{S}, (x,\xi) \mapsto h(-x,\xi)\right\rangle. \qquad (1.2.2)$$

Just as in the automorphic function environment, the linear span of the Hecke operators T_N^{dist} with $N \geq 1$ makes up an algebra, which is generated, as such, by the operators T_p^{dist} with p prime. Automorphic distributions which are left invariant, or change to their negatives, under T_{-1}^{dist}, are said to be of even or odd type: this answers the question whether such (globally even) distributions are separately even or odd with respect to each of the two variables x and ξ.

In what follows, we consider characters χ on $\mathbb{Q}^\times$, by which we mean homomorphisms from $\mathbb{Q}^\times$ to $\mathbb{C}^\times$: we do not assume these to be unitary, but tempered, in the sense that, for some $C \geq 0$, one has

$$\left|\chi\left(\frac{m}{n}\right)\right| \leq |mn|^C \text{ for every fraction } \frac{m}{n}. \tag{1.2.3}$$

In the usual way, an entire function f is said, below, to be polynomially bounded in vertical strips if, given a segment $[a,\,b] \subset \mathbb{R}$, one has $|f(s)| \leq C(1 + |\mathrm{Im}\ s|)^N$ for some pair C, N and all s with $a \leq \mathrm{Re}\ s \leq b$.

Theorem 1.2.2. *Given a tempered character χ on $\mathbb{Q}^\times$ and $\lambda \in \mathbb{R}$, the (even) distribution $\mathfrak{N} = \mathfrak{N}_{\chi,i\lambda} \in \mathcal{S}'(\mathbb{R}^2)$ defined by the equation*

$$\langle \mathfrak{N},\,h\rangle = \frac{1}{4}\sum_{m,\,n\neq 0}\chi\left(\frac{m}{n}\right)\int_{-\infty}^{\infty}|t|^{-1-i\lambda}\left(\mathcal{F}_1^{-1}h\right)\left(\frac{m}{t},\,nt\right)dt\,,\quad h \in \mathcal{S}(\mathbb{R}^2), \tag{1.2.4}$$

satisfies the identity $\langle \mathfrak{N},\,h \circ \left(\begin{smallmatrix}1 & 1\\ 0 & 1\end{smallmatrix}\right)\rangle = \langle \mathfrak{N},\,h\rangle$ for every function $h \in \mathcal{S}(\mathbb{R}^2)$. Also, it is homogeneous of degree $-1 - i\lambda$. Set $\chi(-1) = (-1)^\varepsilon$ with $\varepsilon = 0$ or 1, and define

$$\psi_1(s) = \sum_{m\geq 1}\chi(m)\,m^{-s} = \prod_p\left(1 - \chi(p)\,p^{-s}\right)^{-1},\quad \psi_2(s) = \sum_{n\geq 1}(\chi(n))^{-1}n^{-s}, \tag{1.2.5}$$

two convergent series for $\mathrm{Re}\ s$ large enough. Also, define

$$L(s,\mathfrak{N}) = \psi_1\left(s + \frac{i\lambda}{2}\right)\psi_2\left(s - \frac{i\lambda}{2}\right) \tag{1.2.6}$$

and assume that the function $s \mapsto L(s,\mathfrak{N})$ extends as an entire function of s, polynomially bounded in vertical strips. Then, the distribution $\mathfrak{N}$ admits a decomposition into bihomogeneous components, given as

$$\mathfrak{N} = \frac{1}{8i\pi}\int_{\mathrm{Re}\ \rho=1}B_\varepsilon\left(\frac{\rho-i\lambda}{2}\right)L\left(\frac{2-\rho}{2},\mathfrak{N}\right)\mathrm{hom}_{\rho,-i\lambda}^{(\varepsilon)}\,d\rho: \tag{1.2.7}$$

one can also replace the line $\mathrm{Re}\ \rho = 1$ by any line $\mathrm{Re}\ \rho = c > 1$. It is Γ-invariant, i.e., a modular distribution, if and only if the function

$$L^\natural(s,\mathfrak{N}) = \frac{1}{2}B_\varepsilon\left(\frac{2-i\lambda}{2} - s\right)L(s,\mathfrak{N}) \tag{1.2.8}$$

satisfies the functional equation

$$L^\natural(s,\mathfrak{N}) = (-1)^\varepsilon\,L^\natural(1-s,\mathfrak{N}). \tag{1.2.9}$$

If such is the case, $\mathfrak{N}$ is of necessity a Hecke distribution, by which is meant that it is an eigendistribution of the operator T_N^{dist} for every $N = 1, 2, \dots$ and for $N = -1$.

Anticipating the relation, to be made explicit in Chapter 2, between the automorphic theories available in the plane and in the hyperbolic half-plane, let us make some points clear right now.

Remarks 1.2.1. (i) In Section 2.1, we shall define a linear map from automorphic distributions in the plane to automorphic functions in the hyperbolic half-plane: then, it will be shown that every Hecke eigenform $\mathcal{N}$, an eigenfunction of Δ for the eigenvalue $\frac{1+\lambda^2}{4}$, is the image of some Hecke distribution. The distribution $\mathfrak{N}$ contains more information than the Hecke eigenform $\mathcal{N}$, the knowledge of which only determines λ^2, not λ. The L-function of $\mathcal{N}$, as defined in a usual way, will be seen to coincide with the function $L(\,.\,, \mathfrak{N})$ as defined in (1.2.6), Note, on the other hand, that while $L^{\natural}(\,.\,, \mathfrak{N})$ is well-defined by (1.2.8), one could not substitute $\mathcal{N}$ for $\mathfrak{N}$ there, since this definition depends on λ, not only λ^2. Finally, the functional equations of the function $L^{\natural}(\,.\,, \mathfrak{N})$ and of the more classical function

$$L^*(s, \mathfrak{N}) = \pi^{-s}\, \Gamma\left(\frac{s+\varepsilon}{2} + \frac{i\lambda}{4}\right) \Gamma\left(\frac{s+\varepsilon}{2} - \frac{i\lambda}{4}\right) L(s, \mathfrak{N}) \tag{1.2.10}$$

are identical, since

$$L^{\natural}(s, \mathfrak{N}) = \frac{(-i)^{\varepsilon} \pi^{\frac{1-i\lambda}{2}}}{2\,\Gamma\left(\frac{s+\varepsilon}{2} - \frac{i\lambda}{4}\right) \Gamma\left(\frac{1-s+\varepsilon}{2} - \frac{i\lambda}{4}\right)}\, L^*(s, \mathfrak{N}) \tag{1.2.11}$$

and the factor of proportionality is invariant under the change of s to $1-s$. Still, the function $L^{\natural}(s, \mathfrak{N})$ contains slightly more information than the function $L^*(s, \mathfrak{N})$ (the only one available in the modular form environment), and one may simplify (1.2.7) as

$$\mathfrak{N} = \frac{1}{4i\pi} \int_{\text{Re } \rho=1} L^{\natural}\left(\frac{2-\rho}{2}, \mathfrak{N}\right) \hom_{\rho, -i\lambda}^{(\varepsilon)} d\rho. \tag{1.2.12}$$

(ii) On globally even distributions, the partial Fourier transformation $\mathcal{F}_1$ relates to the symplectic Fourier transformation by the equation

$$\left(\mathcal{F}_1^{-1} \mathcal{F}^{\text{symp}} h\right)(\xi, \eta) = \left(\mathcal{F}_1^{-1} h\right)(\eta, \xi):$$

it follows that

$$\mathcal{F}^{\text{symp}}\, \mathfrak{N}_{\chi, i\lambda} = \mathfrak{N}_{\chi^{-1}, -i\lambda}. \tag{1.2.13}$$

As a consequence of this equation, together with (1.2.5), (1.2.6), one has $L(s, \mathcal{F}^{\text{symp}} \mathfrak{N}) = L(s, \mathfrak{N})$ for every Hecke distribution $\mathfrak{N}$. Proposition 2.1.1 will give a better explanation of the fact.

(iii) The pair (λ, ε) is uniquely determined by $\mathfrak{N}$, but χ is not: indeed, as will be seen below, splitting the set of primes into two disjoint sets and changing the function $p \mapsto \chi(p)$ to $p \mapsto (\chi(p))^{-1} p^{i\lambda}$ on one of the two sets (then extending the modified version of χ as a character) does not change the distribution $\mathfrak{N}$.

(iv) As will be seen later as a consequence of (2.1.28) and of the fact that, in this equation, b_p is real, one has for every prime p either $|\chi(p)| = 1$ or $\bar{\chi}(p) = \chi(p) \, p^{-i\lambda}$ (whether the first condition always holds is the Ramanujan-Petersson conjecture): then, it follows from Remark (iii) that, if one sets $\chi_1(q) = (\bar{\chi}(q))^{-1}$ for $q \in \mathbb{Q}^\times$, one has $\mathfrak{N}_{\chi_1,i\lambda} = \mathfrak{N}_{\chi,i\lambda}$.

(v) From (1.1.38), the Eisenstein distribution $\frac{1}{2}\mathfrak{E}_{i\lambda}$ has, up to two extra terms, a totally similar Fourier decomposition, taking this time $\chi = 1$.

Proof of Theorem 1.2.2. One has

$$\left(\mathcal{F}_1^{-1} \left(h \circ \left(\begin{smallmatrix} 1 & 1 \\ 0 & 1 \end{smallmatrix} \right) \right) \right) (s, t) = e^{-2i\pi st} \left(\mathcal{F}_1^{-1} h \right) (s, t), \qquad (1.2.14)$$

from which the invariance of $\mathfrak{N}$ under $\left(\begin{smallmatrix} 1 & 1 \\ 0 & 1 \end{smallmatrix} \right)$ follows. However, invariance under $\left(\begin{smallmatrix} 0 & 1 \\ -1 & 0 \end{smallmatrix} \right)$ will necessitate a condition, similar to the ones occurring in so-called "converse theorems". Before analyzing it, let us observe, starting from the identity $\left(r^{2i\pi\mathcal{E}} h \right) (x, \xi) = r \, h(rx, r\xi)$, that, for $r > 0$, one has

$$\left(\mathcal{F}_1^{-1} \left(r^{2i\pi\mathcal{E}} h \right) \right) \left(\frac{m}{t}, nt \right) = \left(\mathcal{F}_1^{-1} h \right) \left(\frac{m}{rt}, rnt \right); \qquad (1.2.15)$$

after a change of variable $t \mapsto r^{-1} t$ in the integral (1.2.4) defining $\langle \mathfrak{N}, r^{2i\pi\mathcal{E}} h \rangle$, one obtains that this coincides with $r^{i\lambda} \langle \mathfrak{N}, h \rangle$: in other words, $\mathfrak{N}$ is homogeneous of degree $-1 - i\lambda$. Another obvious point is the identity $\mathfrak{N}(-x, \xi) = \chi(-1) \, \mathfrak{N}(x, \xi)$.

We now turn to the question of decomposing $\mathfrak{N}$ into bihomogeneous components. One may rewrite (1.2.4) as

$$\langle \mathfrak{N}, h \rangle = \frac{1}{4} \sum_{k \in \mathbb{Z}^\times} \phi(k) \int_{-\infty}^{\infty} |t|^{-1-i\lambda} \left(\mathcal{F}_1^{-1} h \right) \left(\frac{k}{t}, t \right) dt, \qquad (1.2.16)$$

setting

$$\phi(k) = \sum_{mn=k} \chi \left(\frac{m}{n} \right) |n|^{i\lambda}, \quad k \in \mathbb{Z}^\times \qquad (1.2.17)$$

(note that $\phi(1) = 2$). Permuting m and n, one sees that ϕ is unchanged if χ is changed to χ_1, with $\chi_1(s) = \chi \left(s^{-1} \right) |s|^{i\lambda}$: this justifies a remark made above, since we could also split the set of primes into two subsets and perform the change $\chi \mapsto \chi_1$ only on rational numbers which are products of powers of primes of the first category, leaving the other factor unchanged.

The case when $\varepsilon = 0$ can be treated in a way quite similar to the one used toward the decomposition into bihomogeneous components of the distribution $\frac{1}{2}\mathfrak{E}_\nu^{\mathrm{main}}$, as done in the proof of Theorem 1.1.8: only, we must replace ν by $i\lambda$ and

$\sigma_\nu(k)$ by $\frac{1}{2}\phi(k)$. Then, $\sum_{n\geq 1}\sigma_\nu(n)\,n^{\frac{\rho-\nu-2}{2}}$, as computed in (1.1.50), must be replaced by

$$\frac{1}{2}\sum_{k\geq 1}\phi(k)\,k^{\frac{\rho-i\lambda-2}{2}} = \psi_1\left(\frac{2-\rho+i\lambda}{2}\right)\psi_2\left(\frac{2-\rho-i\lambda}{2}\right) \tag{1.2.18}$$

(a convergent series when $-\mathrm{Re}\,\rho$ is large because the character χ is tempered), and (1.1.51) becomes now

$$\mathfrak{N}(x,\xi)$$
$$= \frac{1}{8i\pi}\int_{\mathrm{Re}\,\rho=c}\pi^{\frac{\rho-i\lambda-1}{2}}\frac{\Gamma(\frac{2-\rho+i\lambda}{4})}{\Gamma(\frac{\rho-i\lambda}{4})}\psi_1(\frac{2-\rho+i\lambda}{2})\psi_2(\frac{2-\rho-i\lambda}{2})\,\mathrm{hom}^{(0)}_{\rho,-i\lambda}(x,\xi)\,d\rho$$
$$= \frac{1}{8i\pi}\int_{\mathrm{Re}\,\rho=c}B_0\left(\frac{\rho-i\lambda}{2}\right)L(\frac{2-\rho}{2},\mathfrak{N})\,\mathrm{hom}^{(0)}_{\rho,-i\lambda}(x,\xi)\,d\rho. \tag{1.2.19}$$

The change of contour from $\mathrm{Re}\,\rho=c$ (with $-c$ large) to $\mathrm{Re}\,\rho=1$ does not require, this time, computing any residue, since the function $L(\,\boldsymbol{\cdot}\,,\mathfrak{N})$ is entire, while the factor $B_0(\frac{\rho-i\lambda}{2})$ kills the poles of the distribution $\mathrm{hom}^{(0)}_{\rho,-i\lambda}$ at $\rho=i\lambda,i\lambda-4,\dots$: we shall examine later the change of contour to a line $\mathrm{Re}\,\rho=c>1$.

Some changes are necessary when $\varepsilon=1$, replacing the study of the distribution $\mathfrak{S}_n$ in (1.1.41) by that of the distribution such that

$$\langle\mathfrak{S}_n^-,h\rangle = \frac{1}{2}\int_{-\infty}^{\infty}|t|^{-i\lambda-1}\left[(\mathcal{F}_1^{-1}h)\left(\frac{n}{t},t\right)-(\mathcal{F}_1^{-1}h)\left(-\frac{n}{t},t\right)\right]dt$$
$$= i\int_{-\infty}^{\infty}|\xi|^{-i\lambda-1}\,d\xi\int_{-\infty}^{\infty}h(x,\xi)\sin\frac{2\pi nx}{\xi}\,dx, \tag{1.2.20}$$

in other words

$$\mathfrak{S}_n^-(x,\xi) = i\,|\xi|^{-1-i\lambda}\sin\frac{2\pi nx}{\xi}: \tag{1.2.21}$$

the only difference with the preceding case is that one must apply (1.1.18) in place of (1.1.17). The decomposition (1.2.7) is proved, whether $\varepsilon=0$ or 1.

Since

$$\mathrm{hom}^{(\varepsilon)}_{\rho,\nu}(-\xi,x) = (-1)^\varepsilon\,\mathrm{hom}^{(\varepsilon)}_{2-\rho,\nu}(x,\xi), \tag{1.2.22}$$

that the functional equation (1.2.9) is equivalent to the identity $\mathfrak{N}(-\xi,x)=\mathfrak{N}(x,\xi)$ follows from the decomposition.

Whether $\varepsilon=0$ or 1, we verify now that the line $\mathrm{Re}\,\rho=1$ can be changed to $\mathrm{Re}\,\rho=c>1$. The poles of $B_\varepsilon(\frac{\rho-i\lambda}{2})$ to be taken care of are simple, at the points $2+2\varepsilon+i\lambda,6+2\varepsilon+i\lambda,\dots$: but let us recall [4, p. 107] that not only the function $L(s,\mathfrak{N})$ is entire, but so is the function $L^*(s,\mathfrak{N})$ in (1.2.10), one of the two extra factors defining it being $\Gamma(\frac{s+\varepsilon}{2}+\frac{i\lambda}{4})$. Dividing by this factor, evaluated at $s=\frac{2-\rho}{2}$, precisely kills the poles of $B_\varepsilon(\frac{\rho-i\lambda}{2})$ under consideration. On the other hand, the poles of $\hom_{2-\rho,-i\lambda}^{(\varepsilon)}$ to be taken care of are at $\rho = 2 + 2\varepsilon - i\lambda, 6 + 2\varepsilon - i\lambda, \dots$: they are killed with the help of the other factor $(\Gamma(\frac{s+\varepsilon}{2} - \frac{i\lambda}{4}))^{-1}$ present if using the entire function $L^*(s, \mathfrak{N})$ in place of $L(s, \mathfrak{N})$.

What remains to be done is to show that $\mathfrak{N}$ changes, for every integer $N \geq 1$, to a multiple, under the operator T_N^{dist}. For $N \geq 1$, one has (with $d > 0$)

$$\left(\mathcal{F}_1^{-1}\left((x, \xi) \mapsto N^{-\frac{1}{2}} h\left(\frac{dx - b\xi}{\sqrt{N}}, \frac{a\xi}{\sqrt{N}}\right)\right)\right)(s, t)$$

$$= d^{-1} \left(\mathcal{F}_1^{-1} h\right)\left(\frac{\sqrt{N}}{d} s, \frac{at}{\sqrt{N}}\right) \exp\left(2i\pi \frac{bst}{d}\right), \quad (1.2.23)$$

so that

$$\langle T_N^{\text{dist}} \mathfrak{N}, h \rangle = \frac{1}{4} \sum_{m, n \neq 0} \chi\left(\frac{m}{n}\right)$$

$$\sum_{\substack{ad=N, d>0 \\ b \bmod d}} d^{-1} \int_{-\infty}^{\infty} |t|^{-1-i\lambda} e^{2i\pi \frac{bmn}{d}} \left(\mathcal{F}_1^{-1} h\right)\left(\frac{\sqrt{N}}{d} \frac{m}{t}, \frac{ant}{\sqrt{N}}\right) dt. \quad (1.2.24)$$

After a change of variable $t \mapsto \frac{\sqrt{N}}{an} t$, one finds

$$\langle T_N^{\text{dist}} \mathfrak{N}, h \rangle = \frac{1}{4} \sum_{m, n \neq 0} \chi\left(\frac{m}{n}\right) \sum_{\substack{ad=N, d>0 \\ b \bmod d}} d^{-1} \left|\frac{an}{\sqrt{N}}\right|^{i\lambda}$$

$$e^{2i\pi \frac{bmn}{d}} \int_{-\infty}^{\infty} |t|^{-1-i\lambda} \left(\mathcal{F}_1^{-1} h\right)\left(\frac{a}{d} \frac{mn}{t}, t\right) dt. \quad (1.2.25)$$

It is sufficient to examine further the case when N coincides with a prime number p: then, one can have ($a = 1, d = p, b \bmod p$) or ($a = p, d = 1, b = 0$). In the first case, one has $\sum_b e^{2i\pi \frac{bmn}{d}} = 0$ if p does not divide mn, and the same sum is p if $p|mn$. In the second case, this sum reduces to 1. Hence, (1.2.25) simplifies as

$$\langle T_p^{\text{dist}} \mathfrak{N}, h \rangle = \frac{1}{4} \sum_{k \in \mathbb{Z}^\times} a_k \int_{-\infty}^{\infty} |t|^{-1-i\lambda} \left(\mathcal{F}_1^{-1} h\right)\left(\frac{k}{t}, t\right) dt, \quad (1.2.26)$$

with

$$a_k = \sum_{mn=pk} \left|\frac{n}{\sqrt{p}}\right|^{i\lambda} \chi\left(\frac{m}{n}\right) + \sum_{pmn=k} |n\sqrt{p}|^{i\lambda} \chi\left(\frac{m}{n}\right)$$

$$= p^{-\frac{i\lambda}{2}} \phi(pk) + p^{\frac{i\lambda}{2}} \phi\left(\frac{k}{p}\right), \quad (1.2.27)$$

with the convention that $\phi(k) = 0$ unless $k \in \mathbb{Z}^\times$. Given any integer $k \neq 0$, one has

$$\{(m, n): mn = pk\} = \{(pm_1, n): m_1 n = k\} \cup \{(m, pn_1): mn_1 = k\}, \quad (1.2.28)$$

not a disjoint union if $p \mid k$: the two sets intersect along the set $\{(pm_1, pn_1): m_1 n_1 = \frac{k}{p}\}$, which leads to the equation

$$\phi(pk) = \left[\chi(p) + p^{i\lambda} \chi(p^{-1}) \right] \phi(k) - p^{i\lambda} \phi\left(\frac{k}{p} \right), \quad k \in \mathbb{Z}^\times. \quad (1.2.29)$$

It follows that

$$p^{-\frac{i\lambda}{2}} \phi(pk) + p^{\frac{i\lambda}{2}} \phi\left(\frac{k}{p} \right) = \left[p^{-\frac{i\lambda}{2}} \chi(p) + p^{\frac{i\lambda}{2}} \chi(p^{-1}) \right] \phi(k). \quad (1.2.30)$$

Coupling this equation with the pair of equations (1.2.26), (1.2.27), one obtains the equation

$$T_p^{\text{dist}} \mathfrak{N} = \left[\chi(p) p^{-\frac{i\lambda}{2}} + \chi(p^{-1}) p^{\frac{i\lambda}{2}} \right] \mathfrak{N}, \quad (1.2.31)$$

proving that $\mathfrak{N}$ is indeed a Hecke eigenform if it is automorphic (i.e., invariant under $\left(\begin{smallmatrix} 0 & 1 \\ -1 & 0 \end{smallmatrix} \right)$). This completes the proof of Theorem 1.2.2. $\qquad \square$

Let us sum up the main results of the chapter just concluded as follows. Modular distributions, either of Eisenstein or of Hecke type, can be defined in the plane without appealing to a previously defined notion of non-holomorphic modular form: the relation between the two species of notions will be treated in the next chapter. Each modular distribution admits two types of expansions. First, the "Fourier series" expansion, to wit ((1.1.38) and (1.2.4))

$$\frac{1}{2} \mathfrak{E}_\nu(x, \xi) = \frac{1}{2} \zeta(-\nu) |\xi|^{-\nu - 1}$$

$$+ \frac{1}{2} \zeta(1 - \nu) |x|^{-\nu} \delta(\xi) + \frac{1}{2} \sum_{k \neq 0} \sigma_\nu(|k|) |\xi|^{-1-\nu} \exp\left(2i\pi \frac{kx}{\xi} \right),$$

$$\mathfrak{N}(x, \xi) = \frac{1}{4} \sum_{m, n \neq 0} |n|^{i\lambda} \chi\left(\frac{m}{n} \right) |\xi|^{-1-i\lambda} \exp\left(2i\pi \frac{mnx}{\xi} \right). \quad (1.2.32)$$

On the other hand, according to Theorem 1.1.8 and Theorem 1.2.2 (or (1.2.12)), every modular distribution admits a continuous expansion (up to the addition of a few special terms in the case of Eisenstein distributions) into bihomogeneous functions, the coefficient of which is provided by the L-function relative to the given modular distribution.

Let us set $h_{\nu, q}(x, \xi) = |\xi|^{-1-\nu} e^{2i\pi \frac{qx}{\xi}}$, a symbol the analysis of which (with an emphasis on sharp products of such) will keep us busy throughout Chapter 4. The functions $h_{\nu,q}$ with $q \in \mathbb{Z}^\times$ are the basic terms of expansions (1.2.32) into Fourier series. They are linked to bihomogeneous functions by the pair of formulas

$$\mathrm{hom}_{\rho,\nu}^{(\varepsilon)} = B_\varepsilon \left(\frac{2-\rho-\nu}{2} \right) \int_{-\infty}^{\infty} |q|_\varepsilon^{\frac{-\rho-\nu}{2}} h_{\nu-1,q} dq, \quad \mathrm{Re}\ \nu > 0,\ 0 < \mathrm{Re}\ \frac{\rho+\nu}{2} < 1,$$

(1.2.33)

as it follows from (1.1.6), while, from Lemma 1.1.1, if $q \neq 0$, $a > 0$ and $x\xi \neq 0$,

$$h_{\nu,q}(x,\xi) = \frac{1}{4i\pi} \sum_{\varepsilon=0,1} (-1)^\varepsilon \int_{\mathrm{Re}\ \mu=a} B_\varepsilon(1-\mu) |q|_\varepsilon^{-\mu} \mathrm{hom}_{\nu-2\mu+2,-\nu}^{(\varepsilon)}(x,\xi)\, d\mu,$$

(1.2.34)

which can be rewritten in the following form, more immediately comparable to (1.1.46) and (1.2.7): if $q \neq 0$, $b < \mathrm{Re}\ \nu + 2$ and $x\xi \neq 0$,

$$h_{\nu,q}(x,\xi) = \frac{1}{8i\pi} \sum_{\varepsilon=0,1} (-1)^\varepsilon \int_{\mathrm{Re}\ \rho=b} B_\varepsilon(\frac{\rho-\nu}{2}) |q|_\varepsilon^{\frac{\rho-\nu-2}{2}} \mathrm{hom}_{\rho,-\nu}^{(\varepsilon)}(x,\xi)\, d\rho. \quad (1.2.35)$$

Chapter 2

From the plane to the half-plane

This chapter provides a dictionary from automorphic distribution theory (in the plane) to automorphic function theory (in the hyperbolic half-plane). More precisely, one defines, with the help of the so-called dual Radon transformation, a linear operator $\Theta = (\Theta_0, \Theta_1)$ from automorphic distributions to pairs of automorphic functions: a two-component operator is needed because two distributions in the plane which are images of each other under the symplectic Fourier transformation have the same image under Θ_0. We show that the Θ_0-transforms of Eisenstein, or Hecke, distributions are Eisenstein, or Maass-Hecke modular forms, and that the notions of L-functions defined in the two environments are fully coherent.

We also transfer (non-automorphic) bihomogeneous functions, which leads to further decompositions of Eisenstein or Maass-Hecke modular forms. There, new functions $F_{\rho,\nu}^{(\varepsilon)}$ or, with a different normalization, $\Psi_{\rho,\nu}^{(\varepsilon)}$, show up: ν enters the (generalized) eigenvalue $\frac{1-\nu^2}{4}$ relative to the modular Laplacian Δ, while ρ is an eigenvalue associated to the operator

$$\mathrm{Eul}^{\Pi} = \frac{1}{i\pi} \left(z\frac{\partial}{\partial z} + \bar{z}\frac{\partial}{\partial \bar{z}} \right). \tag{2.0.1}$$

The functions $F_{\rho,\nu}^{(\varepsilon)}$ are much more complicated than the distributions in the plane they originate from: the quite simple spectral-theoretic role of L-functions in automorphic distribution theory does not stay so simple in the automorphic function environment.

On the other hand, these functions led in [39, chap. 4] to the construction of a new class of automorphic functions in the hyperbolic half-plane with interesting singularities on the set of lines congruent to the line $(0, i\infty)$, a task briefly implemented in the last section of this chapter for the sake of completeness: only the case of functions invariant under the symmetry $z \mapsto -\bar{z}$ had been considered in the given reference. Disregarding completely Sections 2.2 and 2.3 would not harm understanding the rest of the book. But Section 2.1 and the notation, relating Hecke distributions and Hecke eigenforms, introduced there, is essential for the sequel.

2.1   Modular distributions and non-holomorphic modular forms

With the exception of Theorem 2.1.2 below (a converse of Theorem 1.2.2), all non classical facts in this section needed in this book have been detailed in [34] and, with a refreshed proof, in Sections 2.1 and 3.1 of [39]. Before dealing with the automorphic situation, we relate general analysis on the plane to that on the hyperbolic half-plane Π.

Recall that the standard Iwasawa decomposition NAK of $G = SL(2, \mathbb{R})$ involves the subgroup $K = SO(2)$ and the subgroups N and A consisting respectively of all matrices $\left(\begin{smallmatrix} 1 & b \\ 0 & 1 \end{smallmatrix}\right)$ with $b \in \mathbb{R}$ and $\left(\begin{smallmatrix} a & 0 \\ 0 & a^{-1} \end{smallmatrix}\right)$ with $a > 0$; one considers also the group M consisting of the two matrices $\pm \left(\begin{smallmatrix} 1 & 0 \\ 0 & 1 \end{smallmatrix}\right)$. The homogeneous space G/K can be identified with the half-plane Π with base-point i since K is the subgroup of G leaving this point fixed. The generic point (x, ξ) of $\mathbb{R}^2 \backslash \{0\}$, regarded as the left column of the matrix $g = \left(\begin{smallmatrix} x & b \\ \xi & d \end{smallmatrix}\right)$, can be identified with the class gN: further dividing by M, we may regard G/MN as the quotient of the former space by the equivalence $(x, \xi) \sim (-x, -\xi)$, so that functions on G/MN become exactly even functions in $\mathbb{R}^2$.

The dual Radon transform V^* — a concept which can be defined and studied in considerable generality [11] — is the map from continuous even functions in $\mathbb{R}^2$ to functions on Π defined as

$$(V^* h)(g \cdot i) = \int_K h((gk) \cdot \left(\begin{smallmatrix} 1 \\ 0 \end{smallmatrix}\right)) \, dk \qquad (2.1.1)$$

or, making the choice $g = \begin{pmatrix} y^{\frac{1}{2}} & y^{-\frac{1}{2}} x \\ 0 & y^{-\frac{1}{2}} \end{pmatrix}$,

$$(V^* h)(x + iy) = \frac{1}{2\pi} \int_0^{2\pi} h\left(\pm \begin{pmatrix} y^{\frac{1}{2}} \cos \frac{\theta}{2} - x\, y^{-\frac{1}{2}} \sin \frac{\theta}{2} \\ -y^{-\frac{1}{2}} \sin \frac{\theta}{2} \end{pmatrix} \right) d\theta. \qquad (2.1.2)$$

In other words, $(V^* h)(i) = \langle d\sigma_i, h \rangle$ if $d\sigma_i$ is the rotation-invariant measure on the unit circle with total mass 1; more generally, $(V^* h)(z) = \langle d\sigma_z, h \rangle$ if $d\sigma_z$ is the measure supported in the ellipse $\{(x, \xi) : \frac{|x - z\xi|^2}{\operatorname{Im} z} = 1\}$ (one irritant is that one cannot use simultaneously the coordinates x, ξ in $\mathbb{R}^2$ and $x + iy$ in Π), invariant under the group of linear transformations preserving this ellipse and with total mass 1.

The Radon transform V, which will be useful later as well, works in the other direction, from functions f on Π to even functions on $\mathbb{R}^2$: it is defined by the equation

$$(Vf)(g \cdot (\begin{smallmatrix} 1 \\ 0 \end{smallmatrix})) = \int_N f((gn) \cdot i) \, dn, \tag{2.1.3}$$

with $dn = db$ if $n = (\begin{smallmatrix} 1 & b \\ 0 & 1 \end{smallmatrix})$. Recall that the hyperbolic distance d on Π is $SL(2, \mathbb{R})$-invariant and characterized as such by its special case $\cosh d(z, i) = \frac{1+|z|^2}{2 \operatorname{Im} z}$. The integral is convergent, yielding a continuous function Vf if, say,

$$|f(z)| \leq C \, (\cosh d(i, z))^{-\frac{1}{2}-\varepsilon}$$

for some $\varepsilon > 0$: indeed, when $g = (\begin{smallmatrix} a & 0 \\ 0 & a^{-1} \end{smallmatrix})$, it is immediate that $2 \cosh d(i, (gn) \cdot i)$ $= a^{-2} + a^2 (1 + b^2)$, a formula which remains true if g is replaced by kg with $k \in K$. Explicitly, completing if $x \neq 0$ the column $(\begin{smallmatrix} x \\ \xi \end{smallmatrix})$ into the matrix $(\begin{smallmatrix} x & 0 \\ \xi & x^{-1} \end{smallmatrix})$, one has

$$(Vf)(\pm (\begin{smallmatrix} x \\ \xi \end{smallmatrix})) = \frac{1}{\pi} \int_{-\infty}^{\infty} f \left(\frac{x^2 (i + b)}{x\xi (i + b) + 1} \right) db, \quad x \neq 0. \tag{2.1.4}$$

One pair of transformations, working in the same direction as V^*, will be of considerable interest in this book. It is the pair (Θ_0, Θ_1) of maps from even functions, or even tempered distributions on $\mathbb{R}^2$, to functions on Π, defined by the equations

$$(\Theta_0 \mathfrak{S})(z) = \left\langle \mathfrak{S}, (x, \xi) \mapsto 2 \exp\left(-2\pi \frac{|x - z\,\xi|^2}{\operatorname{Im} z} \right) \right\rangle, \quad \Theta_1 \mathfrak{S} = \Theta_0 \, (2i\pi \mathcal{E} \mathfrak{S}). \tag{2.1.5}$$

This pair of operators has a useful interpretation in terms of pseudodifferential analysis and of the canonical set of coherent states of the metaplectic representation, and the same is true of its adjoint: we shall come back to it in the next chapter. What we need to know about is the (immediate) covariance of this pair of maps, to wit the pair of relations

$$\Theta_\kappa \, (\mathfrak{S} \circ g) = (\Theta_\kappa \mathfrak{S}) \circ g, \quad g \in G, \quad \kappa = 0, 1, \tag{2.1.6}$$

in which $g = (\begin{smallmatrix} a & b \\ c & d \end{smallmatrix}) \in G = SL(2, \mathbb{R})$ acts on Π by means of the equation $(\begin{smallmatrix} a & b \\ c & d \end{smallmatrix}) \cdot z = \frac{az+b}{cz+d}$, on $\mathbb{R}^2$ by means of the equation $(\begin{smallmatrix} a & b \\ c & d \end{smallmatrix}) \cdot (\begin{smallmatrix} x \\ \xi \end{smallmatrix}) = (\begin{smallmatrix} ax+b\xi \\ cx+d\xi \end{smallmatrix})$. Also, recalling that we have already defined, in (1.1.19), the Euler operator $2i\pi\mathcal{E}$ on $\mathbb{R}^2$, we need the fundamental transfer property expressed by the equation

$$\Theta_\kappa \, (\pi^2 \mathcal{E}^2 \mathfrak{S}) = \left(\Delta - \frac{1}{4} \right) \Theta_\kappa \, \mathfrak{S}, \tag{2.1.7}$$

with $\Delta = (z - \bar{z})^2 \frac{\partial^2}{\partial z \partial \bar{z}} = -y^2 \left(\frac{\partial^2}{\partial x^2} + \frac{\partial^2}{\partial y^2} \right)$ if $z = x + iy$. Setting $\rho = \frac{|x - z\,\xi|^2}{\operatorname{Im} z}$, this identity can be written as

$$-\left(\rho \frac{d}{d\rho} + \frac{1}{2} \right)^2 k(\rho) = \left(\Delta - \frac{1}{4} \right) k(\rho) \tag{2.1.8}$$

for every C^2 function k: the calculation of the right-hand side presents no difficulty since, taking advantage of the invariance of both operators involved under the appropriate actions of G, one may assume that $(x, \xi) = (1, 0)$, so that $\rho = (\text{Im } z)^{-1}$.

The operator $\mathcal{E}$ is essentially self-adjoint on $L^2(\mathbb{R}^2)$ (i.e., it admits a unique self-adjoint extension) if given the initial domain $C_0^\infty(\mathbb{R}^2 \backslash \{0\})$. This makes it possible to define, in the spectral-theoretic sense, functions of $\mathcal{E}$. The map Θ_0 connects to the dual Radon transformation by the equation

$$\Theta_0 = V^* (2\pi)^{\frac{1}{2} - i\pi \mathcal{E}} \, \Gamma\left(\frac{1}{2} + i\pi \mathcal{E}\right). \tag{2.1.9}$$

To prove this, one first decomposes the function $h \in \mathcal{S}_{\text{even}}(\mathbb{R}^2)$ the two sides of (2.1.9) are to be tested on into homogeneous components $h_{i\lambda}$, as will be done in (3.2.1), after which, performing a change of variable in Euler's integral formula for the Gamma function (details are given in [39, p.52] if so desired), one obtains

$$(\Theta_0 \, h_{i\lambda})(z) = (2\pi)^{\frac{i\lambda - 3}{2}} \Gamma(\frac{1 - i\lambda}{2}) \int_{\mathbb{R}^2} h(x, \xi) \left(\frac{|x - z\xi|^2}{\text{Im } z}\right)^{\frac{i\lambda - 1}{2}} dx \, d\xi. \tag{2.1.10}$$

On the other hand, the operator $(2\pi)^{\frac{1}{2} - i\pi \mathcal{E}} \Gamma\left(\frac{1}{2} + i\pi \mathcal{E}\right)$ acts on $h_{i\lambda}$ as multiplication by the scalar $(2\pi)^{\frac{1 + i\lambda}{2}} \Gamma(\frac{1 - i\lambda}{2})$. Temporarily denoting z as $z = x' + iy$, one writes, with the help of (3.2.1) and (2.1.2),

$$(V^* h_{i\lambda})(z) = \frac{1}{4\pi^2} \int_0^{2\pi} d\theta \int_0^\infty t^{i\lambda} h\left(\begin{pmatrix} t\left(y^{\frac{1}{2}} \cos\frac{\theta}{2} - x'y^{-\frac{1}{2}} \sin\frac{\theta}{2}\right) \\ t\left(-y^{-\frac{1}{2}} \sin\frac{\theta}{2}\right) \end{pmatrix}\right) dt : \tag{2.1.11}$$

then, one performs the change of variables defined by the equations

$$x = t\left(y^{\frac{1}{2}} \cos\frac{\theta}{2} - x'y^{-\frac{1}{2}} \sin\frac{\theta}{2}\right), \quad \xi = t\left(-y^{-\frac{1}{2}} \sin\frac{\theta}{2}\right), \tag{2.1.12}$$

so that $t^2 = \frac{|x - z\xi|^2}{\text{Im } z}$ and $dx \, d\xi = t \, dt \, d\theta$. The identity (2.1.9) follows.

Let $\mathcal{G}$ be the rescaled version of the symplectic Fourier transformation (1.1.29) defined on $L^1(\mathbb{R}^2)$ (next on $\mathcal{S}'(\mathbb{R}^2)$) as $\mathcal{G} = 2^{2i\pi\mathcal{E}} \mathcal{F}^{\text{symp}} = 2^{i\pi\mathcal{E}} \mathcal{F}^{\text{symp}} 2^{-i\pi\mathcal{E}}$, i.e.,

$$(\mathcal{G}h)(x, \xi) = 2 \int_{\mathbb{R}^2} h(y, \eta) \, e^{4i\pi(x\eta - y\xi)} \, dy \, d\eta. \tag{2.1.13}$$

One then has the identities

$$\Theta_0 \, (\mathcal{G} \, \mathfrak{S}) = \Theta_0 \, \mathfrak{S}, \quad \Theta_1 \, (\mathcal{G} \, \mathfrak{S}) = -\Theta_1 \, \mathfrak{S}. \tag{2.1.14}$$

If $\mathfrak{S} \in \mathcal{S}'_{\text{even}}(\mathbb{R}^2)$, the image of $\mathfrak{S}$ under Θ_0 (resp. Θ_1) characterizes the part of $\mathfrak{S}$ invariant (resp. changing to its negative) under $\mathcal{G}$. The proof of this fact, fundamental for our purposes, is to be postponed to the end of Section 3.1, after we have related the transforms Θ_0 and Θ_1 to pseudodifferential analysis:

We can now consider the automorphic situation, recalling that a tempered distribution $\mathfrak{S}$ is automorphic if it is invariant under the action of the group $\Gamma = SL(2, \mathbb{Z})$ by linear changes of coordinates. Because of the covariance formula (2.1.6), its Θ-transform will consist of a pair of automorphic functions in Π. A modular distribution is an automorphic distribution homogeneous of some degree $-1 - \nu$. As a consequence of (2.1.7), its Θ_0-transform is a (possibly generalized) eigenfunction of Δ for the eigenvalue $\frac{1-\nu^2}{4}$, in other words a non-holomorphic modular form; so is its Θ_1-transform, but no novel information is carried by it if ν (not only ν^2) is known. Note, in view of the first equation (2.1.14) and of the identity $\mathcal{G}(2i\pi\mathcal{E}) = (-2i\pi\mathcal{E})\mathcal{G}$, that two modular distributions, one the image of the other under $\mathcal{G}$, have the same Θ_0-transform, and that their degrees of homogeneity are then $-1 - \nu$ and $-1 + \nu$ for some ν.

We make all this explicit, for which we need to give a crash course on automorphic function theory in Π, limiting ourselves to what is absolutely needed in the sequel. Very nice presentations of this theory (accessible to non-experts, including the present author) are to be found in [4, 14, 16] and elsewhere. The first thing to recall is that it is useful to complete the Riemann zeta function $\zeta(s) = \sum_{n \geq 1} n^{-s}$ (a convergent series if Re $s > 1$) as the function

$$\zeta^*(s) = \pi^{-\frac{s}{2}} \Gamma\left(\frac{s}{2}\right) \zeta(s), \tag{2.1.15}$$

which extends as a meromorphic function in the entire plane, the only poles of which lie at 1 and 0 and are simple: moreover, it satisfies the fundamental functional equation $\zeta^*(s) = \zeta^*(1 - s)$. A great bulk of non-holomorphic modular form theory is made up of the so-called Eisenstein series. If Re $\nu < -1$, the series

$$E_{\frac{1-\nu}{2}}(z) = \frac{1}{2} \sum_{\substack{m,n \in \mathbb{Z} \\ (m,n)=1}} \left(\frac{|mz - n|^2}{\text{Im } z}\right)^{\frac{\nu-1}{2}} \tag{2.1.16}$$

(where (m, n) denotes the g.c.d. of the pair m, n) is convergent, and its sum is a non-holomorphic modular form for the eigenvalue $\frac{1-\nu^2}{4}$. It is periodic of period 1, and the function

$$E^*_{\frac{1-\nu}{2}}(z) = \zeta^*(1 - \nu)\, E_{\frac{1-\nu}{2}}(z) \tag{2.1.17}$$

admits the Fourier series expansion (with respect to $x = $ Re z)

$$E^*_{\frac{1-\nu}{2}}(x + iy)$$
$$= \zeta^*(1 - \nu)\, y^{\frac{1-\nu}{2}} + \zeta^*(1 + \nu)\, y^{\frac{1+\nu}{2}} + 2 y^{\frac{1}{2}} \sum_{k \neq 0} |k|^{-\frac{\nu}{2}}\, \sigma_\nu(|k|)\, K_{\frac{\nu}{2}}(2\pi\, |k|\, y)\, e^{2i\pi kx},$$

$$\tag{2.1.18}$$

with $\sigma_\nu(|k|) = \sum_{1 \le d | k} d^\nu$: this provides its analytic continuation as a function of ν.

An easy separation of variables relative to the operator Δ shows that every generalized eigenfunction of Δ for the eigenvalue $\frac{1-\nu^2}{4}$, periodic of period 1 with respect to $x = \operatorname{Re} z$, admits a Fourier expansion of the kind

$$f(x + iy) = a_+ \, y^{\frac{1-\nu}{2}} + a_- \, y^{\frac{1+\nu}{2}} + y^{\frac{1}{2}} \sum_{k \ne 0} b_k \, e^{2i\pi k x} \, K_{\frac{\nu}{2}}(2\pi \, |k| \, y), \qquad (2.1.19)$$

unless it is far from being bounded, as $y \to \infty$, by some power of $1+y$ (in which case one would have to substitute for $K_{\frac{\nu}{2}}$ another linear combination of the functions $I_{\pm \frac{\nu}{2}}$).

We now introduce the *standard* fundamental domain D of Γ, consisting of all points $z \in \Pi$ with $-\frac{1}{2} < \operatorname{Re} z < \frac{1}{2}$ and $|z| > 1$: it satisfies the property that no two distinct points of D are congruent under Γ (i.e., the images of each other under some transformation in Γ), while every point of Π is congruent to at least one point in the topological closure of D. Outside a set of measure zero, an automorphic function is then characterized by its restriction to D: using in Π the invariant measure $dm(x + iy) = y^{-2} dx \, dy$, one may then introduce the Hilbert space, denoted as $L^2(\Gamma \backslash \Pi)$, which is just $L^2(D, dm)$ in terms of these restrictions. Standard Hilbert space techniques then show that there exists an at most — an adverb which can be dispensed with thanks to Selberg's trace formula — countable set of linearly independent modular forms (the so-called Maass forms), which satisfy the property (not shared by Eisenstein series) that they are rapidly decreasing as $y = \operatorname{Im} z \to \infty$, in a way uniform with respect to $x = \operatorname{Re} z$: in other words, the first two coefficients a_+ and a_- are zero. As can be seen, ν must be pure imaginary: one usually sets $\nu = i\lambda$ with, say, $\lambda > 0$, or $\nu = i\lambda_r$ since the possible λ's make up a sequence going to ∞. It is not known whether (in the case of Γ, the only discrete group under consideration here) there may exist linearly independent Maass forms corresponding to the same eigenvalue $\frac{1+\lambda_r^2}{4}$.

A much clearer picture emerges after one has introduced the so-called Hecke operators T_N, $N \ge 1$, defined by the equation

$$(T_N f)(z) = N^{-\frac{1}{2}} \sum_{\substack{ad = N, \, d > 0 \\ b \bmod d}} f\left(\frac{az + b}{d}\right): \qquad (2.1.20)$$

they can be shown to commute pairwise, while commuting with Δ and with the parity operator T_{-1} defined by the equation $(T_{-1} f)(z) = f(-\bar{z})$. One has the fundamental formal relation between Dirichlet series

$$\sum_{N \ge 1} N^{-s} T_N = \prod_{p \text{ prime}} \left(1 - p^{-s} T_p + p^{-2s}\right)^{-1}, \qquad (2.1.21)$$

a compact form of an infinite set of polynomial relations among the Hecke operators. Maass forms which are joint eigenfunctions of all Hecke operators (including T_{-1}) are called Hecke eigenforms. Just as Δ, the Hecke operators are self-adjoint in the space $L^2(\Gamma\backslash\Pi)$. Consider a true eigenvalue $\frac{1+\lambda_r^2}{4}$ of Δ, to wit one for which some Maass forms do exist. Then, standard Hilbert space methods (the theory of commuting families of compact self-adjoint operators) show that there exists a finite family $(\mathcal{M}_{r,\ell})_{1\leq\ell\leq\kappa}$, where κ is an r-dependent finite number, of Hecke eigenforms making up an orthonormal basis of the eigenspace of Δ corresponding to the given eigenvalue. Another normalization of Hecke eigenforms (to be referred to as Hecke's normalization) is very useful: it is the one for which one substitutes for $\mathcal{M}_{r,\ell}$ the proportional Hecke eigenform $\mathcal{N}_{r,\ell}$ such that the coefficient b_1 from its Fourier expansion (2.1.19) is 1: then, one has the collection of identities $T_N\mathcal{N}_{r,\ell} = b_N\,\mathcal{N}_{r,\ell}$. Again, a self-adjointness argument shows that all coefficients b_k of such a Hecke eigenform must be real numbers.

The spectral theorem relative to a certain natural self-adjoint realization of the operator Δ in $L^2(\Gamma\backslash\Pi)$, together with the collection of Hecke operators, makes it possible to show that every automorphic function $f \in L^2(\Gamma\backslash\Pi)$ admits a so-called Roelcke-Selberg expansion, to wit a decomposition of the kind

$$f(z) = \Phi^0 + \frac{1}{8\pi}\int_{-\infty}^{\infty}\Phi(\lambda)\,E_{\frac{1-i\lambda}{2}}(z)\,d\lambda + \sum_{r\geq 1}\sum_{\ell}\Phi^{r,\ell}\,\mathcal{M}_{r,\ell}(z). \qquad (2.1.22)$$

In Chapter 5, we shall prove an analogous expansion for a large class of automorphic distributions. What is much harder to prove, at least in the case of automorphic functions invariant under the map $z \mapsto -\bar{z}$ (it requires using the so-called Selberg's trace formula), is that there does exist an infinite sequence $\left(\frac{1+\lambda_r^2}{4}\right)_{r\geq 1}$ of true eigenvalues of Δ. Note that Eisenstein series, as defined for general values of ν by their expansion (2.1.18), can never be *true* eigenfunctions of Δ, in that they never lie in $L^2(\Gamma\backslash\Delta)$ (as seen by an application of Hadamard's theorem that $\zeta(s)$ has no zero on the line Re $s = 0$, while one trivially has $\zeta(s) \neq 0$ for Re $s > 1$ in view of the Euler product expansion $(\zeta(s))^{-1} = \prod_{p\,\text{prime}}(1 - p^{-s})$).

The following proposition establishes the link between modular distribution theory and non-holomorphic modular form theory. Before stating it, we define the rescaled version of a tempered distribution $\mathfrak{S}$ as $\mathfrak{S}^{\text{resc}} = 2^{-\frac{1}{2}+i\pi\mathcal{E}}\mathfrak{S}$: the operator $\mathcal{G}$ is the conjugate of the operator $\mathcal{F}^{\text{symp}}$ under the rescaling operator. In particular, $\mathfrak{E}_\nu^{\text{resc}} = 2^{\frac{-1-\nu}{2}}\mathfrak{E}_\nu$. The rescaling operator cannot be dispensed with when interested in the Weyl calculus, since the "most natural" one and two-dimensional Gaussian functions in this context are $x \mapsto 2^{\frac{1}{4}}e^{-\pi x^2}$ and $(x,\xi) \mapsto 2\,e^{-2\pi(x^2+\xi^2)}$: the second one (the symbol of the operator of orthogonal projection on the first one) is the rescaled version of $2\,e^{-\pi(x^2+\xi^2)}$. There are also reasons of elementary algebraic number theory leading to the same two types of normalization of Gaussian functions [20, p. 282].

Proposition 2.1.1. *For every $\nu \in \mathbb{C}$, $\nu \neq \pm 1$, one has $\Theta_0 (\mathfrak{E}_\nu^{\mathrm{resc}}) = E^*_{\frac{1-\nu}{2}}$. Next, let $\mathcal{N}$ be a cusp-form with the Fourier expansion*

$$\mathcal{N}(x+iy) = y^{\frac{1}{2}} \sum_{k \neq 0} b_k \, K_{\frac{i\lambda}{2}}(2\pi |k| y) \, e^{2i\pi kx} : \qquad (2.1.23)$$

this only defines the number λ^2 and, choosing $\lambda = \sqrt{\lambda^2}$, we define a pair $(\mathfrak{N}_\pm)$ of distributions in the plane by setting, for $h \in \mathcal{S}(\mathbb{R}^2)$,

$$\langle \mathfrak{N}_\pm, h \rangle = \frac{1}{2} \sum_{k \neq 0} |k|^{\frac{\pm i\lambda}{2}} b_k \int_{-\infty}^{\infty} |t|^{-1\mp i\lambda} \left(\mathcal{F}_1^{-1} h\right) \left(\frac{k}{t}, t\right) dt. \qquad (2.1.24)$$

The distribution $\mathfrak{N}_\pm$ is a modular distribution, homogeneous of degree $-1 \mp i\lambda$. The two distributions are related by the identity $\mathcal{F}^{\mathrm{symp}} \, \mathfrak{N}_\pm = \mathfrak{N}_\mp$. The Θ-transform of the rescaled version $\mathfrak{N}_\pm^{\mathrm{resc}}$ is given by the equation

$$\left(\Theta_0 \, \mathfrak{N}_\pm^{\mathrm{resc}}\right)(z) = \mathcal{N}(z). \qquad (2.1.25)$$

Proof. In view of the (similar) Fourier expansions (1.2.32) and (1.2.4) of Eisenstein and Hecke distributions, the statement reduces to the results of computations involving the functions $h_{\nu,k}(x,\xi) = |\xi|^{-1-\nu} \exp\left(2i\pi \frac{k}{\xi}\right)$: these will be given (in a slightly more general version) in (4.1.2) and (4.6.3). □

Remarks 2.1.1. (i) From (1.1.30), one has $\mathcal{G} \mathfrak{E}_\nu^{\mathrm{resc}} = \mathfrak{E}_{-\nu}^{\mathrm{resc}}$: the invariance of the function $E^*_{\frac{1-\nu}{2}}$ under the change of ν to $-\nu$, which follows then from (2.1.14), can also be seen from the Fourier expansion (2.1.18). In the same way, the modular distributions $\mathfrak{N}_\pm^{\mathrm{resc}}$ are $\mathcal{G}$-related and have the same Θ_0-transform.

(ii) While, as indicated above, we denote as $\left(\frac{1+\lambda_r^2}{4}\right)_{r \geq 1}$ the increasing sequence of true eigenvalues of the automorphic Laplacian and, for each r, we denote as $(\mathcal{N}_{r,\ell})_\ell$ the finite associated set (unique up to permutation) of Hecke eigenforms, normalized in Hecke's way, the following slight change is necessary when dealing with modular distributions: with the same convention about (r, ℓ), and given a Hecke eigenform $\mathcal{N} = \mathcal{N}_{r,\ell}$, we now denote as $\mathfrak{N}_{r,\ell}$ (resp. $\mathfrak{N}_{-r,\ell}$) the modular distribution (a Hecke distribution as will be seen presently) defined as $\mathfrak{N}_+$ (resp. $\mathfrak{N}_-$) by (2.1.24). In other words, a proper "total" set of Hecke distributions will then be the set $(\mathfrak{N}_{r,\ell})_{r,\ell}$ where, this time, the condition on r is $r \neq 0$. We shall always assume that, for $r \geq 1$, λ_r is the positive square root of λ_r^2, and it will be convenient to set $\lambda_{-r} = -\lambda_r$ so that, whether $r \geq 1$ or $r \leq -1$, one should always have

$$\langle \mathfrak{N}_{r,\ell}, h \rangle = \frac{1}{2} \sum_{k \neq 0} |k|^{\frac{i\lambda_r}{2}} b_k \int_{-\infty}^{\infty} |t|^{-1-i\lambda_r} \left(\mathcal{F}_1^{-1} h\right) \left(\frac{k}{t}, t\right) dt. \qquad (2.1.26)$$

A clear understanding of the relation between the collection $(\mathcal{N}_{r,\ell})_{r\geq 1}$ of Hecke eigenforms and the collection $(\mathfrak{N}_{r,\ell})_{r\in\mathbb{Z}^\times}$ of Hecke distributions will be necessary in Chapter 5.

(iii) The coefficient b_1 in the series (2.1.23) for $\mathcal{N}$ is normalized (in Hecke's way, not in any Hilbert sense) to the value 1: making the same choice in the Eisenstein case leads (2.1.18) to considering $\frac{1}{2}E^*_{\frac{1-\nu}{2}}$ as the correctly normalized Eisenstein series. Finally, looking at the first sentence in Proposition 2.1.1, it is $\frac{1}{2}\mathfrak{E}_\nu$ that ought to be considered as normalized in Hecke's way.

We prove now a fact announced in Remark 1.2.1 (i) following Theorem 1.2.2, to some extent a converse of that theorem.

Theorem 2.1.2. *Every Hecke eigenform $\mathcal{N}$ with the Fourier expansion (2.1.23), normalized so that the coefficient b_1 is 1, coincides, for some choice of χ, with the image under Θ_0 of the rescaled version of the Hecke distribution $\mathfrak{N} = \mathfrak{N}_{\chi,i\lambda}$ as defined in Theorem 1.2.2. Setting when $\mathrm{Re}\ s$ is large, as is usual, $L(s,\mathcal{N}) = \sum_{k\geq 1}\frac{b_k}{k^s}$, one has $L(s,\mathcal{N}) = L(s,\mathfrak{N})$. Recall that, with $\varepsilon = 0$ or 1 according to the parity of $\mathcal{N}$ under the map $z \mapsto -\bar{z}$, one sets*

$$L^*(s,\mathcal{N}) = \pi^{-s}\Gamma\left(\frac{s+\varepsilon}{2} + \frac{i\lambda}{4}\right)\Gamma\left(\frac{s+\varepsilon}{2} - \frac{i\lambda}{4}\right)L(s,\mathcal{N}), \qquad (2.1.27)$$

obtaining as a result the identity $L^*(s,\mathcal{N}) = (-1)^\varepsilon L^*(1-s,\mathcal{N})$.

Proof. In this direction, we start from a Hecke eigenform $\mathcal{N}$, normalized in the way indicated and, defining $\mathfrak{N} = \mathfrak{N}_+$ according to Proposition 2.1.1, our problem is showing that $\mathfrak{N}$ coincides for some choice of $(\chi, i\lambda)$ with the Hecke distribution $\mathfrak{N}_{\chi,i\lambda}$ as defined by means of Theorem 1.2.2. First define $\varepsilon = 0$ or 1 according to the choice made in the statement of the theorem: then, $b_{-k} = (-1)^\varepsilon b_k$. On the other hand, for every prime p, let θ_p be any of the two roots of the equation

$$\theta_p^2 - b_p\,\theta_p + 1 = 0. \qquad (2.1.28)$$

Denote as σ the collection of data

$$\sigma = \{\varepsilon,\ i\lambda,\ (\theta_p)_{p\,\mathrm{prime}}\}, \qquad (2.1.29)$$

where λ is any of the two square roots of λ^2. To each such set σ of spectral data, one associates in a one-to-one way the pair $(\chi, i\lambda)$, where the character χ on $\mathbb{Q}^\times$ is defined by the set of conditions

$$\chi(p) = p^{\frac{i\lambda}{2}}\theta_p, \quad \chi(\pm 1) = (-1)^\varepsilon : \qquad (2.1.30)$$

it is quite well-known that χ, so defined, is a tempered character. The correspondence $\sigma \mapsto \mathfrak{N}_{\chi,i\lambda}$, or $(\chi, i\lambda) \mapsto \mathfrak{N}_{\chi,i\lambda}$, introduced in Theorem 1.2.2, is not one-to-one because, besides ε and λ, only the set of sums $\theta_p + \theta_p^{-1}$ is needed to define $\mathfrak{N}_{\chi,i\lambda}$, as it follows from (1.2.16), (1.2.17).

We now prove that, with $\mathfrak{N}$ as defined above, one has the identity (1.2.4), here recalled:

$$\langle \mathfrak{N}, h \rangle = \frac{1}{4} \sum_{m,\,n \neq 0} \chi\left(\frac{m}{n}\right) \int_{-\infty}^{\infty} |t|^{-1-i\lambda} \left(\mathcal{F}_1^{-1} h\right) \left(\frac{m}{t}, nt\right) dt, \quad h \in \mathcal{S}(\mathbb{R}^2).$$

$$(2.1.31)$$

To do so, let us rewrite the set of Fourier coefficients b_k in terms first of σ, next of the pair $(i\lambda, \chi)$, relying on Hecke's theory: the computation reduces to that of b_k for $k \geq 1$. One has $T_k \mathcal{N} = b_k \mathcal{N}$. On the other hand, one has the formal identity (2.1.21) between Dirichlet series: applying the operator there to $\mathcal{N}$ and using (2.1.28), one has

$$\sum_{k \geq 1} k^{-s} b_k = \prod_p \left(1 - p^{-s} \theta_p\right)^{-1} \left(1 - p^{-s} \theta_p^{-1}\right)^{-1}$$

$$= \prod_p \left(1 + p^{-s} \theta_p + p^{-2s} \theta_p^2 + \dots\right) \left(1 + p^{-s} \theta_p^{-1} + p^{-2s} \theta_p^{-2} + \dots\right). \quad (2.1.32)$$

The number b_k, which is the coefficient of k^{-s} in the right-hand side, can thus be written, if $k = \prod_p p^{j_p}$, as

$$b_k = \prod_p \left(\sum_{\substack{r_p,\, s_p \geq 0 \\ r_p + s_p = j_p}} \theta_p^{r_p - s_p} \right) = \sum_{r+s=j} \prod_p \theta_p^{r_p - s_p}, \quad k \geq 1 \quad (2.1.33)$$

if, in the last expression, one sets $j = (j_2, j_3, j_5, \dots)$ and one considers similarly defined vectors r and s with non-negative coordinates, indexed by the set of primes. To each pair (r, s), associate the pair (m, n) of positive integers $m = \prod_p p^{r_p}$, $n = \prod_p p^{s_p}$, so that $k = mn$, an arbitrary decomposition of $k \geq 1$ as a product of two integers ≥ 1. Defining the character θ on $\mathbb{Q}_+^\times$ by the equation $\theta(p) = \theta_p$ for all p, one can thus write (2.1.33) as

$$b_k = \sum_{\substack{m,\,n \geq 1 \\ mn=k}} \theta(m)\,\theta(n)^{-1} = \sum_{\substack{m,\,n \geq 1 \\ mn=k}} \theta\left(\frac{m}{n}\right), \quad k \geq 1. \quad (2.1.34)$$

Then, for any $k \in \mathbb{Z}^\times$,

$$|k|^{\frac{i\lambda}{2}} b_k = |k|_\varepsilon^{\frac{i\lambda}{2}} b_{|k|} = \sum_{\substack{m,\,n \geq 1 \\ mn=|k|}} (\text{sign } k)^\varepsilon |mn|^{\frac{i\lambda}{2}} \theta\left(\left|\frac{m}{n}\right|\right)$$

$$= \frac{1}{4} \sum_{\substack{m,\,n \neq 0 \\ mn=k}} |m|_\varepsilon^{\frac{i\lambda}{2}} \theta(|m|) \times |n|^{i\lambda} \left[|n|_\varepsilon^{\frac{i\lambda}{2}} \theta(|n|)\right]^{-1}$$

$$= \frac{1}{4} \sum_{\substack{m, n \neq 0 \\ mn = k}} |n|^{i\lambda} \chi\left(\frac{m}{n}\right). \tag{2.1.35}$$

Hence,

$$\langle \mathfrak{N}, h \rangle = \frac{1}{4} \sum_{m, n \neq 0} \chi\left(\frac{m}{n}\right) |n|^{i\lambda} \int_{-\infty}^{\infty} |t|^{-1-i\lambda} \left(\mathcal{F}_1^{-1} h\right) \left(\frac{mn}{t}, t\right) dt : \tag{2.1.36}$$

performing the change of variable $t \mapsto nt$, we are done. $\qquad\square$

Remarks 2.1.2. (i) Recall that the Hecke eigenform $\mathcal{N}$ is said to satisfy the Ramanujan-Petersson conjecture if $|\theta_p| = 1$ for every prime p, in other words if χ is unitary.

(ii) That the Hecke operator T_N (acting on non-holomorphic modular forms), as defined in (2.1.20), is the transfer under Θ_0 of the operator T_N^{dist} (acting on automorphic distributions) in (1.2.1) is easy: nothing more than relating the two actions (by linear or fractional-linear transformations) of the group G is needed.

(iii) Eisenstein series are of course not cusp-forms: however, Eisenstein distributions can be recovered in the same way, with the exception of the first two terms of its decomposition (1.1.38). We start from the Fourier series expansion (2.1.18), both sides of which have been multiplied by $\frac{1}{2}$ for proper normalization, so that the coefficient b_1 taken from this expansion should be 1. Following the construction in the proof which precedes, one sees that $\varepsilon = 0$ and that, for $k \geq 1$, one has $b_k = k^{-\frac{\nu}{2}} \sigma_\nu(k)$. In particular, for any prime p, one has $b_p = p^{-\frac{\nu}{2}} (1 + p^\nu)$, so that $\theta_p = p^{-\frac{\nu}{2}}$ is one solution of equation (2.1.28). Any corresponding character χ is trivial on p. This leads to the main part of the expansion (1.1.38) of the Eisenstein distribution $\frac{1}{2} \mathfrak{E}_\nu$.

Even though it is natural to put less emphasis on Hilbert space methods in the automorphic distribution environment than in the automorphic function (in Π) environment, there is a perfectly natural Hilbert space $L^2(\Gamma \backslash \mathbb{R}^2)$, despite the fact that there is no fundamental domain for the action by linear changes of coordinates of Γ in $\mathbb{R}^2$ (most orbits are dense). This will be recalled at the end of Section 3.2. The decomposition of rather general automorphic distributions into their homogeneous components (the analogue of the Roelcke-Selberg theory) will be treated in Section 5.1.

2.2 Bihomogeneous functions and joint eigenfunctions of $(\Delta, \mathrm{Eul}^{\Pi})$

N.B. The present section and the one which follows are not required for further reading.

However, the question of how bihomogeneous functions in the plane transfer to the half-plane is a natural one. Theorems 2.2.4 and 2.2.5 below are the analogues, in the half-plane, of the decomposition formulas (1.1.46) and (1.2.7) of Eisenstein or Hecke distributions into bihomogeneous components. On the other hand, Theorem 2.2.3 can be stated as the fact that $\Psi_{\rho,\nu}^{(\varepsilon)}$ is *almost* a generalized eigenfunction of Δ, in that it satisfies the required differential equation outside a one-dimensional set.

We first compute the dual Radon transform of $\hom_{\rho,\nu}^{(\varepsilon)}$, a task already performed in [39, section 2.3] in the case when $\varepsilon = 0$. To do so, we need to introduce for $\nu \notin \mathbb{Z}$ and $\rho \pm \nu \notin 2\mathbb{Z}$ the function on $\mathbb{R}\backslash\{0\}$

$$\chi_{\rho,\nu}(t) = 2^{\nu-1}\,\pi^{-\frac{1}{2}}\,\frac{\Gamma(\frac{\nu}{2})}{\Gamma(\frac{2-\rho+\nu}{2})}$$
$$\times \left(\frac{-1-i\,t}{2}\right)_{+}^{\frac{\rho+\nu-2}{2}} {}_2F_1\left(\frac{1-\nu}{2},\frac{2-\rho-\nu}{2};1-\nu;\frac{2}{1+it}\right),\quad (2.2.1)$$

where the power z_+^{α}, for $z \notin\,] - \infty,\,0]$, is defined as $e^{i\alpha\theta}$, θ being the argument of z lying in $]-\pi,\pi[$. It is undefined at 0 since, for a proper definition of the hypergeometric function, we must exclude values of the argument lying on the half-line $[1,\infty[$. Its main property is that the function

$$z \mapsto (\operatorname{Im} z)^{\frac{\rho-1}{2}}\,\chi_{\rho,\nu}\left(\frac{\operatorname{Re} z}{\operatorname{Im} z}\right)\qquad\qquad (2.2.2)$$

is in the complement, in the hyperbolic half-plane, of the line $\operatorname{Re} z = 0$, a generalized eigenvalue of Δ for the eigenvalue $\frac{1-\nu^2}{4}$: this will be detailed below in (2.2.30), (2.2.31). We also set

$$F_{\rho,\nu}^{(\varepsilon)}(z) = \begin{cases} (\operatorname{Im} z)^{\frac{\rho-1}{2}}\,\chi_{\rho,\nu}^{\mathrm{even}}\left(\frac{\operatorname{Re} z}{\operatorname{Im} z}\right) & \text{if } \varepsilon = 0, \\[2mm] (\operatorname{Im} z)^{\frac{\rho-1}{2}}\,\chi_{\rho,\nu}^{\mathrm{odd}}\left(\frac{\operatorname{Re} z}{\operatorname{Im} z}\right) & \text{if } \varepsilon = 1, \end{cases}\qquad (2.2.3)$$

and, with another normalization,

$$\Psi_{\rho,\nu}^{(\varepsilon)}(z) = \pi^{-\frac{1}{2}}\,\Gamma\left(\frac{4-2\varepsilon-\rho-\nu}{4}\right)\Gamma\left(\frac{4-2\varepsilon-\rho+\nu}{4}\right)F_{\rho,\nu}^{(\varepsilon)}(z).\qquad (2.2.4)$$

These functions are obviously generalized eigenfunctions of the operator Eul^{Π} introduced in (2.0.1).

Lemma 2.2.1. *One has*

$$(1+t^2)^{\frac{1-\rho}{2}}\,\chi_{\rho,\nu}^{\mathrm{even}}(t) = \frac{\Gamma(\frac{2+\rho-\nu}{4})\Gamma(\frac{2+\rho+\nu}{4})}{\Gamma(\frac{4-\rho-\nu}{4})\Gamma(\frac{4-\rho+\nu}{4})}\,\chi_{2-\rho,\nu}^{\mathrm{even}}(t),$$

$$(1+t^2)^{\frac{1-\rho}{2}}\,\chi_{\rho,\nu}^{\mathrm{odd}}(t) = \frac{\Gamma(\frac{\rho-\nu}{4})\Gamma(\frac{\rho+\nu}{4})}{\Gamma(\frac{2-\rho-\nu}{4})\Gamma(\frac{2-\rho+\nu}{4})}\,\chi_{2-\rho,\nu}^{\mathrm{odd}}(t).\qquad (2.2.5)$$

Proof. The even case is just equation (2.3.52) in [39]: we (wrongly) omitted the superscript "even" there: the proof of Lemma 2.3.4, from a certain point on, only considered the even part of $\chi_{\rho,\nu}$ (the odd part of this function played no part in that book). Equation (2.3.55) in the given reference can be written as

$$\chi_{2-\rho,\nu}(t) = e^{i\pi\frac{\rho-1}{2}\mathrm{sign}\,t}\,\frac{\Gamma(\frac{2-\rho+\nu}{2})}{\Gamma(\frac{\rho+\nu}{2})}\left(\frac{1+t^2}{4}\right)^{\frac{1-\rho}{2}}\chi_{\rho,\nu}(t). \qquad (2.2.6)$$

Then,

$$\left(\frac{1+t^2}{4}\right)^{\frac{1-\rho}{2}}\chi_{\rho,\nu}^{\mathrm{odd}}(t) = \frac{1}{2}\,\frac{\Gamma(\frac{\rho+\nu}{2})}{\Gamma(\frac{2-\rho+\nu}{2})}\left[e^{i\pi\frac{1-\rho}{2}}\chi_{2-\rho,\nu}(t) - e^{i\pi\frac{\rho-1}{2}}\chi_{2-\rho,\nu}(-t)\right].$$
$$(2.2.7)$$

Using twice the equation [39, (2.3.29)]

$$\chi_{\rho,\nu}(t) = e^{\frac{i\pi(2-\nu-\rho)}{2}}\,\chi_{\rho,\nu}(-t), \qquad t > 0 \qquad (2.2.8)$$

and the equation

$$\frac{1}{2}\left(e^{i\pi\frac{1-\rho}{2}} - e^{i\pi\frac{\nu-1}{2}}\right) = \frac{i\pi\,e^{\frac{i\pi(\nu-\rho)}{4}}}{\Gamma(\frac{2+\rho+\nu}{4})\Gamma(\frac{2-\rho-\nu}{4})}, \qquad (2.2.9)$$

one arrives at the second equation (2.2.5). $\qquad\square$

It follows that

$$|z|^{1-\rho}\,F_{\rho,\nu}^{(\varepsilon)}(z) = \frac{\Gamma(\frac{2-2\varepsilon+\rho-\nu}{4})\Gamma(\frac{2-2\varepsilon+\rho+\nu}{4})}{\Gamma(\frac{4-2\varepsilon-\rho-\nu}{4})\Gamma(\frac{4-2\varepsilon-\rho+\nu}{4})}\,F_{2-\rho,\nu}^{(\varepsilon)}(z), \qquad (2.2.10)$$

hence

$$F_{\rho,\nu}^{(\varepsilon)}(z) = (-1)^{\varepsilon}\,\frac{\Gamma(\frac{2-2\varepsilon+\rho-\nu}{4})\Gamma(\frac{2-2\varepsilon+\rho+\nu}{4})}{\Gamma(\frac{4-2\varepsilon-\rho-\nu}{4})\Gamma(\frac{4-2\varepsilon-\rho+\nu}{4})}\,F_{2-\rho,\nu}^{(\varepsilon)}(-z^{-1}); \qquad (2.2.11)$$

in other words

$$\Psi_{\rho,\nu}^{(\varepsilon)}(z) = (-1)^{\varepsilon}\,\Psi_{2-\rho,\nu}^{(\varepsilon)}(-z^{-1}). \qquad (2.2.12)$$

Another symmetry worth mentioning concerns the function $\mathrm{hom}_{\rho,\nu}^{(\varepsilon)}$: one has

$$\mathcal{F}^{\mathrm{symp}}\,\mathrm{hom}_{\rho,\nu}^{(\varepsilon)} = \pi^{-\nu}\,\frac{\Gamma(\frac{2-\rho+\nu+2\varepsilon}{4})\Gamma(\frac{\rho+\nu+2\varepsilon}{4})}{\Gamma(\frac{\rho-\nu+2\varepsilon}{4})\Gamma(\frac{2-\rho-\nu+2\varepsilon}{4})}\,\mathrm{hom}_{\rho,-\nu}^{(\varepsilon)}, \qquad (2.2.13)$$

or

$$\mathcal{F}^{\mathrm{symp}}\left(B_{\varepsilon}(\frac{\rho+\nu}{2})\,\mathrm{hom}_{\rho,\nu}^{(\varepsilon)}\right) = B_{\varepsilon}(\frac{\rho-\nu}{2})\,\mathrm{hom}_{\rho,-\nu}^{(\varepsilon)}. \qquad (2.2.14)$$

Theorem 2.2.2. *Assume* $\operatorname{Re} \nu > -1 + |\operatorname{Re} \rho - 1|$ *and* $\nu \notin \mathbb{Z}$, $\rho \pm \nu \notin 2\mathbb{Z}$. *Then,
one has, denoting also the even and odd parts of* $\chi_{\rho,\nu}$ *as* $\chi_{\rho,\nu}^{(\varepsilon)}$ *with* $\varepsilon = 0$ *or* 1,

$$\left(V^* \, hom_{\rho,\nu}^{(\varepsilon)} \right)(z) = i^\varepsilon \, (\operatorname{Im} z)^{\frac{\rho-1}{2}}$$

$$\times \; 2^{\frac{\rho-\nu}{2}} \, \pi^{-1} \frac{\Gamma(\frac{2-\rho+\nu}{2}) \, \Gamma(\frac{2\varepsilon+\rho+\nu}{4}) \, \Gamma(\frac{4-2\varepsilon-\rho-\nu}{4})}{\Gamma(\frac{\nu+1}{2})}$$

$$\times \left[\chi_{\rho,-\nu}^{(\varepsilon)} \left(\frac{\operatorname{Re} z}{\operatorname{Im} z} \right) + \chi_{\rho,\nu}^{(\varepsilon)} \left(\frac{\operatorname{Re} z}{\operatorname{Im} z} \right) \right], \qquad (2.2.15)$$

or

$$V^* \, hom_{\rho,\nu}^{(\varepsilon)} = i^\varepsilon \, 2^{\frac{\rho-\nu}{2}} \, \pi^{-1} \frac{\Gamma(\frac{2-\rho+\nu}{2}) \, \Gamma(\frac{2\varepsilon+\rho+\nu}{4}) \, \Gamma(\frac{4-2\varepsilon-\rho-\nu}{4})}{\Gamma(\frac{\nu+1}{2})} \left[F_{\rho,-\nu}^{(\varepsilon)} + F_{\rho,\nu}^{(\varepsilon)} \right].$$

$$(2.2.16)$$

Proof. In the case when $\varepsilon = 0$, this is [39, p. 70], and we consider now the case
when $\varepsilon = 1$. One has, if $\operatorname{Re} \nu > \max(\operatorname{Re} \rho - 2, -\operatorname{Re} \rho) = -1 + |\operatorname{Re} \rho - 1|$,

$$(V^* \, hom_{\rho,\nu}^{(1)})(x + iy) = \frac{1}{2\pi} \int_0^{2\pi} \left| y^{-\frac{1}{2}} \sin \frac{\theta}{2} \right|_1^{\frac{\nu-1}{2}} \left| y^{\frac{1}{2}} \cos \frac{\theta}{2} - x \, y^{-\frac{1}{2}} \sin \frac{\theta}{2} \right|_1^{\frac{\rho+\nu-2}{2}} d\theta$$

$$= y^{\frac{\rho-1}{2}} \times \frac{1}{2\pi} \int_0^{2\pi} \left| \sin \frac{\theta}{2} \right|_1^{\frac{\nu-1}{2}} \left| \cos \frac{\theta}{2} - \frac{x}{y} \sin \frac{\theta}{2} \right|_1^{\frac{\rho+\nu-2}{2}} d\theta.$$

$$(2.2.17)$$

It is immediate (compare [39, p. 71]) that, as $\frac{x}{y} \to +\infty$, one has the equivalent

$$(V^* \, hom_{\rho,\nu}^{(1)})(x + iy) \sim -\pi^{-\frac{1}{2}} \frac{\Gamma(\frac{\nu}{2})}{\Gamma(\frac{\nu+1}{2})} \, y^{\frac{\rho-1}{2}} \left| \frac{x}{y} \right|^{\frac{\rho+\nu-2}{2}}. \qquad (2.2.18)$$

On the other hand, $y^{\frac{1-\rho}{2}} (V^* \, hom_{\rho,\nu}^{(1)})(x + iy)$ is an odd function of x, or of $t = \frac{x}{y}$.
To see this, starting from the definition

$$(V^* h)(g.i) = \int_K h \left((gk).\left({\textstyle \frac{1}{0}} \right) \right) dk, \qquad (2.2.19)$$

only note that $\left(\begin{smallmatrix} 1 & 0 \\ 0 & -1 \end{smallmatrix} \right) \left(\begin{smallmatrix} y^{\frac{1}{2}} & xy^{-\frac{1}{2}} \\ 0 & y^{-\frac{1}{2}} \end{smallmatrix} \right) \left(\begin{smallmatrix} 1 & 0 \\ 0 & -1 \end{smallmatrix} \right) = \left(\begin{smallmatrix} y^{\frac{1}{2}} & -xy^{-\frac{1}{2}} \\ 0 & y^{-\frac{1}{2}} \end{smallmatrix} \right)$ and $\left(\begin{smallmatrix} 1 & 0 \\ 0 & -1 \end{smallmatrix} \right) K \left(\begin{smallmatrix} 1 & 0 \\ 0 & -1 \end{smallmatrix} \right) = K$.

To prove Theorem 2.2.2, one may assume, using analytic continuation, that
$\operatorname{Re} \nu > 0$, in which case an equivalent of $\chi_{\rho,\nu}^{\text{odd}}(t) + \chi_{\rho,-\nu}^{\text{odd}}(t)$ as $t \to \infty$ reduces to
an equivalent of the first term. With $C = 2^{\nu-1} \, \pi^{-\frac{1}{2}} \frac{\Gamma(\frac{\nu}{2})}{\Gamma(\frac{2-\rho+\nu}{2})}$, one has

$$\chi_{\rho,\nu}^{\text{odd}}(t) \sim C \, 2^{\frac{-\rho-\nu}{2}} \, t^{\frac{\rho+\nu-2}{2}} \left[e^{-\frac{i\pi}{4}(\rho+\nu-2)} - e^{\frac{i\pi}{4}(\rho+\nu-2)} \right]$$

$$= i \, 2^{\frac{\nu-\rho}{2}} \, \pi^{\frac{1}{2}} \frac{\Gamma(\frac{\nu}{2})}{\Gamma(\frac{2-\rho+\nu}{2}) \Gamma(\frac{2+\rho+\nu}{4}) \Gamma(\frac{2-\rho-\nu}{4})} \, t^{\frac{\rho+\nu-2}{2}}. \qquad (2.2.20)$$

Now, the function $(V^* hom^{(1)}_{\rho,\nu})(x + iy)$ is C^∞ in Π because it is a generalized eigenfunction of Δ, and it must coincide with a linear combination of the two functions $(\text{Im } z)^{\frac{\rho-1}{2}} \chi^{\text{odd}}_{\rho,\pm\nu} \left(\frac{\text{Re } z}{\text{Im } z}\right)$: so that the jumps on the line $x = 0$ should cancel, only the sum of the two functions qualifies, up to multiplication by a constant. Using the estimates (2.2.18) and (2.2.20), one obtains the theorem. □

We compute now the image of the function $\Psi^{(\varepsilon)}_{\rho,\nu}$ under the operator $\Delta - \frac{1-\nu^2}{4}$, expecting to find a distribution supported in the hyperbolic line from 0 to $i\infty$. We denote as $\delta_{(0,i\infty)}$ the measure $\frac{dy}{y}$ on this line, with $z = x + iy$, and we define the distribution $\delta'_{(0,i\infty)} = y\frac{\partial}{\partial x} \delta_{(0,i\infty)}$ by the equation

$$\langle \delta'_{(0,i\infty)}, h \rangle = -\langle \delta_{(0,i\infty)}, y\frac{\partial h}{\partial x} \rangle, \quad h \in C^\infty_0(\Pi). \tag{2.2.21}$$

Theorem 2.2.3. *Set, with $\varepsilon = 0$ or 1,*

$$G^{(\varepsilon)}(\rho,\nu) = 2^{2-\varepsilon}\pi \frac{\Gamma(\frac{\nu}{2})\Gamma(\frac{2-\nu}{2})}{\Gamma(\frac{2+2\varepsilon-\rho+\nu}{4})\Gamma(\frac{2+2\varepsilon-\rho-\nu}{4})\Gamma(\frac{2\varepsilon+\rho+\nu}{4})\Gamma(\frac{2\varepsilon+\rho-\nu}{4})}. \tag{2.2.22}$$

Then, one has

$$\left(\Delta - \frac{1-\nu^2}{4}\right) \Psi^{(0)}_{\rho,\nu} = G^{(0)}(\rho,\nu) (\text{Im } z)^{\frac{\rho-1}{2}} \delta_{(0,i\infty)},$$

$$\left(\Delta - \frac{1-\nu^2}{4}\right) \Psi^{(1)}_{\rho,\nu} = -G^{(1)}(\rho,\nu) (\text{Im } z)^{\frac{\rho-1}{2}} \delta'_{(0,i\infty)}. \tag{2.2.23}$$

Proof. It was proved in [39, p. 73] that □

$$\left(\Delta - \frac{1-\nu^2}{4}\right) F^{(0)}_{\rho,\nu} = C(\rho,\nu) (\text{Im } z)^{\frac{\rho-1}{2}} \delta_{(0,i\infty)}, \tag{2.2.24}$$

with

$$C(\rho,\nu) = 2^{2-\rho}\pi^{\frac{1}{2}} \frac{\Gamma(\frac{\nu}{2})\Gamma(\frac{2-\nu}{2})}{\Gamma(\frac{2-\rho+\nu}{2})\Gamma(\frac{2-\rho-\nu}{2})\Gamma(\frac{\rho+\nu}{4})\Gamma(\frac{\rho-\nu}{4})}: \tag{2.2.25}$$

this covers the case when $\varepsilon = 0$, in view of (2.2.4) and a new application of the duplication formula.

To cover the case when $\varepsilon = 1$, we must start with a computation of the jump

$$D(\rho,\nu) = \chi_{\rho,\nu}(0^+) - \chi_{\rho,\nu}(0^-). \tag{2.2.26}$$

It is based on the equation [22, p. 48]

$$(-z)^b_+ \,_2F_1(a, b; c; z) = \frac{\Gamma(c)\Gamma(b-a)}{\Gamma(b)\Gamma(c-a)}(-z)^{b-a}_+ \,_2F_1(a, a-c+1; a-b+1; \frac{1}{z})$$
$$+ \frac{\Gamma(c)\Gamma(a-b)}{\Gamma(a)\Gamma(c-b)} \,_2F_1(b, b-c+1; b-a+1; \frac{1}{z}), \tag{2.2.27}$$

applied with $z = \frac{2}{1+it}$ and $a = \frac{1-\nu}{2}$, $b = \frac{2-\rho-\nu}{2}$, $c = 1 - \nu$. One obtains

$$D(\rho, \nu) = 2^{\nu + \frac{1}{2} - \frac{\rho}{2}} i \, \pi^{-\frac{1}{2}} \sin \frac{\pi(1-\rho)}{2}$$

$$\times \frac{\Gamma(\frac{\nu}{2})}{\Gamma(\frac{2-\rho+\nu}{2})} \frac{\Gamma(1-\nu)\Gamma(\frac{1-\rho}{2})}{\Gamma(\frac{2-\rho-\nu}{2})\Gamma(\frac{1-\nu}{2})} {}_2F_1 \left(\frac{1-\nu}{2}, \frac{1+\nu}{2}; \frac{\rho+1}{2}; \frac{1}{2} \right): \quad (2.2.28)$$

using [22, p. 41] to obtain the required value of the hypergeometric function,

$$D(\rho, \nu) = 2^{1-\rho} \pi^{\frac{1}{2}} i \, \frac{\Gamma(\frac{\nu}{2})\Gamma(\frac{2-\nu}{2})}{\Gamma(\frac{2-\rho+\nu}{2})\Gamma(\frac{2-\rho-\nu}{2})\Gamma(\frac{2+\rho+\nu}{4})\Gamma(\frac{2+\rho-\nu}{4})}. \quad (2.2.29)$$

Now, given a function χ on the real line, one has [39, p. 65]

$$\Delta \left(z \mapsto (\text{Im } z)^{\frac{\rho-1}{2}} \chi \left(\frac{\text{Re } z}{\text{Im } z} \right) \right) = (\text{Im } z)^{\frac{\rho-1}{2}} (\mathcal{D}\chi) \left(\frac{\text{Re } z}{\text{Im } z} \right) \quad (2.2.30)$$

with

$$\mathcal{D} = -(1+t^2) \frac{d^2}{dt^2} + (\rho - 3) \frac{d}{dt} - \frac{(\rho-1)(\rho-3)}{4}: \quad (2.2.31)$$

the fact that $\left(\Delta - \frac{1-\nu^2}{4} \right) F_{\rho,\nu}^{(\varepsilon)} = 0$ in the complementary of the hyperbolic line $\text{Re } z = 0$ actually originated from the ordinary differential equation

$$\left(\mathcal{D} - \frac{1-\nu^2}{4} \right) \chi_{\rho,\nu} = 0$$

outside 0. In the distribution sense on the real line, one then has

$$\left(\mathcal{D} - \frac{1-\nu^2}{4} \right) \chi_{\rho,\nu}^{\text{odd}} = -D(\rho, \nu) \frac{d^2}{dt^2} \text{char}(t \geq 0)$$

$$= -D(\rho, \nu) \, \delta', \quad (2.2.32)$$

and

$$\left(\Delta - \frac{1-\nu^2}{4} \right) F_{\rho,\nu}^{(1)} = -D(\rho, \nu) (\text{Im } z)^{\frac{\rho-1}{2}} \delta'_{(0, i\infty)}, \quad (2.2.33)$$

from which the second equation (2.2.23) follows.

Note that both coefficients $G_{\rho,\nu}^{(\varepsilon)}$ change to their negatives under the change $\nu \to -\nu$, and are invariant under the change $\rho \to 2-\rho$: the second property was to be expected in view of (2.2.12), of the invariance of Δ under the map $z \mapsto -z^{-1}$, finally in view of the fact that $\delta_{(0, i\infty)}$ is invariant under the map $z \mapsto -z^{-1}$, while $\delta'_{(0, i\infty)}$ changes to its negative.

We reproduce from [39, p. 75], assuming $\text{Re } \nu < 0$, the equation

$$\left[\left(\Delta - \frac{1-\nu^2}{4} \right)^{-1} (\text{Im } z)^{\frac{\rho-1}{2}} \delta_{(0, i\infty)} \right] (z) = \frac{1}{C(\rho, \nu)} F_{\rho,\nu}^{(0)}, \quad (2.2.34)$$

more precise than (2.2.24) since it singles out the function $F^{(0)}_{\rho,\nu}$ from the pair $F^{(0)}_{\rho,\pm\nu}$: it required a somewhat unusual expression of the resolvent of Δ. The same method makes it possible to obtain the equation

$$\left[\left(\Delta - \frac{1-\nu^2}{4}\right)^{-1}(\text{Im }z)^{\frac{\rho-1}{2}}\delta'_{(0,i\infty)}\right](z) = -\frac{1}{D(\rho,\nu)}F^{(1)}_{\rho,\nu}. \qquad (2.2.35)$$

We now combine the results of Sections 2 and 3 on one hand, of the present section on the other hand, to obtain the decompositions of Eisenstein series $E^*_{\frac{1-\nu}{2}}$ or of Hecke eigenforms $\mathcal{N}$ into joint generalized eigenfunctions of the pair of operators $(\Delta, \text{Eul}^{\text{II}})$ in $L^2(\Pi)$: not in $L^2(\Gamma\backslash\Pi)$, of course, since the second operator is not defined, even formally, in that space.

Theorem 2.2.4. *One has, for* $\nu \neq 0$,

$$\frac{1}{2}E^*_{\frac{1-\nu}{2}}(z) = \frac{1}{2}\zeta^*(\nu)\left[(\text{Im }z)^{\frac{1-\nu}{2}} + (\text{Im }(-z^{-1}))^{\frac{1-\nu}{2}}\right]$$

$$+ \frac{1}{2}\zeta^*(-\nu)\left[(\text{Im }z)^{\frac{1+\nu}{2}} + (\text{Im }(-z^{-1}))^{\frac{1+\nu}{2}}\right]$$

$$+ \frac{1}{8i\pi}\int_{\text{Re }\rho=1}\zeta^*\left(\frac{\rho-\nu}{2}\right)\zeta^*\left(\frac{\rho+\nu}{2}\right)\left[\Psi^{(0)}_{\rho,-\nu}(z) + \Psi^{(0)}_{\rho,\nu}(z)\right]d\rho,$$

$$(2.2.36)$$

with $\Psi^{(0)}_{\rho,\pm\nu}$ *as defined in* (2.2.4).

Proof. One has

$$E^*_{\frac{1-\nu}{2}} = \Theta_0\mathfrak{E}^{\text{resc}}_\nu = V^* \cdot \pi^{\frac{1}{2}-i\pi\mathcal{E}}\Gamma(\frac{1}{2}+i\pi\mathcal{E})\,\mathfrak{E}_\nu = \pi^{\frac{1+\nu}{2}}\Gamma(\frac{1-\nu}{2})V^*\mathfrak{E}_\nu. \qquad (2.2.37)$$

Consider first the term $\frac{1}{2}\zeta(-\nu)|\xi|^{-\nu-1}$ of the decomposition (1.1.46). From [39, p. 50], one has

$$V^*\left((x,\xi)\mapsto|\xi|^{-\nu-1}\right)(z) = \frac{1}{2\pi}\int_0^{2\pi}\left|(\text{Im }z)^{-\frac{1}{2}}\sin\frac{\theta}{2}\right|^{-\nu-1}d\theta$$

$$= \pi^{-\frac{1}{2}}\frac{\Gamma(-\frac{\nu}{2})}{\Gamma(\frac{1-\nu}{2})}(\text{Im }z)^{\frac{1+\nu}{2}}. \qquad (2.2.38)$$

Then, the image of the function $(x,\xi)\mapsto|x|^{-\nu-1}$ under V^* is the transform of the function on the right-hand side of the last equation under the fractional-linear transformation associated to the matrix $\begin{pmatrix}0&1\\-1&0\end{pmatrix}$. These calculations lead to the second term of the expansion (2.2.36) to be proved. In the same way, the equation

$$V^*\left((x,\xi)\mapsto|x|^{-\nu}\delta(\xi)\right)(z) = \frac{1}{2\pi}\int_{-\pi}^{\pi}(\text{Im }z)^{-\frac{\nu}{2}}\delta\left((\text{Im }z)^{-\frac{1}{2}}\sin\frac{\theta}{2}\right)d\theta$$

$$= \pi^{-1}(\text{Im }z)^{\frac{1-\nu}{2}} \qquad (2.2.39)$$

leads to the first term of the expansion.

Now, still using the equation $\Theta_0 = V^* . (2\pi)^{\frac{1}{2}-i\pi\mathcal{E}}\Gamma(\frac{1}{2} + i\pi\mathcal{E})$, we rewrite (2.2.15) as

$$\left(\Theta_0 \hom_{\rho,-\nu}^{(0)}\right)(z)$$

$$= 2^{\frac{\rho+1}{2}+\nu}\pi^{\frac{-1+\nu}{2}}\Gamma(\frac{2-\rho-\nu}{2})\Gamma(\frac{\rho-\nu}{4})\Gamma(\frac{4-\rho+\nu}{4})\left[F_{\rho,\nu}^{(0)}(z) + F_{\rho,-\nu}^{(0)}(z)\right].$$
(2.2.40)

Using then (1.1.46), one obtains that the main term of $\frac{1}{2}E_{\frac{1-\nu}{2}}^*(z)$ is

$$\frac{1}{8\,i\pi}\int_{\text{Re }\rho=1} 2^{\frac{\rho+\nu}{2}}\pi^{\frac{-1+\nu}{2}}\zeta(\frac{2-\rho-\nu}{2})\zeta(\frac{\rho-\nu}{2})$$

$$\Gamma(\frac{2-\rho-\nu}{2})\Gamma(\frac{\rho-\nu}{4})\Gamma(\frac{4-\rho+\nu}{4})\left[F_{\rho,\nu}^{(0)}(z) + F_{\rho,-\nu}^{(0)}(z)\right]d\rho, \quad (2.2.41)$$

an expression which can easily be transformed into the one which occurs in (2.2.36) by an application of the duplication formula. □

Theorem 2.2.5. *Let $\mathcal{N}$ be a Hecke eigenform such that $\Delta\mathcal{N} = \frac{1+\lambda^2}{4}\mathcal{N}$, of even or odd type according to whether $\varepsilon = 0$ or 1. One has the decomposition*

$$\mathcal{N} = \frac{1}{8i\pi}\int_{\text{Re }\rho=1} L^*(\frac{\rho}{2}, \mathcal{N})\left[\Psi_{\rho,i\lambda}^{(\varepsilon)}(z) + \Psi_{\rho,-i\lambda}^{(\varepsilon)}(z)\right]d\rho. \quad (2.2.42)$$

Proof. Let $\mathfrak{N}$ be the Hecke distribution, homogeneous of degree $-1 - i\lambda$ for some choice of a square root of λ^2, such that $\mathcal{N} = \Theta_0\mathfrak{N}^{\text{resc}}$. Rewriting (1.2.7) as

$$\mathfrak{N} = \frac{(-i)^\varepsilon}{8\,i\pi}\int_{\text{Re }\rho=1}\frac{\pi^{\frac{1-i\lambda}{2}}}{\Gamma(\frac{\varepsilon}{2} + \frac{\rho-i\lambda}{4})\Gamma(\frac{\varepsilon}{2} + \frac{2-\rho-i\lambda}{4})}L^*(\frac{\rho}{2}, \mathcal{N})\hom_{\rho,-i\lambda}^{(\varepsilon)}d\rho. \quad (2.2.43)$$

Then,

$$\mathcal{N}(z) = \pi^{\frac{1+i\lambda}{2}}\Gamma(\frac{1-i\lambda}{2})(V^*\mathfrak{N})(z). \quad (2.2.44)$$

Using (2.2.43), (2.2.15) and, again, the duplication formula, one obtains the decomposition (2.2.42). □

2.3   A class of automorphic functions

N.B. This section is not required for further reading.

This short section is meant "for completeness" only. In the last section, we decomposed Eisenstein series or Hecke eigenforms corresponding to the (possibly generalized) eigenvalue $\frac{1-\nu^2}{4}$ of Δ and with a parity under the map $z \mapsto -\bar{z}$ characterized by $\varepsilon = 0$ or 1, as continuous superpositions, for Re $\rho = 1$, of the functions $F^{(\varepsilon)}_{\rho,\pm\nu}$. Our program in this section is different. Starting from one function $F^{(\varepsilon)}_{\rho,\nu}$, we wish to analyze the function defined, when possible, as

$$f^{(\varepsilon)}_{\rho,\nu} = \frac{1}{2} \sum_{g \in \Gamma} F^{(\varepsilon)}_{\rho,\nu} \circ g. \tag{2.3.1}$$

In the case when $\varepsilon = 0$, this was done in [39, chapter 4] and proved to be a lengthy task. We here consider the case when $\varepsilon = 1$, a much easier one since an automorphic function changing to its negative under the map $z \mapsto -\bar{z}$ can only be a series of Hecke eigenforms and does not involve any Eisenstein series in its decomposition: the continuous part of the Roelcke-Selberg decomposition was by far the hardest part when dealing with $f^{(0)}_{\rho,\nu}$. It will be handy to refer to the part of the discrete spectrum of Δ for which there exist (Maass) eigenfunctions invariant (resp. changing to their negatives) under the map $z \mapsto -\bar{z}$ as the even (resp. odd) part of the discrete spectrum of Δ. The two parts may not be disjoint, for all we know, though there is some suspicion that they are.

Whether $\varepsilon = 0$ or 1, we make the standing assumptions that $0 < \text{Re } \rho < 2$, that $\nu \notin \mathbb{Z}$ and $\rho \pm \nu \notin 2\mathbb{Z}$: the last two assumptions make the definition (2.2.1) of $\chi_{\rho,\nu}$, hence that (2.2.3) of $F^{(\varepsilon)}_{\rho,\nu}$, possible. Let us briefly recall the main results obtained when $\varepsilon = 0$. The series for $f^{(0)}_{\rho,\nu}$ is convergent if Re $\nu < -1 - |\text{Re } \rho - 1|$ and can be continued analytically to the domain obtained from the half-plane Re $\nu < 1 - |\text{Re } \rho - 1|$ by removing the following values of ν: the non-trivial zeros of the zeta function, and the points $i\lambda_r$ with $\frac{1+\lambda_r^2}{4}$ in the even part of the discrete spectrum of Δ. If Re $\nu < 0$ and $\rho \neq 1$, one has the Roelcke-Selberg expansion

$$[C(\rho,\nu)]^{-1} f^{(0)}_{\rho,\nu} = \frac{4}{\nu^2 - \rho^2} E_{\frac{1+\rho}{2}} + \frac{4}{\nu^2 - (2-\rho)^2} E_{\frac{3-\rho}{2}}$$

$$+ \frac{1}{2\pi} \int_{-\infty}^{\infty} \frac{1}{\nu^2 + \lambda^2} \frac{\zeta^*(\frac{\rho-i\lambda}{2}) \zeta^*(\frac{\rho+i\lambda}{2})}{\zeta^*(1+i\lambda)} E_{\frac{1-i\lambda}{2}} \, d\lambda$$

$$+ \sum_{r,\ell \text{ even}} \frac{2}{\nu^2 + \lambda_r^2} L^*(\frac{\rho}{2}, \overline{\mathcal{M}}_{r,\ell}) \mathcal{M}_{r,\ell}. \tag{2.3.2}$$

If Re $\nu < 0$ and $\rho = 1$, the equation remains true after one has replaced the linear combination of Eisenstein series making up the first two terms of the right-hand side of (2.3.2) by a linear combination of a constant and of a (not quite homogeneous) automorphic substitute $E_1^\flat$, obtained by a limiting process, for the non-existent Eisenstein series E_1. The coefficient $C(\rho,\nu)$ is that defined in (2.2.25), and the subscript "r, ℓ even" indicates that only Hecke eigenforms of even type (relative to $z \mapsto -\bar{z}$) are to be taken into account.

One of the major points of the analysis was the study of the continuation [39, Theorem 3.6.2] of the Dirichlet series in two variables defined, when Re $s >$ 1, Re $t > 1$, as

$$\zeta_k(s,t) = \frac{1}{4} \sum_{\substack{m_1 m_2 \neq 0 \\ (m_1, m_2) = 1}} |m_1|^{-s} |m_2|^{-t} \exp\left(2i\pi k \frac{\overline{m_2}}{m_1}\right), \tag{2.3.3}$$

where $\overline{m_2}$ is the number mod m_1 defined by the condition $m_2 \overline{m_2} \equiv 1 \bmod m_1$: we also defined the function $\zeta_k^-(s,t)$ by the same formula, in which the extra factor sign $(m_1 m_2)$ has been added on the right-hand side. One has

$$\zeta_0(s,t) = \frac{\zeta(s)\,\zeta(t)}{\zeta(s+t)} \tag{2.3.4}$$

and, for $k \neq 0$, the main features of the two functions under consideration are given as follows [34, p. 101, 108, 112] to be completed by [34, p. 160]:

Theorem 2.3.1. *For $k \neq 0$, the function $\zeta_k(s,t)$ extends as a meromorphic function for Re $s > 0$, Re $t > 0$, $|\mathrm{Re}\,(s-t)| < 1$, $s \neq 1$, $t \neq 1$, holomorphic outside the set of points (s,t) with $s + t = 1 - i\lambda_r$, $\frac{1+\lambda_r^2}{4}$ in the even part of the discrete spectrum of Δ, or $s + t = \omega$, a non-trivial zero of the zeta function. So far as the function $\zeta_k^-(s,t)$ is concerned, the same result holds, with the difference that it is now the odd part of the discrete spectrum of Δ that must be considered, and the zeros of zeta do not enter the picture any more.*

The given references also give the residue of a function such as

$$\mu \mapsto \zeta_k(\frac{1-\nu-\mu}{2}, \frac{1+\nu-\mu}{2}) \quad \text{or} \quad \zeta_k^-(\frac{1-\nu-\mu}{2}, \frac{1+\nu-\mu}{2}) \tag{2.3.5}$$

at $\mu = i\lambda_r$. However, we shall only take advantage, here, of the version just given in Theorem 2.3.1, more precisely of the part dealing with ζ_k^- of that theorem. The proof of Theorem 2.3.1 is quite lengthy, and based on the spectral decomposition of the pointwise product of two Eisenstein series in the case when $\varepsilon = 0$ (this part was somewhat simplified in [39, section 3.6]), on that of the Poisson bracket of two Eisenstein series when $\varepsilon = 1$: these results are related to the main subject of the present book. The role of the function ζ_k in the proof of (2.3.2) is that it provides the continuation of the coefficients of a Fourier series expansion of the main part of the function $z \mapsto f_{\rho,\nu}^{(0)}(-z^{-1})$.

Theorem 2.3.2. *Keeping the standing assumptions relative to (ρ, ν) given in the beginning of this section, the series for $f_{\rho,\nu}^{(1)}$ is convergent if Re $\nu < -1 - |\mathrm{Re}\,\rho - 1|$ and can be continued analytically, as a function of ν, to the domain obtained from the half-plane Re $\nu < 1 - |\mathrm{Re}\,\rho - 1|$ by removing the points $\nu = i\lambda_r$ with $\frac{1+\lambda_r^2}{4}$ in* *the odd part of the discrete spectrum of Δ. If* $\operatorname{Re} \nu < 0$, *one has the Roelcke-Selberg expansion*

$$[D(\rho,\nu)]^{-1} f_{\rho,\nu}^{(1)} = \sum_{r,\ell \text{ odd}} \frac{4i}{\nu^2 + \lambda_r^2} L^*(\frac{\rho}{2}, \overline{\mathcal{M}_{r,\ell}}) \mathcal{M}_{r,\ell}, \qquad (2.3.6)$$

with $D(\rho,\nu)$ *as defined in* (2.2.28).

Proof. The possibility of extending the function $\nu \mapsto f_{\rho,\nu}^{(1)}$ to the domain indicated is proved exactly like the corresponding fact for $f_{\rho,\nu}^{(0)}$ in [34, section 4.3]: since the core of the proof is an application of Theorem 2.3.1, the non-trivial zeros of zeta do not enter the picture in the case when $\varepsilon = 1$. Then, the function $f_{\rho,\nu}^{(1)}$ obtained is automorphic and changes to its negative under the map $z \mapsto -\bar{z}$, so that it can be written as

$$f_{\rho,\nu}^{(1)} = \sum_{r,\ell \text{ odd}} \left(\mathcal{M}_{r,\ell} \mid f_{\rho,\nu}^{(1)} \right)_{L^2(\Gamma \backslash \Pi)} \mathcal{M}_{r,\ell}. \qquad (2.3.7)$$

Recalling (2.2.35), we obtain

$$\left(\Delta - \frac{1-\nu^2}{4} \right) \left[(D(\rho,\nu))^{-1} f_{\rho,\nu}^{(1)} \right] = \frac{1}{2} \sum_{g \in \Gamma} \mathfrak{s}_\rho \circ g, \qquad (2.3.8)$$

where $\mathfrak{s}_\rho$ is the distribution $-(\operatorname{Im} z)^{\frac{\rho-1}{2}} \delta'_{0,i\infty}$ supported in the hyperbolic line $(0, i\infty)$ from 0 to $i\infty$: we restrict it, in what follows, to the space of functions in Π invariant under the map $z \mapsto -z^{-1}$, the $\frac{\partial}{\partial x}$-derivative of which is rapidly decreasing at infinity on $(0, i\infty)$. The formula then remains true if one replaces $\mathfrak{s}_\rho$ by its symmetrized version

$$\mathfrak{s}_\rho^{\text{sym}} = \frac{1}{2} \left[\mathfrak{s}_\rho + \mathfrak{s}_\rho \circ \left(\begin{smallmatrix} 0 & 1 \\ -1 & 0 \end{smallmatrix} \right) \right] = \frac{1}{2} \left[(\operatorname{Im} z)^{\frac{3-\rho}{2}} - (\operatorname{Im} z)^{\frac{\rho-1}{2}} \right] \delta'_{0,i\infty} : \qquad (2.3.9)$$

to prove this latter formula, observe that, if $\mathcal{M}$ satisfies the conditions just listed, one has

$$\mathcal{M}(x + iy) = \mathcal{M}\left(\frac{-x+iy}{x^2+y^2} \right), \quad \text{so that} \quad \frac{\partial \mathcal{M}}{\partial x}(iy) = -\frac{1}{y^2} \frac{\partial \mathcal{M}}{\partial x}\left(\frac{i}{y} \right). \qquad (2.3.10)$$

Let Σ be the union of the (locally finite) collection of g-transforms, with $g \in \Gamma$, of the line $(0, i\infty)$. Since there are 4 elements of Γ preserving this line, the distribution $ds_\Sigma^{(\rho,1)} = \frac{1}{4} \sum_{g \in \Gamma} \mathfrak{s}_\rho^{\text{sym}} \circ g$ is the only Γ-invariant distribution in Π supported in Σ and coinciding with $\mathfrak{s}_\rho^{\text{sym}}$ when tested on a function supported in a sufficiently small neighbourhood of $(0, i\infty)$. Then, one writes

$$
\frac{\nu^2 + \lambda_r^2}{4} \left(\mathcal{M}_{r,\ell} \mid f_{\rho,\nu}^{(1)} \right)_{L^2(\Gamma\backslash\Pi)} = \left(\left(\Delta - \frac{1-\bar{\nu}^2}{4} \right) \mathcal{M}_{r,\ell} \mid f_{\rho,\nu}^{(1)} \right)_{L^2(\Gamma\backslash\Pi)}
$$

$$
= \left(\mathcal{M}_{r,\ell} \mid \left(\Delta - \frac{1-\nu^2}{4} \right) f_{\rho,\nu}^{(1)} \right)_{\Gamma\backslash\Pi}
$$

$$
= D(\rho,\nu) \langle \frac{1}{2} \sum_{g\in\Gamma} \mathfrak{s}_\rho^{\mathrm{sym}} \circ g, \overline{\mathcal{M}_{r,\ell}} \rangle_{\Gamma\backslash\Pi}. \qquad (2.3.11)
$$

Since the line $(0, i\infty)$ intersects the usual fundamental domain along the half-line from i to $i\infty$ only, one has, using (2.3.10) again,

$$
\langle \frac{1}{2} \sum_{g\in\Gamma} \mathfrak{s}_\rho^{\mathrm{sym}} \circ g, \overline{\mathcal{M}_{r,\ell}} \rangle_{\Gamma\backslash\Pi} = \int_1^\infty \left(y^{\frac{\rho-1}{2}} - y^{\frac{3-\rho}{2}} \right) \frac{\partial \overline{\mathcal{M}_{r,\ell}}}{\partial x}(iy)\, dy
$$

$$
= \int_0^\infty y^{\frac{\rho-1}{2}} \frac{\partial \overline{\mathcal{M}_{r,\ell}}}{\partial x}(iy)\, dy. \qquad (2.3.12)
$$

As

$$
\mathcal{M}_{r,\ell}(x + iy) = y^{\frac{1}{2}} \sum_{k\neq 0} b_k\, e^{2i\pi kx}\, K_{\frac{i\lambda_r}{2}}(2\pi |k| y), \qquad (2.3.13)
$$

one has, if $\mathrm{Re}\, s$ is large, with the help of [22, p. 91], the well-known equation [4, p. 107]

$$
\int_0^\infty y^s \frac{\partial \mathcal{M}_{r,\ell}}{\partial x}(iy)\, dy = 4i\pi \sum_{k\geq 1} k\, b_k \int_0^\infty y^{s+\frac{1}{2}}\, K_{\frac{i\lambda_r}{2}}(2\pi ky)\, dy
$$

$$
= i\, \pi^{-s-\frac{1}{2}} \Gamma\left(\frac{s}{2} + \frac{3+i\lambda_r}{4} \right) \Gamma\left(\frac{s}{2} + \frac{3-i\lambda_r}{4} \right) L(s+\frac{1}{2}, \mathcal{M}_{r,\ell})
$$

$$
= i\, L^*(s + \frac{1}{2}, \mathcal{M}_{r,\ell}). \qquad (2.3.14)
$$

Hence, using also the fact that $\overline{\mathcal{M}_{r,\ell}} = -\mathcal{M}_{r,\ell}$,

$$
\left(\mathcal{M}_{r,\ell} \mid f_{\rho,\nu}^{(1)} \right) = \frac{4i\, D(\rho,\nu)}{\nu^2 + \lambda_r^2}\, L^*\left(\frac{\rho}{2}, \overline{\mathcal{M}_{r,\ell}} \right), \qquad (2.3.15)
$$

which proves Theorem 2.3.2.

Corollary 2.3.3. *The distribution* $ds_\Sigma^{(\rho,1)}$, *which is Γ-invariant and supported in Σ, admits the series expansion, convergent in the space of distributions in Π,*

$$
ds_\Sigma^{(\rho,1)} = \frac{i}{2} \sum_{r,\ell\ \mathrm{odd}} L^*(\frac{\rho}{2}, \overline{\mathcal{M}_{r,\ell}})\, \mathcal{M}_{r,\ell}. \qquad (2.3.16)
$$

Remark 2.3.1. This corollary is only a rephrasing of (2.3.14), but the following comparison is useful. We gave in [39, section 4.7] the spectral expansion (involving a continuous part too) of the measure $ds_{\Sigma}^{(\rho)}$ built in the same way from $\delta_{0,i\infty}$ in place of $\delta'_{0,i\infty}$. In both cases, ν has disappeared from the object under consideration, while the coefficients of the decomposition are given in terms of the restriction of zeta to the line Re $s = \frac{a}{2}$ with $a = \text{Re } \rho$ (when $\varepsilon = 0$) and (in both cases) of the L-functions of Hecke eigenforms at $\frac{\rho}{2}$. One can then interpret the zeros of zeta or of L-functions of Maass-Hecke type on any given line Re $s = \frac{a}{2}$ as points where a spectral density vanishes, or where a Hecke eigenform is missing from a certain decomposition. Needless to say, this is nothing more than an aesthetic satisfaction. □

Chapter 3

A short introduction to the Weyl calculus

Pseudodifferential operator theory, or analysis, a.k.a. the symbolic calculus of operators, started its development half a century ago as an alternative to representing operators by their integral kernels. It soon became one of the major tools of partial differential equations. Their domain of applications in the present book, however, is certainly aside from the main trend, and new methods had to be developed as a consequence. They will, most of the time, be unknown to practitioners of pseudodifferential analysis for P.D.E. purposes: this is especially true for what concerns the composition formula, or the properties linked to the use of the Euler operator in the plane.

Several symbolic calculi are available at present: in the major part of the book, we shall be concerned only with the most important one, to wit the Weyl calculus. Section 2.1 contains an exposition of the first properties of this calculus: newcomers to pseudodifferential analysis should concentrate on the covariance properties (3.1.10) and (3.1.11), and on the so-called coherent state theory, explained in Proposition 3.1.4. Equation (3.1.19) shows that the maps Θ_0 and Θ_1 (2.1.5) on which we relied to link automorphic function theory (in Π) to automorphic distribution theory (in $\mathbb{R}^2$) has a very natural interpretation in terms of the Weyl calculus: it is not limited to the automorphic environment either.

With the help of some spectral theory and of calculations involving Bessel functions, we shall construct, in Proposition 3.2.5, some associates of the dual Radon transformation: these will make it possible eventually, in Chapter 5, to move in the other direction, from modular forms to modular distributions. We shall then be able to take benefit from an important result of the automorphic function theory, the so-called Roelcke-Selberg decomposition theorem, and to transfer it to the automorphic distribution environment.

Section 3.3 deals with the sharp composition of symbols, by which is meant the operation on symbols that corresponds under the Weyl calculus to the composition of operators, when the latter one is well-defined. The formula recalled there (it was initiated in [34] for purposes identical to the present ones) is based on the decomposition of symbols as integral superpositions of their homogeneous components: it bears no relation to the composition formula used in view of P.D.E. applications, which is always a variant of the one giving the sharp composition of polynomials, as quoted in (7.1.2).

We shall then see, in the same section, that if coupled with the decomposition of symbols into homogeneous components in a proper way, the anti-commutator (as physicists say) and commutator of two operators reduce in a way to the pointwise product and Poisson bracket of functions in the hyperbolic half-plane.

3.1 An introduction to the Weyl calculus limited to essentials

We specialize here in the one-dimensional case of the Weyl calculus, even though not much would be needed to cover a part of the program we have in mind in the n-dimensional case: we would just have to replace $SL(2, \mathbb{R})$ by the symplectic group $\text{Sp}(n, \mathbb{R})$ so as to define the metaplectic representation [42], at the same time replacing the upper half-plane by the complex tube over the cone of positive-definite symmetric matrices. But the arithmetic calculations needed in connection with modular form theory would be quite another matter.

Definition 3.1.1. The one-dimensional Weyl calculus associates with a function $h \in \mathcal{S}(\mathbb{R}^2)$, the Schwartz space of C^∞ functions on $\mathbb{R}^2$ rapidly decreasing at infinity, the linear endomorphism $\text{Op}(h)$ of $\mathcal{S}(\mathbb{R})$ defined by the equation

$$(\text{Op}(h)\, u)\, (x) = \int_{\mathbb{R}^2} h(\frac{x+y}{2}, \eta)\, e^{2i\pi(x-y)\eta}\, u(y)\, dy\, d\eta : \qquad (3.1.1)$$

the operator $\text{Op}(h)$ is called the pseudodifferential operator with (Weyl) symbol h. Given two functions $u, v \in \mathcal{S}(\mathbb{R})$, one defines their Wigner function as the function $W(v, u)$ on $\mathbb{R}^2$ such that

$$W(v, u)\, (x, \xi) = 2 \int_{-\infty}^{\infty} \bar{v}(x+t)\, u(x-t)\, e^{4i\pi t\xi}\, dt. \qquad (3.1.2)$$

Remark 3.1.1. On $L^2(\mathbb{R})$, we define the scalar product

$$(v \,|\, u) = \int_{-\infty}^{\infty} \bar{v}(x)\, u(x)\, dx \qquad (3.1.3)$$

as an object antilinear with respect to the variable on the left side. This choice is the one generally made by physicists, whereas mathematicians usually make the other choice. It is really a question of taste: note that if making the "mathematicians' choice", it is necessary, for coherence, to denote as $W(u, v)$ the function denoted as $W(v, u)$ according to our choice. The adjoint of the operator $\mathrm{Op}(h)$ is $\mathrm{Op}(\bar{h})$, an immediate but important fact. Incidentally, note that straight brackets $\langle\,,\,\rangle$, as opposed to curly brackets $(\,|\,)$, will always denote bilinear pairings (such as the ones associated to a duality between two linear spaces).

Proposition 3.1.2. *The Wigner function $W(v, u)$ is the one which makes the formula*

$$(v \,|\, \mathrm{Op}(h)\, u) = \int_{\mathbb{R}^2} h(x, \xi)\, W(v, u)(x, \xi)\, dx\, d\xi \qquad (3.1.4)$$

valid for every symbol $h \in \mathcal{S}(\mathbb{R}^2)$: it is also the symbol of the operator $w \mapsto (v|w)u$. If $h \in \mathcal{S}(\mathbb{R}^2)$, the operator $\mathrm{Op}(f)$ extends linearly as a continuous operator from $\mathcal{S}'(\mathbb{R})$ (the space of tempered distributions on the line) to $\mathcal{S}(\mathbb{R})$. In the other direction, it is possible in a unique way to extend the map $h \mapsto \mathrm{Op}(h)$ as a continuous linear map from $\mathcal{S}'(\mathbb{R}^2)$ to the space of weakly continuous linear maps from $\mathcal{S}(\mathbb{R})$ to $\mathcal{S}'(\mathbb{R})$. Finally, if $h \in L^2(\mathbb{R}^2)$, the operator $\mathrm{Op}(h)$ is a Hilbert-Schmidt operator in the space $L^2(\mathbb{R})$. Note its consequence

$$\mathrm{Tr}\left(\mathrm{Op}(h_1)\,\mathrm{Op}(h_2)\right) = \int_{\mathbb{R}^2} h_1(x, \xi)\, h_2(x, \xi)\, dx\, d\xi \qquad (3.1.5)$$

if h_1 and h_2 lie in $L^2(\mathbb{R}^2)$.

Proof. The claims regarding the two roles of the Wigner function are verified by means of immediate calculations. Then, one verifies that if u and v lie in $\mathcal{S}(\mathbb{R})$, the function $W(u, v)$ lies in $\mathcal{S}(\mathbb{R}^2)$: as a consequence (using duality), the operator $\mathrm{Op}(h)$ makes sense, if $h \in \mathcal{S}'(\mathbb{R}^2)$, as a linear operator from $\mathcal{S}(\mathbb{R})$ to $\mathcal{S}'(\mathbb{R})$. We shall often denote tempered distributions in the plane, especially automorphic ones, as $\mathfrak{S}$ rather than h. One verifies also that if u and v lie in $\mathcal{S}'(\mathbb{R})$, the integral (3.1.2) is weakly convergent in $\mathcal{S}'(\mathbb{R}^2)$, i.e., the integral obtained when testing it against a function in $\mathcal{S}(\mathbb{R}^2)$ is convergent: the second "topological" claim of the proposition follows. Finally, the integral kernel of the operator $\mathrm{Op}(f)$ is the function $K(x, y) = \left(\mathcal{F}_2^{-1}h\right)\left(\frac{x+y}{2}, x - y\right)$, and the map from h to K is an isometry of $L^2(\mathbb{R}^2)$: this proves the last assertion. □

Two unitary representations in the space $L^2(\mathbb{R})$ are all-important in pseudodifferential analysis. The first one is defined with the help of the symplectic form $[\,,\,]$ on $\mathbb{R}^2 \times \mathbb{R}^2$, which is the (alternate) bilinear form such that

$$[(y, \eta), (y', \eta')] = -y\eta' + y'\eta. \qquad (3.1.6)$$

Given $(y, \eta) \in \mathbb{R}^2$, consider the unitary transformation $\tau_{y,\eta}$ of $L^2(\mathbb{R})$ defined by the equation

$$(\tau_{y,\eta}\, u)(x) = u(x - y)\, e^{2i\pi(x - \frac{y}{2})\eta}. \qquad (3.1.7)$$

One has the general identity

$$\tau_{y,\eta}\,\tau_{y',\eta'} = e^{i\pi\,[(y,\eta),\,(y',\eta')]}\,\tau_{y+y',\eta+\eta'},\tag{3.1.8}$$

which shows that only the scalar factor $e^{i\pi\,[(y,\eta),\,(y',\eta')]}$ prevents the map $(y,\eta) \mapsto \tau_{y,\eta}$ from being a representation of the additive group $\mathbb{R}^2$ in $L^2(\mathbb{R})$: in other words, this map is a so-called projective unitary representation of $\mathbb{R}^2$ in $L^2(\mathbb{R})$. Replacing $\mathbb{R}^2$ by a central extension, to wit the 3-dimensional group known as the Heisenberg group, one would obtain a genuine representation. However, extra factors such as the one we just came across, of absolute value 1 (to be called, generally, phase factors) will not bother us since they will, in this book, disappear from most formulas of interest.

The other unitary representation of interest for us is even more fundamental for our applications in the present book. It is the so-called metaplectic (projective) representation [42] of the group $G = SL(2,\mathbb{R})$ in $L^2(\mathbb{R})$ defined on generators as follows, starting from the warning that, given $g \in G$, only the unordered pair $(\pm\mathrm{Met}(g))$ is well-defined. If $g = \left(\begin{smallmatrix} a & 0 \\ 0 & a^{-1} \end{smallmatrix}\right)$ with $a > 0$, $\mathrm{Met}(g)$ is (plus or minus) the transformation $u \mapsto v$, with $v(x) = a^{-\frac{1}{2}}u(a^{-1}x)$; if $g = \left(\begin{smallmatrix} 1 & 0 \\ c & 1 \end{smallmatrix}\right)$, the same holds with $v(x) = u(x)e^{i\pi cx^2}$; if $g = \left(\begin{smallmatrix} 0 & 1 \\ -1 & 0 \end{smallmatrix}\right)$, $\mathrm{Met}(g)$ is (plus or minus) the transformation $e^{-\frac{i\pi}{4}}\mathcal{F}$, where the Fourier transformation $\mathcal{F}$ on the real line is defined as

$$(\mathcal{F}u)(x) = \widehat{u}(x) = \int_{-\infty}^{\infty} u(y)\,e^{-2i\pi xy}\,dx.\tag{3.1.9}$$

For every $g \in G$, any of the two transformations $\pm\mathrm{Met}(g)$, besides being a unitary transformation of $L^2(\mathbb{R})$, preserves the space $\mathcal{S}(\mathbb{R})$; also, it extends as a continuous transformation of $\mathcal{S}'(\mathbb{R})$ (we always consider on this space its weak topology of topological dual of $\mathcal{S}(\mathbb{R})$, not that it would really matter). Note that, again, one could replace the representation just defined by a genuine one, in which no factors ± 1 would appear: to that effect, one would just have to replace G by its twofold cover $\widetilde{G}$, called (of course) the metaplectic group ; this group has no faithful linear realization, which implies that it is difficult to give its elements very concrete realizations. The following is fundamental.

Proposition 3.1.3. *The Weyl calculus enjoys the following two covariance properties. For every* $\mathfrak{S} \in \mathcal{S}'(\mathbb{R}^2)$*, one has*

$$\tau_{y,\eta}\,\mathrm{Op}(\mathfrak{S})\,\tau_{y,\eta}^{-1} = \mathrm{Op}\,((x,\xi) \mapsto \mathfrak{S}(x-y,\xi-\eta)), \quad (y,\eta) \in \mathbb{R}^2.\tag{3.1.10}$$

Also, for every $g \in G$*, one has*

$$\mathrm{Met}(\widetilde{g})\,\mathrm{Op}(\mathfrak{S})\,\mathrm{Met}(\widetilde{g})^{-1} = \mathrm{Op}(\mathfrak{S} \circ g^{-1})\tag{3.1.11}$$

if $\widetilde{g}$ *is any of the two elements of* $\widetilde{G}$ *lying above* g*.*

Proof. The simplest way to prove these identities is to prove in both cases their specializations involving particular sets of generators of the group involved. The list has already been given for the metaplectic representation; in the first case, one can take the elements $\tau_{y,0}$ and $\tau_{0,\eta}$. Note that, even if $\tau_{y,\eta}$ had only been defined up to an unknown phase factor (as is traditional, for good reasons, when dealing with projective representations), no indeterminacy would be present anyway on the left-hand side of (3.1.10). □

Let us introduce the "Heisenberg operators" (i.e., the infinitesimal operators of the Heisenberg representation) Q, which is the operator that multiplies functions of x on the line by x, and $P = \frac{1}{2i\pi} \frac{d}{dx}$. The symbols of these operators are respectively (as functions of (x, ξ)) x and ξ. If $A = \text{Op}(h)$ for some $h \in \mathcal{S}(\mathbb{R}^2)$, it is a completely elementary matter, starting from (3.1.1) and using when needed one integration by parts, to obtain the symbols of the operators obtained when composing A, on either side, by Q or P. One obtains the formulas

$$\text{Symb}(QA)(x, \xi) = x\, h - \frac{1}{4i\pi} \frac{\partial h}{\partial \xi}, \quad \text{Symb}(PA)(x, \xi) = \xi\, h + \frac{1}{4i\pi} \frac{\partial h}{\partial x},$$

$$\text{Symb}(AQ)(x, \xi) = x\, h + \frac{1}{4i\pi} \frac{\partial h}{\partial \xi}, \quad \text{Symb}(AP)(x, \xi) = \xi\, h - \frac{1}{4i\pi} \frac{\partial h}{\partial x}. \quad (3.1.12)$$

Easy continuity arguments show that the formulas remain valid for every $h \in \mathcal{S}'(\mathbb{R}^2)$.

The single most important operator in this book is the Euler operator $\mathcal{E}$ introduced in (1.1.19), which commutes with all operators $\mathfrak{S} \mapsto \mathfrak{S} \circ g^{-1}$ with $g \in SL(2, \mathbb{R})$: it is not only formally self-adjoint in $L^2(\mathbb{R}^2)$, it is also essentially self-adjoint with as small an initial domain as the set of C^∞ functions with compact support disjoint from $\{0\}$. If $h \in L^2(\mathbb{R}^2)$ and $t > 0$, one sets (as led to doing by Stone's theorem on one-parameter unitary groups)

$$\left(t^{2i\pi\mathcal{E}} h\right)(x, \xi) = t\, h(tx, t\xi). \quad (3.1.13)$$

The transpose of $\mathcal{E}$ with respect to the duality between $\mathcal{S}'(\mathbb{R}^2)$ and $\mathcal{S}(\mathbb{R}^2)$ is $-\mathcal{E}$, which leads to the action of the operator $t^{2i\pi\mathcal{E}}$ on tempered distributions defined in (1.1.20). The role of $\mathcal{E}$ in pseudodifferential analysis is the following: defining the operation $\text{mad}(P \wedge Q)$ on operators ("mad" stands for "mixed adjoint") by the equation

$$\text{mad}(P \wedge Q)\, A = PAQ - QAP, \quad (3.1.14)$$

the symbol of $\text{mad}(P \wedge Q)\, \text{Op}(\mathfrak{S})$ is $\mathcal{E}\mathfrak{S}$ for every tempered distribution $\mathfrak{S}$. This is an immediate consequence of the equations (3.1.12).

We come now to questions of parity: these will prove to be of great importance. A function on the line is said to be of a definite parity if it is even or odd. An operator with (Weyl) symbol $\mathfrak{S} \in \mathcal{S}'(\mathbb{R}^2)$ preserves the parity, i.e., transforms every even (*resp.* odd) function on the line into an even (*resp.* odd) distribution if and only if $\mathfrak{S}$ is an even distribution, i.e., $\mathfrak{S}(x, \xi) = \mathfrak{S}(-x, -\xi)$: only even distributions in $\mathbb{R}^2$ will have to be considered here. On the other hand, we must deal with functions on the line without definite parity. Recall that the automorphism $\mathcal{G}$ of $\mathcal{S}(\mathbb{R}^2)$ or $\mathcal{S}'(\mathbb{R}^2)$ was defined by the equation (2.1.13), or

$$\mathcal{G} = 2^{2i\pi\mathcal{E}} \mathcal{F}^{\mathrm{symp}} = 2^{-\frac{1}{2} + i\pi\mathcal{E}} \mathcal{F}^{\mathrm{symp}} \left(2^{-\frac{1}{2} + i\pi\mathcal{E}}\right)^{-1}. \tag{3.1.15}$$

Set in the usual way $\overset{\vee}{u}(x) = u(-x)$. Starting from (3.1.1), one obtains with the help of elementary manipulations with the Fourier transform the general identity

$$\mathrm{Op}\,(\mathcal{G}h)\,u = \mathrm{Op}(h)\,\overset{\vee}{u}. \tag{3.1.16}$$

In particular, the symbol of the check operator $\mathrm{ch}\colon u \mapsto \overset{\vee}{u}$ is the distribution $\frac{1}{2}\delta$, half the unit mass at $(0,0)$.

The rescaling operator $2^{-\frac{1}{2} + i\pi\mathcal{E}}\colon \mathfrak{S} \mapsto \mathfrak{S}^{\mathrm{resc}}$ already played a role in Section 2.1, dealing with automorphic distributions in the plane and automorphic functions in the hyperbolic half-plane. It is a small irritant to appear consistently in this book, in view of the link $\mathcal{G}\mathfrak{S}^{\mathrm{resc}} = (\mathcal{F}^{\mathrm{symp}}\mathfrak{S})^{\mathrm{resc}}$, valid for every $\mathfrak{S} \in \mathcal{S}'(\mathbb{R}^2)$: indeed, $\mathcal{F}^{\mathrm{symp}}$ is in some sense the most natural Fourier transformation in the plane, while $\mathcal{G}$ is the one with an important role in the Weyl calculus. This distinction would have been avoided if we had defined the Weyl calculus (3.1.1) in a slightly different way, replacing $e^{2i\pi(x-y)\eta}$ by $2\,e^{4i\pi(x-y)\eta}$, which would amount to "choosing $\frac{1}{2}$ as a Planck constant". But we decided against it, because of the very large set of facts and formulas regarding the Weyl calculus on which we shall depend. Note that many authors introduce in a systematic way a "small" Planck constant in the basic formulas: expansions with respect to it constitute one way to approach the useful (and quite popular) domain of semi-classical analysis.

We introduce now two "sets of coherent states" related to the even and odd parts of the metaplectic representation, and we give an interpretation, in terms of pseudodifferential analysis, of the transforms Θ_0 and Θ_1 (2.1.5).

Proposition 3.1.4. *Given $z \in \Pi$, set*

$$\phi_z^0(x) = 2^{\frac{1}{4}} \left(\mathrm{Im}\,(-z^{-1})\right)^{\frac{1}{4}} \exp \frac{i\pi x^2}{\bar{z}},$$

$$\phi_z^1(x) = 2^{\frac{5}{4}} \pi^{\frac{1}{2}} \left(\mathrm{Im}\,(-z^{-1})\right)^{\frac{3}{4}} x \exp \frac{i\pi x^2}{\bar{z}}. \tag{3.1.17}$$

Given $g = \left(\begin{smallmatrix} a & b \\ c & d \end{smallmatrix}\right) \in SL(2,\mathbb{R})$ and any $\tilde{g}$ lying above g in the metaplectic group, one has for some phase factors ω_0, ω_1 depending on z, $\tilde{g}$ the equations

$$\mathrm{Met}(\tilde{g})\,\phi_z^0 = \omega_0\,\phi_{\frac{az+b}{cz+d}}^0, \quad \mathrm{Met}(\tilde{g})\,\phi_z^1 = \omega_1\,\phi_{\frac{az+b}{cz+d}}^1. \tag{3.1.18}$$

The set $\{\phi_z^0 : z \in \Pi\}$ (resp. $\{\phi_z^1 : z \in \Pi\}$) is total in $L^2_{\text{even}}(\mathbb{R})$ (resp. $L^2_{\text{odd}}(\mathbb{R})$). Any distribution $\mathfrak{S} \in \mathcal{S}'_{\text{even}}(\mathbb{R}^2)$ is characterized by the pair of functions

$$(\Theta_\kappa \mathfrak{S})(z) = (\phi_z^\kappa \,|\, \mathrm{Op}(\mathfrak{S}) \, \phi_z^\kappa), \quad \kappa = 0 \text{ or } 1. \tag{3.1.19}$$

Proof. Let us first apologize for the misprint that occurred in [39, p. 22]: the factor $2^{\frac{5}{4}}$ in the second equation (3.1.17) was unfortunately typed as $2^{\frac{3}{4}}$. Since Met is a projective representation, it suffices to prove the equations (3.1.18) when g belongs to the set of generators of $SL(2, \mathbb{R})$ which we used to define the metaplectic representation: the verification is easy in each case.

In view of (3.1.18) and of the covariance formula (3.1.11), computing, say, the Wigner function $W(\phi_z^0, \phi_z^0)$ is an immediate task, as it can be reduced to the case when $z = i$: note that the phase factor ω_0, not made explicit here, disappears in the process. One obtains

$$W(\phi_z^0, \phi_z^0)(x, \xi) = 2 \exp\left(-\frac{2\pi}{\mathrm{Im}\, z} |x - z\xi|^2\right),$$
$$W(\phi_z^1, \phi_z^1) = -(2i\pi\mathcal{E}) \, W(\phi_z^0, \phi_z^0). \tag{3.1.20}$$

Then, one has

$$(\Theta_0 \mathfrak{S})(z) = \langle \mathfrak{S}, \, W(\phi_z^0, \phi_z^0) \rangle, \quad (\Theta_1 \mathfrak{S})(z) = \langle \mathfrak{S}, \, W(\phi_z^1, \phi_z^1) \rangle, \tag{3.1.21}$$

which confirms that the definition (3.1.19) of Θ_0, Θ_1 coincides with the definition (2.1.5) of this pair of operators.

To prove that, given $\mathfrak{S} \in \mathcal{S}'_{\text{even}}(\mathbb{R}^2)$, the conditions $\Theta_0 \mathfrak{S} = \Theta_1 \mathfrak{S} = 0$ imply $\mathfrak{S} = 0$, we remark first that they imply the conditions $(\phi_w^0 \,|\, \mathrm{Op}(\mathfrak{S}) \, \phi_z^0) = (\phi_w^1 \,|\, \mathrm{Op}(\mathfrak{S}) \, \phi_z^1) = 0$ for every pair w, z of points of Π. This follows from the fact, a consequence of the definition of the two sets of coherent states, that the first scalar product becomes a sesquiholomorphic function (holomorphic with respect to w, antiholomorphic with respect to z) after it has been multiplied by $\left(\mathrm{Im}\, (-w^{-1}) \, \mathrm{Im}\, (-z^{-1})\right)^{-\frac{1}{4}}$, and the same goes so far as the second scalar product is concerned after one has replaced the exponent $-\frac{1}{4}$ by $-\frac{3}{4}$. To conclude the proof, it suffices to remark that the linear space generated by the functions ϕ_z^0 (resp. ϕ_z^1) is dense in $\mathcal{S}_{\text{even}}(\mathbb{R})$ (resp. $\mathcal{S}_{\text{odd}}(\mathbb{R})$). In the odd case (only, for convergence), one can use the polarized version of the easily proved identity

$$(8\pi)^{-1} \int_\Pi |(\phi_z^1 \,|\, u)|^2 \, dm(z) = \|u\|^2, \quad u \in L^2_{\text{odd}}(\mathbb{R}), \tag{3.1.22}$$

to wit

$$u(x) = \frac{1}{8\pi} \int_\Pi (\phi_z^1 \,|\, u) \, \phi_z^1(x) \, dm(z). \tag{3.1.23}$$

In the even case, subtracting first a suitable multiple of the function ϕ_i^0, one can assume that the function $v \in \mathcal{S}_{\text{even}}(\mathbb{R})$ to be approached is the derivative of a function $u \in \mathcal{S}_{\text{odd}}(\mathbb{R})$: writing

$$\frac{d}{dx}\left(x\,e^{\frac{i\pi x^2}{\bar{z}}}\right) = \left[1 - \frac{\bar{z}^2}{2}\frac{d}{d\bar{z}}\right]\left(e^{\frac{i\pi x^2}{\bar{z}}}\right) \tag{3.1.24}$$

and using an integration by parts in the identity (3.1.23), we are done. $\qquad\square$

The map $\Theta = (\Theta_0, \Theta_1)$ intertwines the two actions of $SL(2, \mathbb{R})$ on functions in the plane or in Π. Next, (2.1.7), which states that $\pi^2 \mathcal{E}^2$ transfers under Θ_0 or Θ_1 to $\Delta - \frac{1}{4}$, has the pleasant consequence, to be used throughout this book, that the study of the Laplacian in Π can be replaced by that of the Euler operator in $\mathbb{R}^2$: this remains valid in the automorphic situation.

Our understanding of operators with automorphic symbols, beyond the fact (immediate from the metaplectic covariance) that they can be characterized as those commuting with the maps $u \mapsto v$ with $v = \mathcal{F}u$ or $v(x) = e^{i\pi x^2}u(x)$, is limited, and one of the aims of this book is to make improvements in this direction. After much effort, we shall obtain in Chapter 6 some understanding of the restriction to $\mathcal{S}_{\text{odd}}(\mathbb{R})$ of the product of $\text{Op}(\mathfrak{E}_\nu)$ by its adjoint, assuming that $|\text{Re }\nu| < \frac{1}{2}$. This will lead again, in Proposition 6.4.3, to a natural appearance of Eisenstein distributions the parameters of which are critical zeros of zeta.

Eisenstein distributions with trivial zeros of zeta (if so wished: recall that $\mathcal{F}^{\text{symp}}\mathfrak{E}_\nu = \mathfrak{E}_{-\nu}$) as parameters show up in the identity [40, section 10]

$$2\sum_{m\geq 1}\frac{(-q^2)^m}{m!}\frac{\pi^{2m+\frac{1}{2}}}{\Gamma(m+\frac{1}{2})\zeta(2m+1)}\,(u\,|\,\text{Op}(\mathfrak{E}_{2m})u)$$

$$= \sum_{(j,k)=1}'\left[\left|\left(\psi_{j,k}^{(q)}\,|\,u\right)\right|^2 - \left|\left(\psi_{j,k}^0\,|\,u\right)\right|^2\right], \tag{3.1.25}$$

where the sign $\sum'$ means that pairs j, k and $-j, -k$ must be associated before the summation is implemented, and

$$\psi_{j,k}^{(q)}(x) = \begin{cases} |j|^{-\frac{1}{2}}e^{\frac{i\pi k}{j}x^2}e^{\frac{2i\pi q}{j}x} & \text{if } j \neq 0, \\ \delta(x+\frac{q}{k}) & \text{if } j = 0,\ k = \pm 1: \end{cases} \tag{3.1.26}$$

this characterizes the operator with symbol $\mathfrak{E}_{2m}$ as a Taylor coefficient of some explicit hermitian form of arithmetic interest, depending on some parameter q.

Another class of arithmetic symbols is obtained as the result of applying to the Dirac comb any partial product of the Euler expansion of the operator $(\zeta(2i\pi\mathcal{E}))^{-1}$. The associated operators are fully understood [40, prop. 8.4] and consist of "projections" onto finite-dimensional spaces of discrete measures of arithmetic interest on the line: these spaces generalize the notion of modular form of holomorphic type, of weight $\frac{1}{2}$ or $\frac{3}{2}$.

3.2 Spectral decompositions in $L^2(\mathbb{R}^2)$ and $L^2(\Pi)$

The main formula of this section is (3.2.32) below, which reduces the problem of finding the spectral decomposition of functions in $L^2(\Pi)$ relative to Δ to the simpler one of decomposing tempered distributions in $\mathbb{R}^2$ into homogeneous components. Applying this formula, in place of the one (3.2.8) involving a Legendre function, leads to simpler calculations and extends in an easier way to more general functions f. Given a function h in the plane, we denote here as $h_{i\lambda}$ what was denoted as h_λ in Section 4 of [34] and, given a function f in the half-plane, we denote here as $f_{\frac{1+\lambda^2}{4}}$ what was denoted there and in [39] as f_λ. Then, the subscripts $i\lambda$ and $\frac{1+\lambda^2}{4}$ indicate generalized eigenvalues relative to $-2i\pi\mathcal{E}$ in the first case, to Δ in the second.

With the help of a Mellin transformation or, after an exponential change of variable, a Fourier transformation, every function $h \in \mathcal{S}_{\text{even}}(\mathbb{R}^2)$ can be decomposed into homogeneous functions of degrees $-1 - i\lambda$, $\lambda \in \mathbb{R}$, according to the equation

$$h = \int_{-\infty}^{\infty} h_{i\lambda} \, d\lambda \quad \text{with} \quad h_{i\lambda}(x,\xi) = \frac{1}{2\pi} \int_0^\infty t^{i\lambda} h(tx, t\xi) \, dt : \qquad (3.2.1)$$

the function $h_{i\lambda}$ in the plane is characterized by the function $h_{i\lambda}^\flat$ on the real line such that $h_{i\lambda}^\flat(s) = h_{i\lambda}(s,1)$, since

$$h_{i\lambda}(x,\xi) = |\xi|^{-1-i\lambda} h_{i\lambda}^\flat\left(\frac{x}{\xi}\right). \qquad (3.2.2)$$

Both functions $i\lambda \mapsto h_{i\lambda}$ or $h_{i\lambda}^\flat$ extend as analytic functions of $\nu \in \mathbb{C}$, $\text{Re } \nu > -1$.

Given $h \in \mathcal{S}_{\text{even}}(\mathbb{R}^2)$, the functions $h_{i\lambda}^\flat$ will not, generally, lie in $\mathcal{S}(\mathbb{R})$. Set, for $h \in \mathcal{S}_{\text{even}}(\mathbb{R}^2)$ and $g = \left(\begin{smallmatrix} a & b \\ c & d \end{smallmatrix}\right) \in SL(2,\mathbb{R})$, $(\pi(g)h)(x,\xi) = h(dx - b\xi, -cx + a\xi)$: then, one has

$$(\pi(g)h)_{-i\lambda}^\flat(s) = |cs - a|^{-1+i\lambda} h_{i\lambda}^\flat\left(\frac{ds - b}{-cs + a}\right) = \left(\pi_{-i\lambda}(g) h_{i\lambda}^\flat\right)(s), \qquad (3.2.3)$$

recognizing in the second equation a definition of the representation $\pi_{-i\lambda}$ from the principal series of $SL(2,\mathbb{R})$. Now, for $h \in \mathcal{S}(\mathbb{R}^2)$, $u = h_{-i\lambda}^\flat$ does not belong to $\mathcal{S}(\mathbb{R})$, but it is a C^∞ vector of the unitary representation $\pi_{-i\lambda}$: in particular, one has $|s|^{1-i\lambda} u(s) \to u_\infty$ as $|s| \to \infty$, where $u_\infty = \frac{1}{2\pi} \int_0^\infty t^{-i\lambda} h(t,0) \, dt$ is a generally nonzero number. Details can be found if so desired in [34, sections 2,3].

Recall that we have already defined (1.1.20) the operator $t^{2i\pi\mathcal{E}}$ (with $t > 0$) acting on $\mathcal{S}(\mathbb{R})$ and, by duality, the operator $t^{-2i\pi\mathcal{E}}$ acting on $\mathcal{S}'(\mathbb{R})$. Rewriting the definition of $h_{i\lambda}$ as $h_{i\lambda} = \frac{1}{2\pi} \int_0^\infty t^{-1+i\lambda+2i\pi\mathcal{E}} h \, dt$, one is led, so as to preserve this definition in the case of tempered distributions, to trying to define the homogeneous components of a distribution $\mathfrak{S} \in \mathcal{S}'_{\text{even}}(\mathbb{R}^2)$ by setting

$\langle \mathfrak{S}_{i\lambda}, h \rangle = \langle \mathfrak{S}, h_{-i\lambda} \rangle$ for $h \in \mathcal{S}(\mathbb{R}^2)$. However, this is not possible in general: a tempered distribution cannot be tested, usually, on the homogeneous function associated to a C^∞ vector of the unitary representation $\pi_{-i\lambda}$ or, more generally, a homogeneous component of a function in $\mathcal{S}(\mathbb{R}^2)$.

What one can do instead is the following. Consider the resolvent of the Euler operator (an essentially self-adjoint operator in $L^2(\mathbb{R}^2)$ if given the initial domain $\mathcal{S}(\mathbb{R}^2)$) defined as

$$((2i\pi\mathcal{E} + \mu)^{-1}h)(x, \xi) = \begin{cases} \int_0^1 t^\mu h(tx, t\xi)\, dt & \text{if } \mathrm{Re}\ \mu > 0, \\ -\int_1^\infty t^\mu h(tx, t\xi)\, dt & \text{if } \mathrm{Re}\ \mu < 0. \end{cases} \tag{3.2.4}$$

If h lies in $\mathcal{S}(\mathbb{R}^2)$, the same will not be true, in general, of the functions $(2i\pi\mathcal{E} + \mu)^{-1}h$, preventing a definition by duality of the same operator on tempered distributions: in general, this function is not rapidly decreasing for $\mathrm{Re}\ \mu > 0$, and not C^∞ for $\mathrm{Re}\ \mu < 0$. However, the space $\mathcal{S}(\mathbb{R}^2)$ is the projective limit of a decreasing sequence of spaces $\mathcal{S}_N(\mathbb{R}^2)$, the Nth space consisting of functions lying in $L^2(\mathbb{R}^2)$ which remain there after having been applied a number $\leq N$ of operators chosen among the operators $\frac{\partial}{\partial x}, \frac{\partial}{\partial \xi}$ and the operators of multiplication by x, ξ. Now, it is true that, given $N = 0, 1, \ldots$, the function $(2i\pi\mathcal{E} + \mu)^{-1}h$ will lie in $\mathcal{S}_N(\mathbb{R})$ for every $h \in \mathcal{S}(\mathbb{R}^2)$ if $|\mathrm{Re}\ \mu|$ large enough. This makes it possible, given $\mathfrak{S} \in \mathcal{S}'(\mathbb{R}^2)$, to define by duality the distributions $(2i\pi\mathcal{E} + \mu)^{-1}\mathfrak{S}$ for $|\mathrm{Re}\ \mu|$ large enough. One can then prove [39, p. 191] the equation, here inserted for clarity only,

$$\mathfrak{S} = \frac{1}{2i\pi} \lim_{A\to\infty} \int_{\delta-iA}^{\delta+iA} \left[(\mu + 2i\pi\mathcal{E})^{-1} - (-\mu + 2i\pi\mathcal{E})^{-1} \right] \mathfrak{S}\, d\mu \tag{3.2.5}$$

for δ large enough, in a way depending on $\mathfrak{S}$. In the case when $\mathfrak{S} \in L^2(\mathbb{R}^2)$, one can let δ go to 0, obtaining as a result the classical Weyl-Kodaira-Titchmarsh formula.

While, as made clear by what precedes, there is no a priori reason why a given tempered distribution should have a decomposition into homogeneous components of degrees lying on the line $-1 + i\mathbb{R}$, we shall consistently come across distributions which do admit such decompositions, even in the automorphic environment. However, in that case, we shall sometimes have to add finitely many homogeneous (Eisenstein) distributions the degrees of which lie off the spectral line: several examples will show up in Chapter 5.

Remark 3.2.1. Given a tempered distribution $\mathfrak{S}$, $\mathfrak{R}_\mu = (\mu - 2i\pi\mathcal{E})^{-1}\mathfrak{S}$ is a well-defined distribution when $|\mathrm{Re}\ \mu|$ is large enough, and one has $\langle \mathfrak{R}_\mu, (\mu + 2i\pi\mathcal{E})h \rangle = \langle \mathfrak{S}, h \rangle$ for every $h \in \mathcal{S}(\mathbb{R}^2)$. But we shall often have to use the operator $j - 2i\pi\mathcal{E}$ for some specific values of j (typically, $j = 0, \pm 1$ or ± 2). Then, $\mathfrak{R}_j$ will not be a meaningful distribution any longer but, as shown by the last equation, it will still be a well-defined continuous linear form on the space of functions lying in the image of $\mathcal{S}(\mathbb{R}^2)$ under $j + 2i\pi\mathcal{E}$. It will be important, especially in Chapters 4 and 5, to distinguish clearly between operators such as $b - 2i\pi\mathcal{E}$, with b real and $|b|$ large (the frequent use of powers $(b - 2i\pi\mathcal{E})^M$ will be referred to, for short, as the "(b, M)-trick" in the second paragraph following (5.3.19)), and operators such as $j - 2i\pi\mathcal{E}$, where $j \in \mathbb{Z}$ and $|j|$ is small. Using the ones of the first kind and their inverses will help transforming some $d\lambda$-integrals over the real line into convergent ones (in the weak sense in $\mathcal{S}'(\mathbb{R}^2)$), improving the growth of the integrand at infinity. Operators of the second kind are essential in the role of killing poles of some Gamma factors: but this will be felt more like a necessary ailment than like a technical help.

We must also consider the question whether, given a (weakly) measurable function $\lambda \mapsto \mathfrak{S}_{i\lambda}$ on the real line, in which $\mathfrak{S}_{i\lambda}$ is an even tempered distribution for almost every λ, homogeneous of degree $-1 - i\lambda$, the integral $\int_{-\infty}^{\infty} \mathfrak{S}_{i\lambda} \, d\lambda$ makes sense as a tempered distribution. In an L^2-frame, an answer is provided by the Plancherel formula

$$\| h \|_{L^2(\mathbb{R}^2)}^2 = 4\pi \int_{-\infty}^{\infty} \| h_{i\lambda}^{\flat} \|_{L^2(\mathbb{R})}^2 \, d\lambda. \tag{3.2.6}$$

Actually, we need results both less precise and more general. Given a measurable family $(\mathfrak{S}_\tau)$, depending on some parameter τ, of tempered distributions and a positive function $\tau \mapsto b(\tau)$, say that $\mathfrak{S}_\tau$ is a $O(b(\tau))$ in $\mathcal{S}'(\mathbb{R}^2)$ if there exists a continuous semi-norm q on $\mathcal{S}(\mathbb{R}^2)$ such that the estimate $|\langle \mathfrak{S}_\tau, h \rangle| \leq b(\tau) q(h)$ holds for every τ and every $h \in \mathcal{S}(\mathbb{R})$. Then, the integral $\int_{-\infty}^{\infty} \mathfrak{S}_{i\lambda} \, d\lambda$ will be meaningful as a tempered distribution in the case when, for some N, the distribution $\mathfrak{S}_{i\lambda}$ is a $O((1 + |\lambda|)^N)$: indeed, the possibility to move the exponent N to a value below -1 is a consequence of the fact that $\mathfrak{S}_{i\lambda} = (1 - i\lambda)^{-1}(1 + 2i\pi\mathcal{E})\mathfrak{S}_{i\lambda}$.

It follows from (2.1.7) that, under the transform Θ_0 or close associates to it, one should be able to link decompositions of even functions in the plane into their homogeneous components, as just examined, to the question of decomposing functions in the hyperbolic half-plane into generalized eigenfunctions of Δ for generalized eigenvalues $\frac{1+\lambda^2}{4}$. The latter problem is taken care of by the classical Mehler formulas valid if, say, $f \in C_0^\infty(\Pi)$,

$$f(z) = \int_0^\infty f_{\frac{1+\lambda^2}{4}}(z) \, \pi \frac{\Gamma(\frac{1+i\lambda}{2})\Gamma(\frac{1-i\lambda}{2})}{\Gamma(\frac{i\lambda}{2})\Gamma(\frac{-i\lambda}{2})} \, d\lambda, \tag{3.2.7}$$

with

$$f_{\frac{1+\lambda^2}{4}}(z) = \frac{1}{4\pi^2} \int_\Pi f(w) \, \mathfrak{P}_{-\frac{1}{2} + \frac{i\lambda}{2}}(\cosh d(z, w)) \, dm(w). \tag{3.2.8}$$

Recall that dm is the canonical invariant measure on Π, $d(z, w)$ is the hyperbolic distance, characterized by the property that $d(g.z, g.w) = d(z, w)$ for every $g \in SL(2, \mathbb{R})$ together with the special case $\cosh d(z, i) = \frac{1+|z|^2}{2 \operatorname{Im} z}$ (or $d(i, iy) = \log y$

if $y > 1$). Finally, $\mathfrak{P}_{-\frac{1}{2}+\frac{i\lambda}{2}}$ is the Legendre function so denoted, and the extra factor in (3.2.7) is $\left|c(\frac{i\lambda}{2})\right|^{-2}$ in terms of Harish-Chandra's c-function for $\Pi = SL(2,\mathbb{R})\backslash SO(2)$. References for the Mehler formulas can be found in [28] or, in a considerably more general setting, in [11].

Recall from (2.1.3) and (2.1.1) the definitions of the Radon transform and its dual. It is convenient to introduce in $L^2_{\text{even}}(\mathbb{R}^2)$ the operator, a function in the spectral-theoretic sense of the Euler operator,

$$T^* = \left(\frac{\pi}{2}\right)^{\frac{1}{2}} \frac{\Gamma(\frac{1}{2} + i\pi\mathcal{E})}{\Gamma(i\pi\mathcal{E})}, \tag{3.2.9}$$

and its adjoint T.

Proposition 3.2.1. *Given $h \in \mathcal{S}_{\text{even}}(\mathbb{R}^2)$, one has for every $\lambda \in \mathbb{R}$ the identity*

$$(V^*T^*h_{i\lambda})(x) = (2\pi)^{-\frac{1}{2}} \frac{\Gamma(\frac{1-i\lambda}{2})}{\Gamma(-\frac{i\lambda}{2})} \int_{-\infty}^{\infty} h^\flat_{i\lambda}(s) \left(\frac{|z-s|^2}{\text{Im } z}\right)^{-\frac{1}{2}+\frac{i\lambda}{2}} ds. \tag{3.2.10}$$

In the other direction, given $f \in C_0^\infty(\Pi)$, one obtains for every $\lambda \in \mathbb{R}$ the identity

$$(TV f)^\flat_{i\lambda}(s) = \frac{1}{2}(2\pi)^{-\frac{3}{2}} \frac{\Gamma(\frac{1+i\lambda}{2})}{\Gamma(\frac{i\lambda}{2})} \int_\Pi \left(\frac{|z-s|^2}{\text{Im } z}\right)^{-\frac{1}{2}-\frac{i\lambda}{2}} f(z)\, dm(z). \tag{3.2.11}$$

Also, one has the identity

$$V^*(V f)_{i\lambda} = f_{\frac{1+\lambda^2}{4}}. \tag{3.2.12}$$

Proof. One has $h_{i\lambda}(x,\xi) = |\xi|^{-1-i\lambda}h^\flat_{i\lambda}\left(\frac{x}{\xi}\right)$. The operator T^* acts as a scalar on $h_{i\lambda}$. Substituting the result in (2.1.2) and setting $s = -y \cotan \frac{\theta}{2} + x$ in the formula obtained, one finds (3.2.10). Next, starting from (2.1.4) and (3.2.2), one has the equation

$$(TV f)^\flat_{i\lambda}(s) = (2\pi)^{-\frac{3}{2}} \frac{\Gamma(\frac{1+i\lambda}{2})}{\Gamma(\frac{i\lambda}{2})} \int_0^\infty t^{i\lambda-2}\, dt \int_{-\infty}^\infty f\left(\frac{s^2(i+b)}{s(i+b)+t^2}\right) db: \tag{3.2.13}$$

performing the change of variable such that

$$z = \frac{s^2(i+b)}{s(i+b)+t^2}, \quad dm(z) = \frac{2\, dt\, db}{t}, \tag{3.2.14}$$

so that $t^2 = \frac{|z-s|^2}{\text{Im } z}$, one obtains (3.2.11).

Coupling the two equations just proved, one has

$$(V^*(V f)_{i\lambda})(z) = \frac{1}{4\pi^3} \int_\Pi f(w)\, dm(w) \int_{-\infty}^\infty \left(\frac{|z-s|^2}{\text{Im } z}\right)^{-\frac{1}{2}+\frac{i\lambda}{2}} \left(\frac{|w-s|^2}{\text{Im } w}\right)^{-\frac{1}{2}-\frac{i\lambda}{2}} ds. \tag{3.2.15}$$

Now,

$$\frac{1}{\pi} \int_{-\infty}^{\infty} \left(\frac{|z-s|^2}{\text{Im } z}\right)^{-\frac{1}{2}+\frac{i\lambda}{2}} \left(\frac{|w-s|^2}{\text{Im } w}\right)^{-\frac{1}{2}-\frac{i\lambda}{2}} ds = \mathfrak{P}_{-\frac{1}{2}+\frac{i\lambda}{2}}(\cosh d(z,w)) \quad (3.2.16)$$

as a consequence of Plancherel's formula together with the identities [22, p. 401]

$$\pi^{-\frac{1}{2}} \int_{-\infty}^{\infty} \left(\frac{|z-s|^2}{\text{Im } z}\right)^{-\frac{1}{2}+\frac{i\lambda}{2}} e^{-2i\pi s\sigma} ds$$

$$= \frac{2\pi^{-\frac{i\lambda}{2}}}{\Gamma(\frac{1-i\lambda}{2})} (\text{Im } z)^{\frac{1}{2}} e^{-2i\pi\sigma \, \text{Re } z} |\sigma|^{-\frac{i\lambda}{2}} K_{\frac{i\lambda}{2}}(2\pi |\sigma| \, \text{Im } z) \quad (3.2.17)$$

and [22, p. 413]

$$\int_0^{\infty} K_{\frac{i\lambda}{2}}(2\pi |\sigma| \, \text{Im } z) K_{\frac{i\lambda}{2}}(2\pi |\sigma| \, \text{Im } w) \cos(2\pi\sigma \, \text{Re }(z-w)) \, d\sigma$$

$$= \frac{1}{8} (\text{Im } z \, \text{Im } w)^{-\frac{1}{2}} \Gamma(\frac{1+i\lambda}{2}) \Gamma(\frac{1-i\lambda}{2}) \mathfrak{P}_{-\frac{1}{2}+\frac{i\lambda}{2}}(\cosh d(z,w)). \quad (3.2.18)$$

The equation (3.2.12) follows if one compares the result of this sequence of identities to (3.2.8). □

Remark 3.2.2. One may rewrite (3.2.11) as

$$(TV f)(x,\xi) = \frac{1}{2}(2\pi)^{-\frac{3}{2}} \int_{\Pi} f(z) \, dm(z) \int_{-\infty}^{\infty} \frac{\Gamma(\frac{1+i\lambda}{2})}{\Gamma(\frac{i\lambda}{2})} \left(\frac{|z-s|^2}{\text{Im } z}\right)^{-\frac{1}{2}-\frac{i\lambda}{2}} d\lambda$$

$$(3.2.19)$$

provided that one understands the second integral (divergent in a pointwise sense) as defining a distribution, to wit the image under the operator $1 + 4\pi^2 \mathcal{E}^2$ of the integral obtained after one has inserted the extra factor $(1+\lambda^2)^{-1}$, which ensures convergence.

The operator Θ_0 from $L^2_{\text{even}}(\mathbb{R}^2)$ to $L^2(\Pi, dm)$ relates (2.1.9) to the composition V^*T^* since

$$\Theta_0 = V^*T^* 2(2\pi)^{-i\pi\mathcal{E}} \Gamma(i\pi\mathcal{E}) = V^*(2\pi)^{\frac{1}{2}-i\pi\mathcal{E}} \Gamma\left(\frac{1}{2}+i\pi\mathcal{E}\right). \quad (3.2.20)$$

Its adjoint $\Theta_0^* = 2(2\pi)^{i\pi\mathcal{E}} \Gamma(-i\pi\mathcal{E}) TV$ is given explicitly as

$$(\Theta_0^* f)(x,\xi) = 2 \int_{\Pi} f(z) \exp\left(-2\pi \frac{|x-z\xi|^2}{\text{Im } z}\right) dm(z). \quad (3.2.21)$$

One has $\Theta_0 \mathcal{G} = \Theta_0$, as can be proved from the integral definition (2.1.13) of $\mathcal{G}$ or, better, from the pseudodifferential interpretations (3.1.19) and (3.1.16) of the two operators involved. On the other hand, from (3.2.21), the function $\Theta_0^* f$ is $\mathcal{G}$-invariant for every function f, say bounded, on Π.

Proposition 3.2.2. *The transformation TV, initially defined on the space of continuous functions on Π with a compact support, extends as an isometry from
$L^2(\Pi)$ onto the subspace $\mathrm{Ran}\,(TV)$ of $L^2_{\mathrm{even}}(\mathbb{R}^2)$ consisting of all functions invari-
ant under the unitary involution $(2\pi)^{-2i\pi\mathcal{E}}\,\frac{\Gamma(i\pi\mathcal{E})}{\Gamma(-i\pi\mathcal{E})}\,\mathcal{G}$. The operator V^*T^* extends
on $\mathrm{Ran}\,(TV)$ as the inverse of TV, and is zero on the subspace $(\mathrm{Ran}\,(TV))^\perp$ of
$L^2_{\mathrm{even}}(\mathbb{R}^2)$ consisting of all functions changing to their negatives under the same
involution. Moreover, the isometry TV intertwines the two actions of G on $L^2(\Pi)$
and $L^2_{\mathrm{even}}(\mathbb{R}^2)$ respectively, and transforms the operator $\Delta - \frac{1}{4}$ on $L^2(\Pi)$ into the
operator $\pi^2\mathcal{E}^2$ on $L^2_{\mathrm{even}}(\mathbb{R}^2)$.*

Proof. The identity $(V^*T^*TV)f = f$ for every continuous function f on Π follows from (3.2.12), (3.2.1) and (3.2.7), the factor T^*T reducing to a scalar in the
course of this computation. That functions in the image of the isometry TV are
invariant under the involution under consideration follows from the $\mathcal{G}$-invariance,
mentioned immediately after (3.2.21), of functions in the image of Θ_0^*. In just the
same way, using this time the fact that Θ_0 kills distributions which change to
their negatives under $\mathcal{G}$, one sees that the operator V^*T^* is zero on the subspace
of $L^2_{\mathrm{even}}(\mathbb{R}^2)$ consisting of all functions changing to their negatives under the in-
volution $(2\pi)^{-2i\pi\mathcal{E}}\,\frac{\Gamma(i\pi\mathcal{E})}{\Gamma(-i\pi\mathcal{E})}\,\mathcal{G}$. The intertwining property and the transfer property
of the operator $\Delta - \frac{1}{4}$ follow from the same properties relative to the transform
Θ_0 (2.1.6), (2.1.7), since T, as a function of $\mathcal{E}$, commutes with the action of G on
$L^2(\mathbb{R}^2)$ and reduces to a scalar on the space of functions with a given degree of
homogeneity.

The more difficult part is proving that the image of $L^2(\Pi)$ under TV is dense
in the space of even functions on $\mathbb{R}^2$, invariant under the involution

$$(2\pi)^{-2i\pi\mathcal{E}}\,\frac{\Gamma(i\pi\mathcal{E})}{\Gamma(-i\pi\mathcal{E})}\,\mathcal{G}.$$

Given $w \in \Pi$, consider the function

$$h^{(w)}(x,\xi) = (2\pi)^{-\frac{1}{2}} \int_{-\infty}^{\infty} \frac{\Gamma(\frac{1+i\mu}{2})}{\Gamma(\frac{i\mu}{2})} \left(\frac{|x-w\xi|^2}{\mathrm{Im}\,w}\right)^{\frac{-1-i\mu}{2}} \omega(\mu)\,d\mu, \qquad (3.2.22)$$

where ω is an even C^∞ function on the line, with compact support. We first
prove that the function $h^{(w)}$ is invariant under the involution under consideration.
Indeed, from (3.1.20) and the consideration of a Gamma integral, one has

$$\left(\frac{|x-w\xi|^2}{\mathrm{Im}\,w}\right)^{\frac{-1-i\mu}{2}} = \frac{(2\pi)^{\frac{1+i\mu}{2}}}{\Gamma(\frac{1+i\mu}{2})} \int_0^{\infty} t^{i\mu} W\left(\phi_w^0, \phi_w^0\right)(tx, t\xi)\,dt. \qquad (3.2.23)$$

Since

$$\mathcal{G}\left[(x,\xi) \mapsto W\left(\phi_w^0, \phi_w^0\right)(tx, t\xi)\right] = t^{-2} W\left(\phi_w^0, \phi_w^0\right)(t^{-1}x, t^{-1}\xi), \qquad (3.2.24)$$

the function

$$(2\pi)^{\frac{-1-i\mu}{2}} \Gamma(\frac{1+i\mu}{2}) \left(\frac{|x - w\xi|^2}{\operatorname{Im} w}\right)^{\frac{-1-i\mu}{2}} \tag{3.2.25}$$

changes under $\mathcal{G}$ to what is obtained if changing μ to $-\mu$. Our first claim is thus a consequence of the parity of the function ω.

Using (3.2.10) and (3.2.16), we obtain

$$\left(V^*T^*h^{(w)}\right)(z) = \frac{1}{2\pi} \int_{-\infty}^{\infty} \omega(\mu)\, d\mu \int_{-\infty}^{\infty} \left(\frac{|s - w|^2}{\operatorname{Im} w}\right)^{\frac{-1-i\mu}{2}} \left(\frac{|z - s|^2}{\operatorname{Im} z}\right)^{\frac{-1+i\mu}{2}} ds$$

$$= \frac{1}{2} \int_{-\infty}^{\infty} \frac{\Gamma(\frac{1-i\mu}{2})}{\Gamma(\frac{-i\mu}{2})} \frac{\Gamma(\frac{1+i\mu}{2})}{\Gamma(\frac{i\mu}{2})} \mathfrak{P}_{\frac{-1+i\mu}{2}}(\cosh\, d(z, w))\, \omega(\mu)\, d\mu, \quad (3.2.26)$$

and we can reduce the interval of integration to $(0, \infty)$, just forgetting the coefficient $\frac{1}{2}$ in front of the integral.

Finally, using (3.2.19) and (3.2.8), one obtains

$$\left(TVV^*T^*h^{(w)}\right)(x, \xi) = \int_0^{\infty} \frac{\Gamma(\frac{1-i\mu}{2})}{\Gamma(\frac{-i\mu}{2})} \frac{\Gamma(\frac{1+i\mu}{2})}{\Gamma(\frac{i\mu}{2})} g_{\frac{1+\mu^2}{4}}(w; x, \xi)\, \omega(\mu)\, d\mu \tag{3.2.27}$$

if

$$g(z; x, \xi) = (2\pi)^{\frac{1}{2}} \int_{-\infty}^{\infty} \frac{\Gamma(\frac{1+i\lambda}{2})}{\Gamma(\frac{i\lambda}{2})} \left(\frac{|x - z\xi|^2}{\operatorname{Im} z}\right)^{-\frac{1}{2} - \frac{i\lambda}{2}} d\lambda \tag{3.2.28}$$

where, as explained in Remark 3.2.2, $g(z; x, \xi)$ is actually a distribution rather than a function with respect to the variables x, ξ. Applying finally (3.2.7) in a weak sense with respect to (x, ξ) (i.e., testing against a function of this pair of variables in $C_0^{\infty}(\mathbb{R}^2)$), one obtains that $TVV^*T^*h^{(w)} = h^{(w)}$.

One concludes the proof, as in the proof of Proposition 3.1.4, with a sesqui-holomorphic argument, which makes it possible to use in place of integral super-positions of the factors $\left(\frac{|x-w\xi|^2}{\operatorname{Im} w}\right)^{\frac{-1-i\mu}{2}}$ integral superpositions of the factors

$$\left(\frac{(x - w_1\xi)(x - \overline{w}_2\xi)}{\frac{w_1 - \overline{w}_2}{2}}\right)^{\frac{-1-i\mu}{2i}}. \tag{3.2.29}$$

□

Since pseudodifferential analysis puts the emphasis on the operators Θ_0 and Θ_1 (the second of which does not deserve a particular study since it is simply the product of the first one by $2i\pi\mathcal{E}$), it is useful to rephrase the propositions that precede as follows

Proposition 3.2.3. *For $f \in C_0^\infty(\Pi)$, one has*

$$f_{\frac{1+\lambda^2}{4}}(z) = \frac{2(2\pi)^{\frac{-3+i\lambda}{2}}}{\Gamma(\frac{1+i\lambda}{2})} \int_{-\infty}^{\infty} (\Theta_0^* f)_{i\lambda}^{\flat}(s) \left(\frac{|s-z|^2}{\text{Im } z}\right)^{\frac{-1+i\lambda}{2}} ds \qquad (3.2.30)$$

and

$$(\Theta_0^* f)_{i\lambda}^{\flat}(s) = (2\pi)^{\frac{-3-i\lambda}{2}} \Gamma(\frac{1+i\lambda}{2}) \int_{\Pi} \left(\frac{|s-z|^2}{\text{Im } z}\right)^{\frac{-1-i\lambda}{2}} f(z) \, dm(z). \qquad (3.2.31)$$

Proof. Using (3.2.20), one may rewrite (3.2.12) as

$$[\Theta_0 (\Theta_0^* f)_{i\lambda}](z) = 2\pi \, \Gamma\left(\frac{1+i\lambda}{2}\right) \Gamma\left(\frac{1+i\lambda}{2}\right) f_{\frac{1+\lambda^2}{4}}(z). \qquad (3.2.32)$$

Setting $h = \Theta_0^* f$, one combines this equation with the equations

$$
\begin{aligned}
(\Theta_0 h_{i\lambda})(z) &= 2 \int_{\mathbb{R}^2} h_{i\lambda}(x,\xi) \exp\left(-2\pi \frac{|x-z\xi|^2}{\text{Im } z}\right) dx \, d\xi \\
&= 2 \int_{\mathbb{R}^2} |\xi|^{-i\lambda} h_{i\lambda}^{\flat}(s) \exp\left(-2\pi \xi^2 \frac{|s-z|^2}{\text{Im } z}\right) ds \, d\xi \\
&= 2(2\pi)^{\frac{-1+i\lambda}{2}} \Gamma(\frac{1-i\lambda}{2}) \int_{-\infty}^{\infty} h_{i\lambda}^{\flat}(s) \left(\frac{|s-z|^2}{\text{Im } z}\right)^{\frac{-1+i\lambda}{2}} ds :
\end{aligned} \qquad (3.2.33)
$$

between the first and second equation in the last sequence, we have used the fact that $h_{i\lambda}$ is homogeneous of degree $-1 - i\lambda$ and set $x = s\xi$. This proves the first formula announced in the proposition. To prove the one that works in the reverse sense, one combines (3.2.20) with the equation (3.2.11).

Even though the proof is complete, one may wonder why the right-hand side of (3.2.30) is an even function of λ. The symmetry leading to this is provided by the (unitary) intertwining operator $\theta_{i\lambda}$ from the representation $\pi_{i\lambda}$ to $\pi_{-i\lambda}$, defined by the equation $\widehat{\theta_{i\lambda} u}(\sigma) = |\sigma|^{-i\lambda} \hat{u}(\sigma)$. It is related [34, p. 28–29] to the symplectic Fourier transformation by the identity (for $h \in \mathcal{S}_{\text{even}}(\mathbb{R}^2)$)

$$(\mathcal{F}^{\text{symp}} h)_{-i\lambda}^{\flat} = \theta_{i\lambda} h_{i\lambda}^{\flat}. \qquad (3.2.34)$$

It follows that, given h and Φ in $\mathcal{S}_{\text{even}}(\mathbb{R}^2)$, one has

$$
\begin{aligned}
\left(\Phi_{i\lambda}^{\flat} \mid h_{i\lambda}^{\flat}\right)_{L^2(\mathbb{R})} &= \left((\mathcal{F}^{\text{symp}}\Phi)_{-i\lambda}^{\flat} \mid (\mathcal{F}^{\text{symp}} h)_{-i\lambda}^{\flat}\right)_{L^2(\mathbb{R})} \\
&= \left((\mathcal{G}\Phi)_{-i\lambda}^{\flat} \mid (\mathcal{G}h)_{-i\lambda}^{\flat}\right)_{L^2(\mathbb{R})}.
\end{aligned} \qquad (3.2.35)
$$

Applying this identity with $\Phi(x,\xi) = 2 \exp\left(-2\pi \frac{|x-z\xi|^2}{\text{Im } x}\right)$ and $h = \Theta_0^* f$, two $\mathcal{G}$-invariant functions, we are done, only noting that

$$\overline{\Phi_{i\lambda}^{\flat}(s)} = (2\pi)^{\frac{-3-i\lambda}{2}} \Gamma(\frac{1-i\lambda}{2}) \left(\frac{|s-z|^2}{\text{Im } z}\right)^{\frac{-1+i\lambda}{2}}. \qquad (3.2.36)$$

$\square$

The following is a consequence of the identity

$$\Theta_0^* \Theta_0 = 2\,\Gamma(-i\pi\mathcal{E})\Gamma(i\pi\mathcal{E})\,[I + \mathcal{G}]\,, \tag{3.2.37}$$

which follows from Proposition 3.2.2.

Proposition 3.2.4. *Recall that an even distribution in the plane invariant under $\mathcal{G}$ is characterized by its Θ_0-transform; if it changes to its negative under $\mathcal{G}$, it is characterized by its Θ_1-transform. If h is $\mathcal{G}$-invariant and lies in the image of $\mathcal{S}_{\mathrm{even}}(\mathbb{R}^2)$ under $2i\pi\mathcal{E}$, one has*

$$\| \Theta_0 h \|_{L^2(\Pi)} = 2 \, \| \, \Gamma(i\pi\mathcal{E})\, h \, \|_{L^2(\mathbb{R}^2)}. \tag{3.2.38}$$

If $h \in \mathcal{S}_{\mathrm{even}}(\mathbb{R}^2)$ changes to its negative under $\mathcal{G}$, one has

$$\| \Theta_1 h \|_{L^2(\Pi)} = 4 \, \| \, \Gamma(1 + i\pi\mathcal{E})\, h \, \|_{L^2(\mathbb{R}^2)}. \tag{3.2.39}$$

Since the Gamma function decreases exponentially at infinity on vertical lines (1.1.8), the transformations Θ_0 and Θ_1 are very far from being (partial) isometries. The dual Radon transformation V^* defined by the equation $(V^*h)(z) = \langle d\sigma_z, h \rangle$ in (2.1.1) is closer to a partial isometry since, in view of (3.2.9) together with the fact that V^*T^* is a partial isometry, one has

$$\|V^*h\|_{L^2(\Pi)} = \left(\frac{\pi}{2}\right)^{-\frac{1}{2}} \|\frac{\Gamma(i\pi\mathcal{E})}{\Gamma(\frac{1}{2} + i\pi\mathcal{E})}\|_{L^2(\mathbb{R}^2)} \tag{3.2.40}$$

if the function $\pi^{i\pi\mathcal{E}}\Gamma(\frac{1}{2} - i\pi\mathcal{E})\, h$ is invariant under the involution $\mathcal{F}^{\mathrm{symp}}$.

Even though a distribution $\mathfrak{S} \in \mathcal{S}'(\mathbb{R}^2)$ is characterized by its (Θ_0, Θ_1)-transform, as will be explained just before Corollary 3.3.2, the partially defined inverse map is extremely far from being continuous in any useful sense. This will be felt, especially, in the automorphic situation: an operator such as Θ_0 is simply "too good". The operator V^* is much better in this respect, in view of the existence of inverse formulas such as (3.2.11): however, as it consists in integrating over ellipses, it cannot be applied to general distributions. We introduce now two families (W_m) and (U_m), with $m = 0, 1, \ldots,$ of variants of V^* and $V^*\mathcal{F}^{\mathrm{symp}}$ with improved properties, but not so good as to destroy their usefulness.

The integral kernels of all the operators from functions on $\mathbb{R}^2$ to functions on Π considered in the proposition below depend only on $\frac{|x-z\xi|^2}{\mathrm{Im}\,z}$. Now, if $g = \left(\begin{smallmatrix} a & b \\ c & d \end{smallmatrix}\right)$, $\left(\begin{smallmatrix} x' \\ \xi' \end{smallmatrix}\right) = g^{-1}\left(\begin{smallmatrix} x \\ \xi \end{smallmatrix}\right)$ and $z' = \frac{az+b}{cz+d}$, one has $\frac{|x'-z\xi'|^2}{\mathrm{Im}\,z} = \frac{|x-z'\xi|^2}{\mathrm{Im}\,z'}$. This expresses that all operators below are covariant under the pair of actions of $SL(2, \mathbb{R})$ on $\mathbb{R}^2$ and Π, by linear or fractional-linear changes of coordinates: it will make it possible to tacitly reduce most calculations to the case $z = i$, in which the quadratic form under consideration is simply $x^2 + \xi^2$. Besides, under any of these maps, the operator $\pi^2\mathcal{E}^2$ transfers to $\Delta - \frac{1}{4}$, as stated in (2.1.7) in relation to Θ_0: the proof is the same in all cases, since it boils down to showing that applying $\pi^2 \mathcal{E}^2$ to any function of $\frac{|x-z\xi|^2}{\operatorname{Im} z}$, considered as a function of (x, ξ), gives the same result as applying it the operator $\Delta - \frac{1}{4}$, when considered as a function of z [39, p. 24].

Proposition 3.2.5. *For $m = 0, 1, \ldots$, define*

$$J_{0,m}(\rho) = \left(\rho^{-1} \frac{d}{d\rho} \right)^m J_0(\rho), \quad \rho > 0, \tag{3.2.41}$$

and, for $m = 1, 2, \ldots$, set

$$g_m(r) = \frac{(-1)^m \pi^{-1}}{2^m (m-1)!} (1 - r^2)^{m-1} \operatorname{char}(r < 1), \quad r > 0. \tag{3.2.42}$$

Let W_m and U_m be the operators from $\mathcal{S}_{\text{even}}(\mathbb{R}^2)$ to functions on Π defined by the equations

$$(W_m h)(z) = \int_{\mathbb{R}^2} h(x, \xi)\, g_m \left(\frac{|x - z\xi|}{(\operatorname{Im} z)^{\frac{1}{2}}} \right) dx\, d\xi, \qquad m = 1, 2, \ldots,$$

$$(U_m h)(z) = \int_{\mathbb{R}^2} h(x, \xi)\, J_{0,m} \left(2\pi \frac{|x - z\xi|}{(\operatorname{Im} z)^{\frac{1}{2}}} \right) dx\, d\xi, \quad m = 0, 1, \ldots, \tag{3.2.43}$$

and complete the first definition by setting $W_0 = V^*$. One has, for every $m = 0, 1, \ldots$, the identity $W_m = U_m \mathcal{F}^{\text{symp}}$. Besides, for every $h \in \mathcal{S}_{\text{even}}(\mathbb{R}^2)$, one has the identities

$$V^* \mathcal{F}^{\text{symp}} h = (-1)^m U_m \left[(1 - 2i\pi\mathcal{E})(3 - 2i\pi\mathcal{E}) \ldots (2m - 1 - 2i\pi\mathcal{E})\, h \right],$$
$$V^* h = (-1)^m W_m \left[(1 + 2i\pi\mathcal{E})(3 + 2i\pi\mathcal{E}) \ldots (2m - 1 + 2i\pi\mathcal{E})\, h \right] : \tag{3.2.44}$$

the polynomial in $\mp i\pi\mathcal{E}$, of Pochhammer's style, on the right-hand sides, should be interpreted as the identity operator when $m = 0$.

Let us take this opportunity to recall the Pochhammer notation, which will come in handy on occasions: given a number a (or an element in any algebra with unit) and $n = 1, 2, \ldots$, one sets $(a)_n = a(a+1) \ldots (a + n - 1)$, completing this by $(a)_0 = 1$.

Proof. In the coordinate $r = \sqrt{x^2 + \xi^2}$, the $\mathcal{F}^{\text{symp}}$-transform of a radial function $h(r)$ is the radial function $r \mapsto 2\pi \int_0^\infty t\, h(t)\, J_0(2\pi rt)\, dt$. If one takes for h the integral kernel of the transformation V^*, to wit $h(r) = \frac{1}{2\pi} \delta(r - 1)$, the $\mathcal{F}^{\text{symp}}$-transform of which is the function $r \mapsto J_0(2\pi r)$, one obtains that the operator $V^* \mathcal{F}^{\text{symp}} = W_0 \mathcal{F}^{\text{symp}}$ coincides with U_0, as claimed.

Since the Bessel function J_0 is analytic and even, the integral kernel of the transformation U_0 is an analytic function of (x, ξ): however, in contrast to the integral kernel of Θ_0, it is not rapidly decreasing at infinity. It is useful to improve (not too much) the growth of this integral kernel at infinity. To this effect, we remark first that the function $J_{0,m}$ is still an even analytic function on the line. Its advantage rests on the asymptotic expansion $J_0(\rho) \sim \sum_{n\geq 0} \rho^{-n-\frac{1}{2}}(a_n \cos \rho + b_n \sin \rho)$, valid [22, p. 139] for some (explicit) coefficients a_n, b_n: applying $\left(\rho^{-1}\frac{d}{d\rho}\right)^m$ substitutes for this expansion a similar one, in which the sum starts from $n = m$. Next, we prove the identity

$$J_0(\rho) = (-1)^m \left(\rho\frac{d}{d\rho} + 2\right)\left(\rho\frac{d}{d\rho} + 4\right) \cdots \left(\rho\frac{d}{d\rho} + 2m\right) J_{0,m}(\rho). \qquad (3.2.45)$$

To check it, we first remark from the differential equation which essentially defines the function J_0 that one has

$$-J_0(\rho) = \left(\frac{d}{d\rho} + \frac{1}{\rho}\right)\frac{d}{d\rho} J_0(\rho) = \left(\rho\frac{d}{d\rho} + 2\right)\rho^{-1}\frac{d}{d\rho} J_0(\rho), \qquad (3.2.46)$$

which is the case $m = 1$ of the claimed identity. A proof by induction will then follow from the identity

$$\left(\rho^{-1}\frac{d}{d\rho}\right)^m \left(\rho\frac{d}{d\rho} + 2\right) = \left(\rho\frac{d}{d\rho} + 2m + 2\right)\left(\rho^{-1}\frac{d}{d\rho}\right)^m, \qquad (3.2.47)$$

proved by testing it on ρ^α (with $\alpha \in \mathbb{C}$ arbitrary) and using the fact that

$$\left(\rho^{-1}\frac{d}{d\rho}\right)^m \rho^\alpha = C(\alpha, m)\,\rho^{\alpha - 2m}$$

for some constant $C(\alpha, m)$.

The two equations (3.2.44) are equivalent to each other, assuming that the equation linking W_m and U_m has already been obtained, and the first follows from (3.2.45), not forgetting that the transpose of $\mathcal{E}$ is $-\mathcal{E}$.

What remains to be proved so as to complete the proof of Proposition 3.2.5 is the validity, for $m = 1, 2, \ldots$, of the identity $W_m = U_m \mathcal{F}^{\text{symp}}$, equivalent to

$$g_m(r) = \lim_{\varepsilon \to 0} g_m^\varepsilon(r), \qquad (3.2.48)$$

with

$$g_m^\varepsilon(r) := 2\pi \int_0^\infty t\, e^{-2\pi\varepsilon t} J_{0,m}(2\pi t)\, J_0(2\pi r t)\, dt, \qquad (3.2.49)$$

which expresses that the $\mathcal{F}^{\text{symp}}$-transform of the integral kernel of U_m is W_m: the case when $m = 0$ has already been treated.

For $m = 0, 1, \ldots$, one has the identity

$$J_{0,m}(\rho) = \frac{(-1)^m \pi^{-\frac{1}{2}}}{2^m \Gamma(m + \frac{1}{2})} \int_{-1}^1 e^{-i\rho\lambda}(1 - \lambda^2)^{m-\frac{1}{2}} d\lambda, \qquad \rho > 0. \qquad (3.2.50)$$

This is to be found, e.g., in [22, p. 79] when $m = 0$. The general case follows by induction, with the help of an integration by parts based on the identity

$$\rho^{-1}\frac{d}{d\rho}(e^{-i\rho\lambda}).(1-\lambda^2)^{m-\frac{1}{2}} = \frac{i}{2m+1}e^{-i\rho\lambda}\rho^{-1}\frac{d}{d\lambda}(1-\lambda^2)^{m+\frac{1}{2}}. \qquad (3.2.51)$$

Assuming from now on that $m \geq 1$, one writes

$$g_m^\varepsilon(r) = \frac{(-1)^m\pi^{\frac{1}{2}}}{2^{m-1}\Gamma(m+\frac{1}{2})}\int_{-1}^1(1-\lambda^2)^{m-\frac{1}{2}}d\lambda\int_0^\infty t\,e^{-2\pi\varepsilon t}J_0(2\pi rt)\,\cos(2\pi\lambda t)\,dt. \qquad (3.2.52)$$

After an integration by parts, one has

$$g_m^\varepsilon(r) = \frac{(-1)^m\pi^{-\frac{1}{2}}}{2^{m-2}\Gamma(m-\frac{1}{2})}\int_0^1\lambda(1-\lambda^2)^{m-\frac{3}{2}}d\lambda\int_0^\infty e^{-2\pi\varepsilon t}J_0(2\pi rt)\,\sin(2\pi\lambda t)\,dt. \qquad (3.2.53)$$

The limit as $\varepsilon \to 0$ of the last integral is [22, p. 425]

$$\frac{1}{2\pi}(\lambda^2-r^2)_+^{-\frac{1}{2}} : = \frac{1}{2\pi}\operatorname{char}(r < \lambda)(\lambda^2-r^2)^{-\frac{1}{2}}. \qquad (3.2.54)$$

It follows that the limit as $\varepsilon \to 0$ of $g_m^\varepsilon(r)$ is

$$\frac{(-1)^m\pi^{-\frac{3}{2}}}{2^{m-1}\Gamma(m-\frac{1}{2})}\int_r^1\lambda(1-\lambda^2)^{m-\frac{3}{2}}(\lambda^2-r^2)_+^{-\frac{1}{2}}d\lambda,$$

where the last integral can be transformed to

$$\frac{1}{2}\int_{r^2}^1(1-t)^{m-\frac{3}{2}}(t-r^2)_+^{-\frac{1}{2}}dt = \frac{1}{2}(1-r^2)_+^{m-1}\int_0^1(1-s)^{m-\frac{3}{2}}s^{-\frac{1}{2}}ds$$

$$= \frac{1}{2}\frac{\pi^{\frac{1}{2}}\Gamma(m-\frac{1}{2})}{(m-1)!}(1-r^2)_+^{m-1}. \qquad (3.2.55)$$

This concludes the proof of Proposition 3.2.5.   □

One may also note its consequence that

$$U_0\left[(1+2i\pi\mathcal{E})(3+2i\pi\mathcal{E})\ldots(2m-1+2i\pi\mathcal{E})\,h\right]$$
$$= (-4)^m\Delta(\Delta+2)(\Delta+6)\ldots(\Delta+m^2-m)\,(U_m h), \qquad (3.2.56)$$

or

$$V^*\left[(1-2i\pi\mathcal{E})(3-2i\pi\mathcal{E})\ldots(2m-1-2i\pi\mathcal{E})\,h\right]$$
$$= (-4)^m\Delta(\Delta+2)(\Delta+6)\ldots(\Delta+m^2-m)\,(W_m h). \qquad (3.2.57)$$

We save for future use the following formula.

Lemma 3.2.6. *For* $m = 1, 2, \ldots$ *and* $\alpha \in \mathbb{R}^\times, \beta > 0$, *set*

$$I_m(\alpha, \beta) = \int_{-\infty}^{\infty} g_m(\beta\sqrt{1+t^2})\, e^{it\alpha} dt, \tag{3.2.58}$$

where the function g_m *has been defined in* (3.2.42). *The integral is zero unless* $\beta < 1$, *in which case one has*

$$I_m(\alpha, \beta) = (-1)^m \pi^{-\frac{1}{2}} \frac{\Gamma(m + \frac{1}{2})}{\Gamma(m)} \left(\frac{1 - \beta^2}{\alpha^2}\right)^{\frac{m}{2}} \beta^{m-1} J_m\left(\frac{|\alpha|}{\beta}\sqrt{1 - \beta^2}\right), \tag{3.2.59}$$

where J_m *is the usual Bessel function so denoted.*

Proof. Recall that $g_m(r) = C(m)\,(1 - r^2)_+^{m-1}$, with $C(m) = \frac{(-1)^m \pi^{-1}}{2^m (m-1)!}$. Assuming that $\beta < 1$, one has, with $\gamma = \sqrt{\beta^{-2} - 1}$,

$$(C(m))^{-1} I_m(\alpha, \beta) = 2 \int_0^\gamma \cos(\alpha t) \left[1 - \frac{1 + t^2}{1 + \gamma^2}\right]^{m-\frac{1}{2}} dt$$

$$= 2\,(1 + \gamma^2)^{\frac{1}{2} - m} \int_0^\gamma (\gamma^2 - t^2)^{m - \frac{1}{2}} \cos(\alpha t)\, dt$$

$$= 2^m \pi^{\frac{1}{2}} \Gamma(m + \frac{1}{2})\,(1 + \gamma^2)^{\frac{1}{2} - m} \left(\frac{\gamma}{|\alpha|}\right)^m J_m(|\alpha|\gamma), \tag{3.2.60}$$

according to [22, p. 401]: (3.2.59) follows.   □

In all that precedes in this section, we have made use of the spaces $L^2(\mathbb{R}^2)$ and $L^2(\Pi)$, not of analogous spaces of automorphic objects such as $L^2(\Gamma \backslash \Pi)$. How to define a Hilbert space $L^2(\Gamma \backslash \mathbb{R}^2)$ is much less obvious since, as already remarked, the action of Γ in $\mathbb{R}^2$ has no fundamental domain. However, there is a way to do so, based on a complete spectral analysis of the Poincaré process (the summation of g-transforms of a given function h, with $g \in \Gamma$), as follows [39, chapter 5].

Whenever a function h lies in the image of $\mathcal{S}_{\text{even}}(\mathbb{R}^2)$ under the operator $\pi^2 \mathcal{E}^2 \left(\frac{1}{4} + \pi^2 \mathcal{E}^2\right)$, the series $\mathfrak{S} = \sum_{g \in \Gamma} h \circ g$ converges weakly in $\mathcal{S}'(\mathbb{R}^2)$, and its sum is an automorphic distribution. Moreover, one may set

$$\|\mathfrak{S}\|_{L^2(\Gamma \backslash \mathbb{R}^2)}^2 := \sum_{g \in \Gamma} \int_{\mathbb{R}^2} (h \circ g)(x, \xi)\, \bar{h}(x, \xi)\, dx\, d\xi \tag{3.2.61}$$

as the right-hand side is positive if $\mathfrak{S} \neq 0$ and depends only on $\mathfrak{S}$, not on h. The proof depends on a complete spectral decomposition of the bilinear operator involved in the summation process, to wit on the formula (in which $f \in \mathcal{S}_{\text{even}}(\mathbb{R}^2)$)

$$\langle \mathfrak{S}, f \rangle = \frac{1}{2\pi} \int_{-\infty}^{\infty} \langle \mathcal{E}_{i\lambda}, h \rangle \langle \mathcal{E}_{-i\lambda}, f \rangle \frac{d\lambda}{\zeta(i\lambda)\zeta(-i\lambda)}$$

$$+ 2 \sum_{r \neq 0} \Gamma\left(\frac{i\lambda_r}{2}\right) \Gamma\left(-\frac{i\lambda_r}{2}\right) \sum_{\ell} \epsilon_{r,\ell}\, \langle \mathfrak{N}_{r,\ell}, h \rangle \langle \mathfrak{N}_{-r,\ell}, f \rangle : \tag{3.2.62}$$

we have set $\epsilon_{r,\ell} = (-1)^\varepsilon$, where $\varepsilon = 0$ or 1 is associated to the Hecke distribution $\mathfrak{N}_{r,\ell}$ as in Theorem 1.2.2, or to $\mathcal{N}_{|r|,\ell}$ as in Theorem 2.1.2.

The norm in the space $L^2(\Gamma\backslash\mathbb{R}^2)$ (the completion of the space of distributions $\mathfrak{S}$ just introduced) relates to the norm in $L^2(\Gamma\backslash\Pi)$ by the pair of formulas

$$\|\Theta_0\mathfrak{S}\|_{L^2(\Gamma\backslash\Pi)} = 2\,\|\Gamma(i\pi\mathcal{E})\,\mathfrak{S}\|_{L^2(\Gamma\backslash\mathbb{R}^2)} \qquad \text{if } \mathcal{G}h = h,$$
$$\|\Theta_1\mathfrak{S}\|_{L^2(\Gamma\backslash\Pi)} = 4\,\|\Gamma(1+i\pi\mathcal{E})\,\mathfrak{S}\|_{L^2(\Gamma\backslash\mathbb{R}^2)} \quad \text{if } \mathcal{G}h = -h. \qquad (3.2.63)$$

Even though the two formulas look identical to the formulas of Proposition 3.2.4, this pair of equations, which deals with automorphic distributions (in the plane) and automorphic functions (in the half-plane), requires a solid 40-page proof.

3.3 The sharp composition of homogeneous functions

There is an elementary integral formula for the sharp composition of two symbols in $\mathcal{S}(\mathbb{R}^2)$: though its usefulness is limited, we recall it since it will help clarify things in a moment. Setting $X = (x,\xi)$, $Y = (y,\eta)$, $Z = (z,\zeta)$ and using the symplectic form introduced in (3.1.6), one has the general formula (cf. e.g. [39, p. 26])

$$(h_1\#h_2)(X) = 4\int_{\mathbb{R}^2\times\mathbb{R}^2} h_1(Y)\,h_2(Z)\,e^{-4i\pi[Y-X,Z-X]}dY\,dZ. \qquad (3.3.1)$$

We recall here the way the sharp composition of symbols (the operation corresponding to the composition of operators) combines with the decompositions of symbols into homogeneous components. We have proved the formula that follows, in various degrees of generality, in more than one place. The shortest proof, at the same time a better explanation of what really goes on here, can be found in [39, section 1.2]: still, it is much too lengthy to be reproduced here. Quoting from Theorem 1.2.2 there, if h^1 is an even symbol homogeneous of degree $-1 - i\lambda_1$, originating from the decomposition of a function in $\mathcal{S}(\mathbb{R}^2)$, and a similar notation goes for h^2, one has, if $h = h^1 \# h^2$, the identity $h = \int_{-\infty}^{\infty} h_{i\lambda}d\lambda$, with

$$h_{i\lambda}^\flat(s) = \frac{1}{4\pi} \sum_{j=0,1} \int_{\mathbb{R}^2} K_{i\lambda_1,i\lambda_2;i\lambda}^{(j)}(s_1,s_2;s)\,\left(h^1\right)_{i\lambda_1}^\flat(s_1)\,\left(h^2\right)_{i\lambda_2}^\flat(s_2)\,ds_1\,ds_2 :$$
$$(3.3.2)$$

the integral kernel is given as

$$K_{i\lambda_1,i\lambda_2;i\lambda}^{(j)}(s_1,s_2;s) = C_{i\lambda_1,i\lambda_2;i\lambda}^{(j)}\,\chi_{i\lambda_1,i\lambda_2;i\lambda}^{(j)}(s_1,s_2;s), \qquad (3.3.3)$$

with

$$C_{i\lambda_1,i\lambda_2;i\lambda}^{(j)} = (-1)^j\,2^{\frac{-1+i(\lambda_1+\lambda_2-\lambda)}{2}}$$
$$\cdot B_j\left(\frac{1+i(-\lambda_1+\lambda_2-\lambda)}{2}\right) B_j\left(\frac{1+i(\lambda_1-\lambda_2-\lambda)}{2}\right) B_j\left(\frac{1+i(\lambda_1+\lambda_2+\lambda)}{2}\right)$$
$$(3.3.4)$$

and

$$\chi^{(j)}_{i\lambda_1, i\lambda_2; i\lambda}(s_1, s_2; s)$$

$$= |s_1 - s_2|_j^{\frac{-1+i(\lambda_1+\lambda_2+\lambda)}{2}} |s_2 - s|_j^{\frac{-1+i(-\lambda_1+\lambda_2-\lambda)}{2}} |s - s_1|_j^{\frac{-1+i(\lambda_1-\lambda_2-\lambda)}{2}}. \qquad (3.3.5)$$

Some observations are necessary. What is really meant here is that if $h^1 = \int_{-\infty}^{\infty} (h^1)_{i\lambda_1} d\lambda_1$ lies in $\mathcal{S}_{even}(\mathbb{R})$ and the same goes with h^2, the symbol $h = h^1 \# h^2$, which lies in $\mathcal{S}_{even}(\mathbb{R})$ too, has a decomposition $h = \int_{\mathbb{R}^2} d\lambda_1 d\lambda_2 \int_{-\infty}^{\infty} h_{i\lambda} d\lambda$, with $h^b_{i\lambda}$ given by the preceding formulas. We shall refer to this way of understanding (3.3.2) as being an identity "in the weak $d\lambda_1 d\lambda_2$-sense". This terminology will make it possible to dispense with writing consistently extra integrals, in already quite complicated equations.

To readers who would like to check this quotation with the given reference, let us indicate that there is now some simplification due to the fact that we deal here with globally even symbols only, so that the three indexes denoted as $\delta_1, \delta_2, \delta$ in [39] are zero: but $j = 0, 1$ survives and, with the notation there, one has $\varepsilon_1 = \varepsilon_2 = \varepsilon = j$. We denote as $K^{(j)}_{i\lambda_1, i\lambda_2; i\lambda}$ what would have been denoted $\left[K^{j,j;j}_{i\lambda_1, i\lambda_2; i\lambda} \right]_j$ there. Also, we denote as $\chi^{(j)}_{i\lambda_1, i\lambda_2; i\lambda}$ what would have been denoted $\chi^{j,j;j}_{i\lambda_1, i\lambda_2; i\lambda}$ there and as $C^{(j)}_{i\lambda_1, i\lambda_2; i\lambda}$ what would have been denoted $C^{j,j;j}_{i\lambda_1, i\lambda_2; i\lambda}$. It is useful to understand the role of the parameter j. From the relation

$$K^{(j)}_{i\lambda_2, i\lambda_1; i\lambda}(s_2, s_1; s) = (-1)^j K^{(j)}_{i\lambda_1, i\lambda_2; i\lambda}(s_1, s_2; s), \qquad (3.3.6)$$

of immediate verification, it follows that, if one keeps only the term with $j = 0$ from the right-hand side of (3.3.2), the result one gets is that corresponding to the decomposition, in place of $h^1 \# h^2$, of the commutative part

$$h^1 \bigtriangleup h^2 = \frac{1}{2} \left(h^1 \# h^2 + h^2 \# h^1 \right) : \qquad (3.3.7)$$

similarly, the term with $j = 1$ corresponds to the decomposition of the anticommutative part $h^1 \bigtriangledown h^2 = \frac{1}{2} \left(h^1 \# h^2 - h^2 \# h^1 \right)$. Note that the formulas for the $\bigtriangleup$ (resp. $\bigtriangledown$) compositions of symbols are obtained from (3.3.1) by simply replacing the exponential $\exp(4i\pi(y\zeta - z\eta))$ there by its cosine (resp. i times its sine) part. One may set also $\bigtriangleup = \underset{0}{\#}$ and $\bigtriangledown = \underset{1}{\#}$ so as to treat the two parts simultaneously.

Then, (3.3.1) yields the pair of identities

$$(h_1 \underset{j}{\#} h_2)(X) = \int_{\mathbb{R}^2 \times \mathbb{R}^2} h_1(Y) \, h_2(Z) \, F_j(4[Y - X, Z - X]) \, dY \, dZ \qquad (3.3.8)$$

with $F_0(t) = \cos t$, $F_1(t) = -i \sin t$. If one considers the map J such that $J(x, \xi) = (-x, \xi)$, it is immediate that

$$[Y - JX, Z - JX] = -[JY - X, JZ - X]. \qquad (3.3.9)$$

Just performing a change of variables, one obtains the following consequence, to be used in Chapter 5: if h_1 and h_2 satisfy the identities $h_1(JY) = (-1)^{\varepsilon_1} h_1(Y)$ and $h_2(JY) = (-1)^{\varepsilon_2} h_2(Y)$, one has

$$(h_1 \underset{j}{\#} h_2)(JX) = (-1)^{\varepsilon_1 + \varepsilon_2 + j}(h_1 \underset{j}{\#} h_2)(X). \tag{3.3.10}$$

There is a link between the sharp composition in the plane and the point-wise product, to be completed with the Poisson bracket, in the half-plane, if one combines the first bilinear operation with the decomposition of symbols into homogeneous components. Given a pair f_1, f_2 of C^∞ functions in the hyperbolic half-plane (for instance modular forms), one can always define their pointwise product and — which is just as important — their Poisson bracket

$$\{f_1, f_2\} = y^2 \left(-\frac{\partial f_1}{\partial y} \frac{\partial f_2}{\partial x} + \frac{\partial f_1}{\partial x} \frac{\partial f_2}{\partial y} \right) \quad (z = x + iy). \tag{3.3.11}$$

It is convenient to set

$$f_1 \underset{j}{\times} f_2 = \begin{cases} f_1 f_2 & \text{if } j = 0, \\ \frac{1}{2}\{f_1, f_2\} & \text{if } j = 1. \end{cases} \tag{3.3.12}$$

We show now that, when coupling the sharp product on $\mathbb{R}^2$ with the decomposition of symbols into homogeneous components (as has been done in (3.3.2) to (3.3.5)), one can, using also the map Θ_0 defined in (2.1.5), express it in terms of the two operations just considered in the half-plane.

Theorem 3.3.1. *Given $h^1, h^2 \in \mathcal{S}_{\text{even}}(\mathbb{R}^2)$, one has*

$$h^1 \# h^2$$

$$= \frac{\pi}{2} \sum_{j=0,1} (-i)^j \int_{-\infty}^{\infty} d\lambda \int_{\mathbb{R}^2} \left(\prod_{\eta_1, \eta_2 = \pm 1} \Gamma\left(\frac{1 + i(\eta_1 \lambda_1 + \eta_2 \lambda_2 + \eta_1 \eta_2 \lambda) + 2j}{4} \right) \right)^{-1}$$

$$\cdot \left[\Theta_0^* \left(\Theta_0(h^1)_{i\lambda_1} \underset{j}{\times} \Theta_0(h^2)_{i\lambda_2} \right) \right]_{i\lambda} d\lambda_1 d\lambda_2. \tag{3.3.13}$$

Proof. The following was proved in [34, p. 71–74] and a version dealing only with the case of globally even symbols (the case of interest here) was proved again in [39, p. 56–57]. Let h^1, h^2 be two even functions in $\mathcal{S}(\mathbb{R}^2)$. For every pair (λ_1, λ_2) of real numbers, one has the identity

$$\left[TV\left((V^*T^*(h^1))_{i\lambda_1}\right) \underset{j}{\times} (V^*T^*(h^2))_{i\lambda_2})\right)\right]^\flat_{i\lambda}(s)$$

$$= 2^{-\frac{9}{2}} \pi^{-2} \frac{1}{\Gamma\left(-\frac{i\lambda_1}{2}\right) \Gamma\left(-\frac{i\lambda_2}{2}\right) \Gamma\left(\frac{i\lambda}{2}\right)}$$

$$\cdot \Gamma\left(\frac{1 - i(\lambda + \lambda_1 + \lambda_2) + 2j}{4}\right) \Gamma\left(\frac{1 + i(\lambda - \lambda_1 + \lambda_2) + 2j}{4}\right)$$

$$\cdot \Gamma\left(\frac{1 + i(\lambda + \lambda_1 - \lambda_2) + 2j}{4}\right) \Gamma\left(\frac{1 + i(\lambda - \lambda_1 - \lambda_2) + 2j}{4}\right)$$

$$\cdot \int_{\mathbb{R}^2} \chi^{(j)}_{i\lambda_1, i\lambda_2; i\lambda}(s_1, s_2; s) \, (h^1)^\flat_{\lambda_1}(s_1) \, (h^2)^\flat_{\lambda_2}(s_2) \, ds_1 \, ds_2. \tag{3.3.14}$$

We shall not write the formula obtained when replacing, on the left-hand side, V^*T^* by Θ_0 and TV by the adjoint Θ_0^* of Θ_0: simply observe, as a consequence of (3.2.20), that the right-hand side must be multiplied by

$$8 \, (2\pi)^{\frac{i(\lambda_1 + \lambda_2 - \lambda)}{2}} \Gamma(-\frac{i\lambda_1}{2}) \Gamma(-\frac{i\lambda_2}{2}) \Gamma(\frac{i\lambda}{2}), \tag{3.3.15}$$

which changes the factor $2^{-\frac{9}{2}} \pi^{-2}$ to $2^{\frac{-3 + i(\lambda_1 + \lambda_2 - \lambda)}{2}} \pi^{\frac{-4 + i(\lambda_1 + \lambda_2 - \lambda)}{2}}$ and kills the product of three Gamma factors in the denominator.

The integral on the last line of (3.3.14) is the same as that in (3.3.2): only the coefficients differ. Making the functions B_j which are factors of (3.3.4) explicit and observing that the product of three Gamma factors obtained upstairs are among the four Gamma factors in (3.3.14), we obtain (3.3.13).

A few hints about the way (3.3.14) was proved in the above-given references, say in the case when $j = 0$, may be useful. Taking benefit of (3.2.10) and (3.2.11), one arrives at an expression of the left-hand side which, up to some Gamma factors, coincides with the integral with respect to $ds_1 \, ds_2$ of the product of $\chi^{(0)}_{i\lambda_1, i\lambda_2; i\lambda}(s_1, s_2; s)$ by the (convergent) integral

$$\int_\Pi \left(\frac{|z|^2}{\operatorname{Im} z}\right)^{-\frac{1}{2} + \frac{i\lambda_1}{2}} \left(\frac{|z|^2 - 1}{\operatorname{Im} z}\right)^{-\frac{1}{2} + \frac{i\lambda_2}{2}} (\operatorname{Im} z)^{\frac{1}{2} + \frac{i\lambda}{2}} \, dm(z). \tag{3.3.16}$$

Using (3.2.17) and Plancherel's formula, one reduces the computation of this integral to the consideration of a Weber-Schafheitlin integral, as will occur in (5.2.11). $\square$

It is one of our aims to extend Theorem 3.3.1 to the case of two factors, each of which will be taken from the Fourier series expansion of an Eisenstein or a Hecke distribution, i.e., will be of the kind $h^k(x, \xi) = |\xi|^{-1 - \nu_k} e^{2i\pi n_k \frac{x}{\xi}}$: doing this will be our task in the next chapter. In Chapter 5, we shall then extend the theorem to the case of two modular distributions.

The first difficulty, when trying to extend Theorem 3.3.1, lies in the presence on the right-hand side of the inverse of a product of 4 Gamma factors: if one neglects (easy to deal with) powers of $|\lambda|$, this function is of the order of $e^{\frac{\pi|\lambda|}{2}}$ as $|\lambda| \to \infty$ (cf. (1.1.8)), which is unsuitable for a $d\lambda$-integration, unless we know enough about the other factor in the integrand. The origin of the problem lies in the fact that a pair of operators such as (Θ_0, Θ_1) is "too good" to have a continuous inverse. It is for this reason that we have introduced the sequence of operators (W_m) in (3.2.43). In Section 5.1, we shall substitute W_m, with m well-chosen, for Θ_0: this will solve the difficulty, especially in the automorphic situation. But it is still true, as proved in Proposition 3.1.4, that a tempered distribution in the plane is fully characterized by its transform under the pair (Θ_0, Θ_1). Not all is lost, as a consequence, if we reformulate Theorem 3.3.1 as follows, with the help of (3.2.32):

Corollary 3.3.2. *Under the assumptions of Theorem 3.3.1, one has*

$$\Theta_0 \left(h^1 \,\#\, h^2 \right) = \pi^2 \sum_{j=0,1} (-i)^j \int_{-\infty}^{\infty} d\lambda \int_{\mathbb{R}^2} \frac{\Gamma(\frac{1+i\lambda}{2})\Gamma(\frac{1-i\lambda}{2})}{\prod_{\eta_1,\eta_2=\pm 1} \Gamma\left(\frac{1+i(\eta_1\lambda_1+\eta_2\lambda_2+\eta_1\eta_2\lambda)+2j}{4} \right)}$$

$$\cdot \left[\Theta_0(h^1)_{i\lambda_1} \underset{j}{\times} \Theta_0(h^2)_{i\lambda_2} \right]_{\frac{1+\lambda^2}{4}} d\lambda_1 d\lambda_2, \quad (3.3.17)$$

and the Θ_1-transform of $h^1 \,\#\, h^2$ is given by the same formula, on the right-hand side of which the extra factor $-i\lambda$ has been inserted.

Proof. Only the last sentence has not yet been justified: it follows from the relation (2.1.5) between Θ_0 and Θ_1, the variable $-i\lambda$ corresponding of course, under the decomposition (3.3.13), to the (generalized) eigenvalue of $2i\pi\mathcal{E}$. □

The advantage of this corollary is that the ratio of Gamma factors on the right-hand side of (3.3.17) is bounded by $(1+|\lambda|)^{1-2j}$: the exponent is unimportant, what matters is our having gotten rid of the exponential factor.

Remark 3.3.1. Combining the (n-dimensional version of the) most elementary composition formula (3.3.1), here rewritten with the help of (3.1.5) as

$$\mathrm{Tr}\left(\mathrm{Op}(h_1)\,\mathrm{Op}(h_2)\,\mathrm{Op}(h)\right)$$

$$= 2^{2n} \int_{\mathbb{R}^{2n}\times\mathbb{R}^{2n}\times\mathbb{R}^{2n}} h_1(Y)\,h_2(Z)\,h(X)\,e^{4i\pi([X,Y]-[Y,Z]+[X,Z])}\,dX\,dY\,dZ \quad (3.3.18)$$

with an elementary case of the composition formula based on decompositions into homogeneous components, one obtains an identity which has attracted some interest lately, though people do not seem to have realized how simple a proof could be obtained from a pseudodifferential argument.

Indeed, with $\ell(x, \xi) = |x|^2 + |\xi|^2$, consider the case when $h = \ell^{\frac{-n-i\lambda}{2}}$ and a similar definition goes for h_1 and h_2 after one has replaced λ by λ_1 or λ_2. The three operators in (3.3.18) are not Hilbert-Schmidt (it just fails), but one can still consider the two sides of (3.3.18) in the weak sense against functions of $\lambda, \ldots$ which extend as holomorphic functions in some strip $|\text{Im } \lambda| < \varepsilon$, rapidly decreasing at infinity. Denoting as $d\sigma$ the Euclidean measure on the unit sphere S^{2n-1} and making the polar changes of variables $X \mapsto rX, \ldots$ with $r > 0$, $X \in S^{2n-1}$, one is left with the consideration of the (weakly convergent) integral

$$2^{2n} \int_0^\infty \int_0^\infty \int_0^\infty (rr_1r_2)^{n-1} r^{-i\lambda} r_1^{-i\lambda_1} r_2^{-i\lambda_2}$$

$$\cdot \exp\left(4i\pi\left(rr_1\left[X,Y\right] - r_1r_2\left[Y,Z\right] + rr_2\left[X,Z\right]\right)\right) dr\, dr_1\, dr_2. \quad (3.3.19)$$

Taking rr_1, r_1r_2, rr_2 as new coordinates, one transforms the integral (3.3.18) to

$$2^{\frac{n}{2}-4} B_0\left(\frac{-n+i(\lambda_1+\lambda_2-\lambda)}{2}\right) B_0\left(\frac{-n+i(-\lambda_1+\lambda_2+\lambda)}{2}\right)$$

$$\cdot B_0\left(\frac{-n+i(\lambda_1-\lambda_2+\lambda)}{2}\right)$$

$$\cdot \int |[Y,Z]|^{\frac{-n+i(\lambda_1+\lambda_2-\lambda)}{2}} |[X,Z]|^{\frac{-n+i(-\lambda_1+\lambda_2+\lambda)}{2}}$$

$$\cdot |[X,Y]|^{\frac{-n+i(\lambda_1-\lambda_2+\lambda)}{2}} d\sigma(X)\, d\sigma(Y)\, d\sigma(Z). \quad (3.3.20)$$

On the other hand, the spectral decomposition of the n-dimensional harmonic oscillator $L = \text{Op}(\pi \ell)$ led to the equation [30]

$$\text{Op}\left(e^{-2\pi s\ell}\right) = (1-s^2)^{-\frac{n}{2}} \left(\frac{1-s}{1+s}\right)^L, \quad (3.3.21)$$

from which it followed [19, p. 986] that the decomposition into homogeneous components of degrees $-n - i\lambda$ of the symbol $g = \ell^{\frac{-n-i\lambda_1}{2}} \# \ell^{\frac{-n-i\lambda_2}{2}}$ is given by the equation

$$g_{i\lambda} = \frac{1}{4}(2\pi)^{\frac{n-2+i(\lambda_1+\lambda_2-\lambda)}{2}} \ell^{\frac{-n-i\lambda}{2}}$$

$$\cdot \frac{\Gamma\left(\frac{n+i(\lambda_1+\lambda_2-\lambda)}{4}\right)\Gamma\left(\frac{n+i(\lambda_1-\lambda_2+\lambda)}{4}\right)\Gamma\left(\frac{n+i(-\lambda_1+\lambda_2+\lambda)}{4}\right)\Gamma\left(\frac{n+i(-\lambda_1-\lambda_2-\lambda)}{4}\right)}{\Gamma\left(\frac{n+i\lambda_1}{2}\right)\Gamma\left(\frac{n+i\lambda_2}{2}\right)\Gamma\left(\frac{n-i\lambda_1}{2}\right)}.$$

$$(3.3.22)$$

Taking the trace (in the weak sense against a nice function of the parameter ν) against the operator with symbol $\ell^{\frac{-n-i\nu}{2}}$ and noting (take the trace against the operator with symbol

$$e^{-\pi\delta\ell} = \frac{1}{4\pi} \int_{-\infty}^\infty \Gamma\left(\frac{n+i\mu}{2}\right) (\pi\delta\ell)^{\frac{-n-i\mu}{2}} d\mu \quad (3.3.23)$$

as an intermediary) that

$$\mathrm{Tr}\left(\left(\mathrm{Op}\left(\ell^{\frac{-n-i\lambda}{2}}\right)\mathrm{Op}\left(\ell^{\frac{-n-i\nu}{2}}\right)\right)\right) = 2\pi\,\omega_{2n-1}\,\delta(\lambda+\nu), \qquad (3.3.24)$$

where ω_{2n-1} is the area of the unit sphere, we obtain an explicit formula for the integral (3.3.20). This formula was given, in various degrees of generality, in [3, 5].

Remark 3.3.2. In the n-dimensional case, there is a great variety of Gaussian functions the exponent of which has a positive-definite real part, parametrized by the complex tube over the cone of positive-definite matrices in $\mathbb{R}^n$, which is a model of the homogeneous space quotient of $\mathrm{Sp}(n,\mathbb{R})$ by its maximal compact subgroup. Each such Gaussian function could serve as a substitute for the function ϕ_z^0 in (3.1.17). Let us consider only, however, the normalized functions in $\mathbb{R}^n$

$$\phi_z(x) = \left(2\,\mathrm{Im}\left(-\frac{1}{z}\right)\right)^{\frac{n}{4}}\exp\left(\frac{i\pi}{\bar{z}}\,|x|^2\right), \qquad (3.3.25)$$

with z in the two-dimensional hyperbolic half-plane. The first Wigner function formula (3.1.20) extends at the sole price of having to replace the factor 2 in front of the right-hand side by 2^n. Computing first the (Gaussian) integral

$$I_z(t) := 2^n \int_{\mathbb{R}^n} \exp\left(-2\pi t(|x|^2 + |\xi|^2)\right)\exp\left(-2\pi\frac{|x - z\xi|^2}{\mathrm{Im}\,z}\right) dx\,d\xi$$

$$= (\mathrm{Im}\,z)^{-\frac{n}{2}}\left[t^2 + \frac{|z|^2 + 1}{\mathrm{Im}\,z}\,t + 1\right]^{-\frac{n}{2}}, \qquad (3.3.26)$$

one obtains with the help of [22, p. 185] the formula

$$\left(\phi_z\,|\,\mathrm{Op}\left(\ell^{\frac{-n-i\lambda}{2}}\right)\phi_z\right) = \frac{(2\pi)^{\frac{n+i\lambda}{2}}}{\Gamma(\frac{n+i\lambda}{2})}\int_0^\infty t^{\frac{n+i\lambda-2}{2}}I_z(t)\,dt$$

$$= \frac{\Gamma(\frac{n+1}{2})\Gamma(\frac{n-i\lambda}{2})}{\Gamma(n)}\left(\frac{\sinh d(i,z)}{4}\right)^{\frac{1-n}{2}}\mathfrak{P}^{\frac{1-n}{2}}_{\frac{-1-i\lambda}{2}}(\cosh d(i,z)) \quad (3.3.27)$$

involving the hyperbolic distance on Π.

In [35, p. 215], this equation was combined with the one-dimensional case of (3.3.22) to give a pseudodifferential proof of a formula, due to Mizony [23], expressing the product of two Legendre functions $\mathfrak{P}_{-\frac{1-i\lambda}{2}}(\delta)$ with the same argument δ but generally different parameters λ as an explicit integral superposition of functions of the same kind. It is very likely that, on the basis of (3.3.27), one should be able to obtain a similar formula for the product of two functions of the kind $(\delta^2 - 1)^{\frac{1-n}{4}}\mathfrak{P}^{\frac{1-n}{2}}_{\frac{-1-i\lambda}{2}}(\delta)$. But this would require pushing somewhat the analysis of the very special case of the n-dimensional Radon transform we have considered, obtaining the generalization of Corollary 3.3.2: we have not done it.

3.4 When the Weyl calculus falls short of doing the job

Only linear operators from $\mathcal{S}(\mathbb{R})$ to $\mathcal{S}'(\mathbb{R})$ have symbols, in the sense of the Weyl calculus. However, we shall come very soon (in Lemma 4.1.2) across operators A with the property that, while both operators PA and AP act from $\mathcal{S}(\mathbb{R})$ to $\mathcal{S}'(\mathbb{R})$, the operator A itself does not. In that case (cf. (3.1.14)), one defines $\mathrm{mad}(P \wedge Q)\, A = (PA)Q - Q(AP)$, as well as the symbol h_1 of this operator: it would coincide with $\mathcal{E}h$ if, after all, A did act from $\mathcal{S}(\mathbb{R})$ to $\mathcal{S}'(\mathbb{R})$ and its symbol were h.

The slight difficulty with this definition is that if $g \in SL(2,\mathbb{R})$ and M_g is one of the two metaplectic unitary transformations lying above g, if PA and AP act from $\mathcal{S}(\mathbb{R})$ to $\mathcal{S}'(\mathbb{R})$, the same does not necessarily hold if one replaces A by $M_g A M_g^{-1}$. Indeed, one has $M_g Q M_g^{-1} = \mathrm{Op}(x \circ g^{-1}) = dQ - bP$ and $M_g P M_g^{-1} = -cQ + aP$. Then,

$$M_g \left(\mathrm{mad}(P \wedge Q)\, A\right) M_g^{-1} = \mathrm{mad}((-cQ + aP) \wedge (dQ - bP))\, (M_g A M_g^{-1}), \quad (3.4.1)$$

if one is defining $\mathrm{mad}((-cQ + aP) \wedge (dQ - bP))\, B = (-cQ + aP)\, B\, (dQ - bP) - (dQ - bP)\, B\, (-cQ + aP)$ under the assumption that both $(-cQ + aP)B$ and $B(-cQ + aP)$ act from $\mathcal{S}(\mathbb{R})$ to $\mathcal{S}'(\mathbb{R})$.

We are thus led to generalizing our definition of $\mathrm{mad}(P \wedge Q)\, A$. We shall consider this operator as a well-defined operator from $\mathcal{S}(\mathbb{R})$ to $\mathcal{S}'(\mathbb{R})$ if, for some choice of $g = \left(\begin{smallmatrix} a & b \\ c & d \end{smallmatrix}\right)$, $A(-cQ + aP)$ and $(-cQ + aP)A$ are operators from $\mathcal{S}(\mathbb{R})$ to $\mathcal{S}'(\mathbb{R})$ (i.e., if A acts from the space $(-cQ + aP)\mathcal{S}(\mathbb{R})$ to $\mathcal{S}'(\mathbb{R})$ and from $\mathcal{S}(\mathbb{R})$ to the dual of the space $(-cQ + aP)\mathcal{S}(\mathbb{R})$), and we define then

$$\mathrm{mad}(P \wedge Q)\, A = (-cQ + aP)\, A\, (dQ - bP) - (dQ - bP)\, A\, (-cQ + aP). \quad (3.4.2)$$

Of course, we have then to show that, if for some $g_1 = \left(\begin{smallmatrix} a_1 & b_1 \\ c_1 & d_1 \end{smallmatrix}\right)$, both $A(-c_1 Q + a_1 P)$ and $(-c_1 Q + a_1 P)A$ act from $\mathcal{S}(\mathbb{R})$ to $\mathcal{S}'(\mathbb{R})$ too, one necessarily has

$$(-cQ + aP)\, A\, (dQ - bP) - (dQ - bP)\, A\, (-cQ + aP)$$
$$= (-c_1 Q + a_1 P)\, A\, (d_1 Q - b_1 P) - (d_1 Q - b_1 P)\, A\, (-c_1 Q + a_1 P). \quad (3.4.3)$$

In the case when the vectors $\left(\begin{smallmatrix} a \\ c \end{smallmatrix}\right)$ and $\left(\begin{smallmatrix} a_1 \\ c_1 \end{smallmatrix}\right)$ are not proportional, this implies that AP, PA, AQ, QA all act from $\mathcal{S}(\mathbb{R})$ to $\mathcal{S}'(\mathbb{R})$: expanding the wedge by bilinearity, one verifies that both sides of (3.4.3) agree with $PAQ - QAP$. If the above vectors are proportional, it is no loss of generality to assume that they are identical (since replacing the vector $\left(\begin{smallmatrix} dQ-bP \\ -cQ+aP \end{smallmatrix}\right)$ by $\left(\begin{smallmatrix} \lambda(dQ-bP) \\ \lambda^{-1}(-cQ+aP) \end{smallmatrix}\right)$ will not change $\mathrm{mad}((-cQ + aP) \wedge (dQ-bP)))$. Then, expanding $(-cQ+aP)\, A\, (dQ-bP) - (dQ-bP)\, A\, (-cQ+aP)$ and its companion (in which g_1 takes the place of g) linearly with respect to the factor, in each term, distinct from $-cQ + aP$, is possible, and leads to (3.4.3).

Our definition of $\mathrm{mad}(P \wedge Q)$ applies now to a class of operators A which is preserved under the map $A \mapsto M_g A M_g^{-1}$. However, it may seem a bit artificial: let us trace the origin of the difficulty. An operator such as $A = A_1 A_2$ in Lemma 4.1.2 below is well-defined on the image of $\mathcal{S}(\mathbb{R})$ under P. This is a one-codimensional subspace of $\mathcal{S}(\mathbb{R})$, which may seem "almost as good", but this is not the case, because it is not preserved under the metaplectic representation. In Chapter 6, we shall interest ourselves in the way a certain operator acts on $\mathcal{S}_{\mathrm{odd}}(\mathbb{R})$ only, a subspace of $\mathcal{S}(\mathbb{R})$ invariant under the metaplectic representation, so that the slight inconvenience just explained will disappear anyway.

The developments that follow in this section are not necessary for further reading, but they will be referred to in a short expository paragraph at the end of Section 6.3 and revisited in Section 7.1.

The inconvenience just alluded to would never show if we had a representation acting on the image of $\mathcal{S}(\mathbb{R})$ under P (or under Q), and an associated pseudodifferential calculus, covariant under this representation. Such a pair exists, and consists of the case $p = 1$ of the following general construction. First, one builds for $p = 0, 1, \dots$ a unitary representation Met_p of $SL(2, \mathbb{R})$ in $L^2(\mathbb{R})$, preserving the space $\mathcal{S}_p(\mathbb{R})$ which is the image of $\mathcal{S}(\mathbb{R})$ under the multiplication by the function $x \mapsto x^p$. Such a representation is defined, if $g = \left(\begin{smallmatrix} a & b \\ c & d \end{smallmatrix} \right)$ with $b > 0$, if u is even and $x > 0$, by the equation

$$(\mathrm{Met}_p(g)\, u)\,(x) = e^{-\frac{i\pi}{2}(p+\frac{1}{2})} \frac{2\pi}{b} \int_0^\infty \sqrt{xy}\, J_{p-\frac{1}{2}} \left(\frac{2\pi xy}{b} \right) e^{i\pi \frac{dx^2 + ay^2}{b}} u(y)\, dy, \tag{3.4.4}$$

while in the case when u is odd, the sole modification to be done on the right-hand side is replacing p by $p + 1$. Then, one builds a symbolic calculus Op^p which is covariant under this representation, the action of elements of $SL(2, \mathbb{R})$ on symbols being the usual one. Needless to say, the pair $(\mathrm{Op}^p, \mathrm{Met}_p)$ consists, when $p = 0$, of the Weyl calculus together with the metaplectic representation. Some more hints regarding this "calculus of level p" will be given in Section 7.1.

Such a construction has been carried in [35, sections 7,9] and, indeed, the general properties of the pseudodifferential calculus Op^p, especially when used with automorphic symbols, improve with p. As will be seen in the next chapter and Chapter 5, the sharp composition, in the Weyl calculus, of any two Hecke distributions is not absolutely meaningful: this is where using $\mathrm{mad}(P \wedge Q)$ can save the situation. But it is possible in the Op^p-calculus if $p \geq 2$: it is even possible, staying within such a calculus, to compose any given number of modular distributions (Eisenstein's $\mathfrak{E}_\nu$ and Hecke, with a bound on $|\mathrm{Re}\,\nu|$ in the first case) provided that p has been chosen large enough.

We shall not follow this path, however, partly because this leads to extremely complicated calculations. We shall stay within the Weyl calculus proper, at the price (which will be fully explained) of having to use repeatedly polynomials in the operator $\mathrm{mad}(P \wedge Q)$ in an appropriate way. What may explain the greater simplicity of the Weyl calculus within the series (Op^p) is that it is the only one which, besides its covariance under the metaplectic representation, enjoys also covariance under a representation of the Heisenberg group (3.1.10).

Chapter 4

Composition of joint eigenfunctions of $\mathcal{E}$ and $\xi\frac{\partial}{\partial x}$

The functions

$$h_{\nu,q}(x,\xi) = |\xi|^{-1-\nu} \exp\left(2i\pi\frac{qx}{\xi}\right),\tag{4.0.1}$$

with $q \in \mathbb{Z}$, are the basic pieces of the decomposition (1.2.32) of modular distributions, such as Eisenstein's and Hecke's, into Fourier series. For this reason, analyzing the sharp composition of two symbols of such a kind (it is preferable not to assume that $q \in \mathbb{Z}$ at this point) is a natural way to approach the question of the sharp composition of two modular distributions. Another possible approach could have relied on the decompositions of Eisenstein or Hecke distributions into bihomogeneous functions, since the coefficients of these decompositions are given in terms of L-functions (a desirable fact) in a direct way. However, extensive calculations, not reproduced in this book, show that the terms of the decomposition of the sharp product of two bihomogeneous functions involve linear combinations of generalized hypergeometric functions $_6F_5$, whereas only standard hypergeometric functions will appear in connection with the method to follow.

Before we compute a sharp product such as $h = h_{\nu_1,q_1} \# h_{\nu_2,q_2}$, we must analyze in which sense it is meaningful. As will be seen in the first section to follow, though h is certainly a tempered distribution when $q_1 + q_2 \neq 0$, one must satisfy oneself, when $q_1 + q_2 = 0$, with defining it as a continuous linear form on the image under $2i\pi\mathcal{E}$ of $\mathcal{S}(\mathbb{R}^2)$. Next, in view of applying the results to series of such products, we must improve the estimates. All this will be done with the help of the following trick: substitute for the symbol h its image under a certain polynomial, of Pochhammer's style, in the operator $2i\pi\mathcal{E}$. Finally, the explicit decomposition of $h_{\nu_1,q_1} \# h_{\nu_2,q_2}$ into homogeneous components will depend on a quite lengthy computation, based on the composition formula provided by the equations (3.3.2) to (3.3.5).

4.1   Estimates of sharp products $h_{\nu_1,q_1} \# h_{\nu_2,q_2}$

If $\Phi \in \mathcal{S}(\mathbb{R}^2)$ and $\Psi = \mathcal{F}_1^{-1}\Phi$, one has if $h_{\nu,q}$ is the function introduced in (4.0.1) the identity $\langle h_{\nu,q}, \Phi \rangle = \int_{-\infty}^{\infty} |\xi|^{-1-\nu} \Psi\left(\frac{q}{\xi}, \xi\right) d\xi$. The function $\xi \mapsto \Psi\left(\frac{q}{\xi}, \xi\right)$ lies in $\mathcal{S}(\mathbb{R})$, whether $q = 0$ or $q \neq 0$: it is, moreover, flat to infinite order at 0 if $q \neq 0$. It follows that the function $h_{\nu,q}$, initially defined as a locally integrable function if Re $\nu < 0$, extends as a tempered distribution for every ν if $q \neq 0$, for $\nu \neq 0, 2, \ldots$ if $q = 0$. This splitting of cases will have consequences to be felt throughout the developments to follow.

The distribution $h_{\nu,q}$ satisfies the pair of (generalized) eigenvalue equations

$$(2i\pi\mathcal{E}) h_{\nu,q} = -\nu h_{\nu,q}, \quad \xi\frac{\partial}{\partial x} h_{\nu,q} = q h_{\nu,q}. \tag{4.1.1}$$

One should note the relations

$$\mathcal{F}^{\text{symp}}\mathfrak{b}_q^{\nu} = \mathfrak{b}_q^{-\nu} \quad \text{if} \quad \mathfrak{b}_q^{\nu}(x,\xi) = |q|^{\frac{\nu}{2}} |\xi|^{-\nu-1} e^{2i\pi q\frac{x}{\xi}}, \quad q \neq 0 \tag{4.1.2}$$

and, for future reference, $\mathcal{G}h_{\nu,q} = 2^{2i\pi\mathcal{E}} \mathcal{F}^{\text{symp}} h_{\nu,q} = 2^{\nu}|q|^{-\nu}h_{-\nu,q}$ if $q \neq 0$. Recall (3.1.14) the equation, valid for every $\mathfrak{S} \in \mathcal{S}'(\mathbb{R}^2)$:

$$\text{mad}(P \wedge Q) \text{Op}(\mathfrak{S}): = P \text{Op}(\mathfrak{S}) Q - Q \text{Op}(\mathfrak{S}) P = \text{Op}(\mathcal{E}\mathfrak{S}). \tag{4.1.3}$$

On the other hand, from (3.1.12),

$$\frac{1}{2}[P^2, \text{Op}(\mathfrak{S})] = \text{Op}\left(\frac{1}{2i\pi}\xi\frac{\partial}{\partial x}\mathfrak{S}\right). \tag{4.1.4}$$

From the latter equation, it follows that the sharp product of two (generalized) eigenfunctions of the operator $\xi\frac{\partial}{\partial x}$ will also be an eigenfunction of the same operator, simply adding the (generalized) eigenvalues. Something similar would hold with the operator $\mathcal{E}^{\natural}$ in view of the eigenvalue equation $\frac{1}{2}[QP + PQ, \text{Op}(\mathfrak{S})] = \text{Op}(\mathcal{E}^{\natural}\mathfrak{S})$, but nothing of the same kind is valid so far as the operator $\mathcal{E}$ is concerned. This is why computing the sharp product of two symbols of the kind $h_{\nu,q}$ will involve an integral with respect to the variable ν.

It is useful to consider, in place of functions $h_{\nu,q}$, more general functions $h(x,\xi) = f(\xi)e^{2i\pi q\frac{x}{\xi}}$, where f is assumed to be an even function on the real line. The operators with such functions for Weyl symbols are easy to describe, but analyzing their composition is much more delicate: this will be our task in this chapter.

Lemma 4.1.1. *Let $q \in \mathbb{R}$. Let $h(x,\xi) = f(\xi)e^{2i\pi q \frac{x}{\xi}}$, and set $g(\xi) = \frac{|\xi|}{2} f(\frac{\xi}{2})$. Assume that f is a continuous (even) function on $\mathbb{R}\backslash\{0\}$, and that $|f(\xi)|$ is bounded, for some pair (C, N), by $C\left(|\xi| + |\xi|^{-1}\right)^N$: if $q = 0$, reinforce the assumption, taking f to be locally integrable near the origin. For $u \in \mathcal{S}(\mathbb{R})$, one has*

$$(\mathcal{F} \operatorname{Op}(h) u)(t) = \begin{cases} \operatorname{char}(t^2 \geq 2q) \sum\limits_{\varepsilon = \pm 1} g(t + \varepsilon\sqrt{t^2 - 2q}) \dfrac{(\mathcal{F}u)(\varepsilon\sqrt{t^2-2q})}{\sqrt{t^2-2q}} & \text{if } q \neq 0, \\ f(t)(\mathcal{F}u)(t) & \text{if } q = 0. \end{cases}$$

(4.1.5)

Let $u \in \mathcal{S}(\mathbb{R})$: if $q < 0$, $\mathcal{F} \operatorname{Op}(h) u$ is continuous on the line and rapidly decreasing at infinity; if $q > 0$, $\mathcal{F} \operatorname{Op}(h) u$ is continuous outside $\pm\sqrt{2q}$, locally summable near these points and rapidly decreasing at infinity.

Proof. The case when $q = 0$ is trivial, and we assume that this is not the case. The first thing to note, generalizing what was said in the beginning of this section, is that the equation

$$\langle h, \Phi \rangle = \int_{-\infty}^{\infty} f(\xi) \left(\mathcal{F}_1^{-1}\Phi\right)\left(\frac{q}{\xi}, \xi\right) d\xi, \quad \Phi \in \mathcal{S}(\mathbb{R}^2),$$

(4.1.6)

gives h a meaning as a tempered distribution, even though f may be far from integrable near 0. We take advantage of the equation $\mathcal{F} \operatorname{Op}(h) \mathcal{F}^{-1} = \operatorname{Op}\left(h \circ \left(\begin{smallmatrix} 0 & -1 \\ 1 & 0 \end{smallmatrix}\right)\right)$ and we write (with $w = \mathcal{F}u$)

$$\left(\operatorname{Op}\left(h \circ \left(\begin{smallmatrix} 0 & -1 \\ 1 & 0 \end{smallmatrix}\right)\right) w\right)(t) = \int_{\mathbb{R}^2} f\left(\frac{t+x}{2}\right) e^{-4i\pi q \frac{\xi}{t+x}} e^{2i\pi(t-x)\xi} w(x) \, dx \, d\xi.$$

(4.1.7)

Performing the change of variable $x \mapsto y = x + \frac{2q}{t+x}$, so that $(t+y)^2 \geq 8q$, one has $x = \frac{1}{2}\left(-t + y + \varepsilon\sqrt{(t+y)^2 - 8q}\right)$ with $\varepsilon = \pm 1$, and

$$\left|\frac{dx}{dy}\right| = \frac{1}{2}\frac{|t+y+\varepsilon\sqrt{(t+y)^2 - 8q}|}{\sqrt{(t+y)^2 - 8q}}.$$

(4.1.8)

The product of exponentials becomes $e^{-2i\pi\xi(t-y)}$, and the calculation is over since $\int e^{2i\pi\xi(t-y)} d\xi = \delta(y - t)$.

The last assertion is a consequence of the fact that the argument $t + \varepsilon\sqrt{t^2 - 2q}$ of g in (4.1.5) is never zero, and can approach zero only when $\varepsilon t \to -\infty$, in which case it has the size of $\frac{1}{|t|}$. □

We come now to the question of defining the sharp product of two functions h^k ($k = 1, 2$) of the type just discussed as a tempered distribution, or the equivalent one of defining the composition $A_1 A_2$ of the associated operators as an operator from $\mathcal{S}(\mathbb{R})$ to $\mathcal{S}'(\mathbb{R})$: as will be seen, in the case when $q_1 + q_2 = 0$, one may have to lower slightly one's expectations.

Lemma 4.1.2. *Let there be given $q_1, q_2 \in \mathbb{R}$, not both zero. Consider two operators $A_k = \mathrm{Op}(h^k)$, $(k = 1, 2)$, with $h^k(x, \xi) = f_k(\xi)e^{2i\pi q_k \frac{x}{\xi}}$. Assume that f_1, f_2 are continuous in $\mathbb{R}\backslash\{0\}$ and that, for some pair (C, N), one has $|f_k(\xi)| \leq C(|\xi| + |\xi|^{-1})^N$: if $q_k = 0$, reinforce the hypothesis about f_k, assuming that this function is locally summable near the origin. Then, if $q_1 + q_2 \neq 0$, or $q_1 > 0$ (or $q_2 < 0$), the operator $A_1 A_2$ acts from $\mathcal{S}(\mathbb{R})$ to $\mathcal{S}'(\mathbb{R})$. If $q_1 + q_2 = 0$ and $q_1 < 0$, the operator $A_1 A_2$ does not generally act from $\mathcal{S}(\mathbb{R})$ to $\mathcal{S}'(\mathbb{R})$ but the operator $\mathrm{mad}(P \wedge Q)(A_1 A_2) = P(A_1 A_2)Q - Q(A_1 A_2)P$, defined as $(PA_1)A_2Q - QA_1(A_2 P)$, does.*

Proof. Let us consider first the case when $q_1 q_2(q_1 + q_2) \neq 0$. Let $u \in \mathcal{S}(\mathbb{R})$. Setting $v = A_2 u$, it follows from Lemma 4.1.1 that $\mathcal{F}v$ is continuous on the line, with the exception of locally summable singularities at $\pm\sqrt{2q_2}$ when $q_2 > 0$, and rapidly decreasing at infinity. Then, the only possible singularities of $\mathcal{F}(A_1 A_2 u) = \mathcal{F}(A_1 v)$ are located at points t such that $t^2 - 2q_1 = 0$ or $2q_2$, i.e., $t = \pm\sqrt{2q_1}$ and $t = \pm\sqrt{2(q_1 + q_2)}$: in view of equation (4.1.5) applied with the operator A_1 (and u replaced by v), and of the properties of $\mathcal{F}v$ just given, the singularities of the second species are obviously locally integrable, and so are those of the first species in view of the equation $\frac{dt}{\sqrt{t^2 - 2q_1}} = \frac{ds}{\sqrt{s^2 + 2q_1}}$ valid in a half-neighbourhood of $t = \pm\sqrt{2q_1}$ if $s = \varepsilon_1 \sqrt{t^2 - 2q_1}$. This gives the operator $A_1 A_2$ a meaning as an operator from $\mathcal{S}(\mathbb{R})$ to $\mathcal{S}'(\mathbb{R})$, and a Weyl symbol in $\mathcal{S}'(\mathbb{R}^2)$: but note that the "intermediary space" in which $A_2 u$ has been found to lie depends on q_2. We shall fix this inconvenience to some extent in Lemma 4.1.5 below. Explicitly, setting $g_k(\xi) = \frac{|\xi|}{2} f_k(\frac{\xi}{2})$, one has

$$\left(\mathcal{F}(A_1 A_2 u)\right)(t)$$

$$= \mathrm{char}(t^2 \geq 2q_1) \sum_{\varepsilon_1} \frac{g_1(t + \varepsilon_1 \sqrt{t^2 - 2q_1})}{\sqrt{t^2 - 2q_1}} \left(\mathcal{F}(A_2 u)\right)(\varepsilon_1 \sqrt{t^2 - 2q_1}), \qquad (4.1.9)$$

and

$$\left(\mathcal{F}(A_1 A_2 u)\right)(t) = \mathrm{char}(t^2 \geq 2q_1)\,\mathrm{char}(t^2 \geq 2q_1 + 2q_2) \sum_{\varepsilon_1} \frac{g_1(t + \varepsilon_1 \sqrt{t^2 - 2q_1})}{\sqrt{t^2 - 2q_1}}$$

$$\cdot \sum_{\varepsilon_2} \frac{g_2(\varepsilon_1 \sqrt{t^2 - 2q_1} + \varepsilon_2 \sqrt{t^2 - 2q_1 - 2q_2})}{\sqrt{t^2 - 2q_1 - 2q_2}} (\mathcal{F}u)(\varepsilon_2 \sqrt{t^2 - 2q_1 - 2q_2}). \quad (4.1.10)$$

When $q_2 = 0$ (and $q_1 \neq 0$), this simplifies to

$$\mathrm{char}(t^2 \geq 2q_1) \sum_{\varepsilon_1} g_1(t + \varepsilon_1 \sqrt{t^2 - 2q_1})\, f_2(\varepsilon_1 \sqrt{t^2 - 2q_1})\, \frac{\left(\mathcal{F}u\right)(\varepsilon_2 \sqrt{t^2 - 2q_1})}{\sqrt{t^2 - 2q_1}}.$$

$$\qquad (4.1.11)$$

Still, the operator $A_1 A_2$ acts from $\mathcal{S}(\mathbb{R})$ to $\mathcal{S}'(\mathbb{R})$ under the strengthened assumption regarding f_2: for, if $q_1 > 0$, the fraction $\frac{f_2(\varepsilon_1 \sqrt{t^2 - 2q_1})}{\sqrt{t^2 - 2q_1}}$ is locally integrable near $\pm\sqrt{2q_1}$ because f_2 is in this case assumed to be locally summable near 0 (change variable, setting again $s = \varepsilon_2 \sqrt{t^2 - 2q_1}$). The case when $q_1 = 0$ and $q_2 \neq 0$ is totally similar to the preceding one, from which it can be derived by transposition.

Things are different in the case when $q_1 + q_2 = 0$ and $q_1 < 0$. For one has in this case (setting $\varepsilon_2 = \varepsilon \operatorname{sign} t$)

$$(\mathcal{F}(A_1 A_2 u))\,(t) = \sum_{\varepsilon_1, \varepsilon = \pm 1} \frac{g_1(t + \varepsilon_1 \sqrt{t^2 - 2q_1})}{\sqrt{t^2 - 2q_1}} \, g_2(\varepsilon t + \varepsilon_1 \sqrt{t^2 - 2q_1}) \, \frac{(\mathcal{F}u)(\varepsilon t)}{|t|},$$

(4.1.12)

and we have to cope with the non locally integrable factor $|t|^{-1}$. However, one obtains an operator from $\mathcal{S}(\mathbb{R})$ to $\mathcal{S}'(\mathbb{R})$ if one multiplies, either on the right or on the left, the operator $A_1 A_2$ by the operator P (do not forget the conjugation by the Fourier transformation): in particular, the operator $P(A_1 A_2)Q - Q(A_1 A_2)P$ will do.   □

Remarks 4.1.1. (i) one can reinterpret the last sentence as asserting that, even if $q_1 + q_2 = 0$, the operator $A_1 A_2$ is well-defined as a weakly continuous operator from the subspace of $\mathcal{S}(\mathbb{R})$ consisting of functions u such that $\int_{-\infty}^{\infty} u(x)\,dx = 0$ to $\mathcal{S}'(\mathbb{R})$, or from $\mathcal{S}(\mathbb{R})$ to the dual of that space.

(ii) even though the operator $A_1 A_2 = \operatorname{Op}(h^1)\operatorname{Op}(h^2)$ does not qualify, in the case when $q_1 + q_2 = 0$, for having a Weyl symbol according to the definition (3.1.1) of the Weyl calculus, which does not extend beyond the case of symbols in $\mathcal{S}'(\mathbb{R}^2)$, the image of such a would-be symbol under the operator $\mathcal{E}$ can be *defined*, taking advantage of (4.1.3), as being the symbol of $P(A_1 A_2)Q - Q(A_1 A_2)P$. In other words, we may define the sharp composition of h^1 and h^2, to be denoted as $\operatorname{Sharp}(h^1, h^2)$ rather than $h^1 \# h^2$ (to avoid taking for granted the generalization of certain facts), as a *quasi-distribution*, by which we mean, in this case, a continuous linear form on the space which is the image under $2i\pi\mathcal{E}$ of $\mathcal{S}(\mathbb{R}^2)$. More general quasi-distributions $\mathfrak{S}$, involving in place of $2i\pi\mathcal{E}$ some polynomials, of Pochhammer's style, in this operator, will have to be used later in this volume. We shall actually provide in most cases a genuine tempered distribution $\mathfrak{T}$ coinciding with $\mathfrak{S}$ as a quasi-distribution of the given type, i.e., when restricted to the image of $\mathcal{S}(\mathbb{R}^2)$ under the specified polynomial in $2i\pi\mathcal{E}$.

Considering the case when $q_1 q_2 (q_1 + q_2) \neq 0$, one sees from (4.1.10) and (4.1.5) that the symbol of $A = A_1 A_2$ is of the kind $(x, \xi) \mapsto f(\xi) e^{2i\pi(q_1 + q_2)\frac{x}{\xi}}$ if, with $g(\xi) = \frac{|\xi|}{2} f(\frac{\xi}{2})$, the identity

$$g(t + \varepsilon\sqrt{t^2 - 2q_1 - 2q_2})$$
$$= \mathrm{char}(t^2 \geq 2q_1) \sum_{\varepsilon_1} \frac{g_1(t + \varepsilon_1\sqrt{t^2 - 2q_1})\, g_2(\varepsilon_1\sqrt{t^2 - 2q_1} + \varepsilon\sqrt{t^2 - 2q_1 - 2q_2})}{\sqrt{t^2 - 2q_1}}$$

$$(4.1.13)$$

holds for $t^2 \geq 2(q_1 + q_2)$. Setting $\xi = t + \varepsilon\sqrt{t^2 - 2q_1 - 2q_2}$, one has

$$t = \frac{\xi}{2} + \frac{q_1 + q_2}{\xi}, \quad \varepsilon\sqrt{t^2 - 2q_1 - 2q_2} = \frac{\xi}{2} - \frac{q_1 + q_2}{\xi}, \tag{4.1.14}$$

which leads the (unique) solution for g, hence for f. This expression, however, would not lead easily to an explicit decomposition of f into homogeneous components, a question we will turn to by another method in Section 4.5.

The following elementary lemma will be used time and again: displaying the two obviously equivalent pairs of inequalities will save (very) minor headaches.

Lemma 4.1.3. *With $\varepsilon_1, \varepsilon_2 = \pm 1$ and $a, b \in \mathbb{R}$, one has*

$$\frac{|a - b|}{\sqrt{t^2 - 2\min(a,b)}} \leq |\varepsilon_1\sqrt{t^2 - 2a} + \varepsilon_2\sqrt{t^2 - 2b}| \leq 2\sqrt{t^2 - 2\min(a,b)},$$

$$\frac{|a - b|}{\sqrt{s^2 + 2\max(a,b)}}| \leq |\varepsilon_1\sqrt{s^2 + 2a} + \varepsilon_2\sqrt{s^2 + 2b}| \leq 2\sqrt{s^2 + 2\max(a,b)},$$

$$(4.1.15)$$

assuming that t^2, or s^2, is large enough for each of the two square roots involved to make sense.

Proof. Let us prove the second one, assuming $a \geq b$ (and, of necessity, $s^2 \geq -2b$), so that $0 \leq \frac{2(a-b)}{s^2 + 2a} \leq 1$. One has

$$\sqrt{s^2 + 2a} - \sqrt{s^2 + 2b} = \sqrt{s^2 + 2a}\left[1 - \left(1 - \frac{2(a-b)}{s^2 + 2a}\right)^{\frac{1}{2}}\right], \tag{4.1.16}$$

and $1 - \sqrt{1 - h} \geq \frac{h}{2}$ if $0 \leq h \leq 1$. $\square$

Lemma 4.1.4. *Keeping the notation and assumptions of Lemma 4.1.2, assume that $q_1 q_2 \neq 0$. If $q_1 + q_2 \neq 0$, or $q_1 > 0$ (or $q_2 < 0$), the operator $A_1 A_2$ is continuous as an operator from the space of functions u the Fourier transform $\mathcal{F}u$ of which is continuous and rapidly decreasing at infinity to the space of functions the Fourier transform of which is summable. Still assuming $q_1 q_2 \neq 0$, but dropping the assumption that $q_1 + q_2 \neq 0$ or $q_1 > 0$, the claim remains valid after one has replaced $A_1 A_2$ by $A_1 A_2 P$, or by $P A_1 A_2$.*

Proof. Set $r = \max(q_1, q_1 + q_2)$, i.e., $r = q_1 + q_2$ if $q_2 > 0$ and $r = q_1$ if $q_2 < 0$. We shall consider separately the cases when $r > 0, r < 0$ and $r = 0$. In the first case, we set $\varepsilon = \varepsilon_2$ if $r = q_1 + q_2$ and $\varepsilon = \varepsilon_1$ if $r = q_1$, and make the change of variable

$$s = \varepsilon\sqrt{t^2 - 2r}, \quad t = \varepsilon'\sqrt{s^2 + 2r}, \quad \frac{dt}{\sqrt{t^2 - 2r}} = \frac{ds}{\sqrt{s^2 + 2r}} \tag{4.1.17}$$

in the integral

$$I := \int_{-\infty}^{\infty} |\mathcal{F}(A_1 A_2 u)(t)|\, dt. \tag{4.1.18}$$

Starting from (4.1.10), we obtain in the case when $r = q_1 + q_2 > 0$,

$$I \le \sum_{\varepsilon,\varepsilon_1} \int_{-\infty}^{\infty} \frac{|g_1(\varepsilon\sqrt{s^2 + 2q_1 + 2q_2} + \varepsilon_1\sqrt{s^2 + 2q_2})|}{\sqrt{s^2 + 2q_2}}$$
$$\cdot |g_2(s + \varepsilon_1\sqrt{s^2 + 2q_2})|\,|(\mathcal{F}u)(s)| \frac{ds}{\sqrt{s^2 + 2q_1 + 2q_2}}, \tag{4.1.19}$$

and in the case when $r = q_1 > 0$,

$$I \le \sum_{\varepsilon,\varepsilon_2} \int_{-\infty}^{\infty} |g_1(s + \varepsilon\sqrt{s^2 + 2q_1})| \frac{|g_2(s + \varepsilon_2\sqrt{s^2 - 2q_2})|}{\sqrt{s^2 - 2q_2}}$$
$$\cdot |(\mathcal{F}u)(\varepsilon_2\sqrt{s^2 - 2q_2})| \frac{ds}{\sqrt{s^2 + 2q_1}}. \tag{4.1.20}$$

The arguments of the functions g_1 and g_2 must be appreciated, in view of the assumption about f_1, f_2 (or g_1, g_2) in Lemma 4.1.2. In the first case, one has $q_1 + q_2 > 0$, $q_2 > 0$ and

$$\frac{q_1}{\sqrt{s^2 + 2q_1 + 2q_2}} \le |\varepsilon\sqrt{s^2 + 2q_1 + 2q_2} + \varepsilon_1\sqrt{s^2 + 2q_2}| \le 2\sqrt{s^2 + 2q_1 + 2q_2},$$
$$\frac{q_2}{\sqrt{s^2 + 2q_2}} \le |s + \varepsilon_1\sqrt{s^2 + 2q_2}| \le 2\sqrt{s^2 + 2q_2}; \tag{4.1.21}$$

in the second case, one has $q_1 > 0$, $q_2 < 0$ and

$$\frac{q_1}{\sqrt{s^2 + 2q_1}} \le |s + \varepsilon\sqrt{s^2 + 2q_1}| \le 2\sqrt{s^2 + 2q_1},$$
$$\frac{|q_2|}{\sqrt{s^2 - 2q_2}} \le |s + \varepsilon_2\sqrt{s^2 - 2q_2}| \le 2\sqrt{s^2 - 2q_2}, \tag{4.1.22}$$

and the result, to wit the fact that the integral I is convergent, follows: note, however, that this result may, or not (it depends on subcases) be uniform with respect to q_1, q_2.

When $r \leq 0$, no change of variable is needed, or possible, and we start from (4.1.10), just forgetting the two needless characteristic functions. Since

$$\sqrt{t^2 - 2q_1 - 2q_2} \geq \max(|t|, \sqrt{t^2 - 2q_1}),$$

the desired result is immediate, estimating the arguments of g_1 and g_2 as done before.

We consider finally the case when $r = 0$, i.e., $q_1 < 0$ and $q_2 = -q_1$. Then, we use (4.1.12), and observe that, though the function $\mathcal{F}(A_1 A_2 u)$ is not summable in general (for $u \in \mathcal{S}(\mathbb{R})$), it becomes so, assuming that $\mathcal{F}u$ is rapidly decreasing at infinity, provided that we multiply it by t (any power of $|t|$ with positive exponent would do, but this would not help). Now, performing such a multiplication amounts to multiplying the (multiplication) operator $\mathcal{F}A_1 A_2 \mathcal{F}^{-1}$, on the left or on the right, by the operator Q, or the (convolution) operator $A_1 A_2$, on the left or on the right, by P. This leads to a proof of the last remaining case of Lemma 4.1.4.   □

In the next lemma, we show that, assuming $q_1 q_2 \neq 0$, one can give each of the two operators $PA_1 A_2$ and $A_1 A_2 P$ a meaning as an operator from $\mathcal{S}(\mathbb{R})$ to $\mathcal{S}'(\mathbb{R})$, the "intermediary" space implied in the composition being independent of q_1, q_2: this is a natural demand when, as is the case in automorphic distribution theory, one is dealing with series (with respect to $q \in \mathbb{Z}^{\times}$) of operators of the type under consideration. It is not possible, however, to choose the same intermediary space for the two operators under consideration. The lemma, meant for clarification only, will not be used in the sequel.

Lemma 4.1.5. *Let $L^{1,\infty}$ be the space of functions $w = w(t)$ on the real line such that, for every M, the function $(1 + |t|)^M w(t)$ is summable; let $L^{\infty,\infty} \subset L^{1,\infty}$ be the space of measurable functions on $\mathbb{R}$, essentially bounded after they have been multiplied by an arbitrary power of $1 + |t|$. Given $q \neq 0$, let $A = \mathrm{Op}(h)$ be the operator considered in Lemma 4.1.1. The operator A sends the space $\mathcal{F}L^{\infty,\infty}$ to $\mathcal{F}L^{1,\infty}$, while the operator AP is an endomorphism of the space $\mathcal{F}L^{\infty,\infty}$, and the operator PA is an endomorphism of $\mathcal{F}L^{1,\infty}$. As a consequence, under the assumptions of Lemma 4.1.1 relative to h^1, h^2, completed by the condition $q_1 q_2 \neq 0$, both operators $A_1 A_2 P$ and $PA_1 A_2$ act from $\mathcal{F}L^{\infty,\infty}$ to $\mathcal{F}L^{1,\infty}$.*

Proof. Set $B = \mathcal{F}A\mathcal{F}^{-1}$. As observed in the proof of Lemma 4.1.4, one has

$$|t + \varepsilon\sqrt{t^2 - 2q}| + |t + \varepsilon\sqrt{t^2 - 2q}|^{-1} \leq \begin{cases} \left(1 + \frac{1}{|q|}\right)\sqrt{t^2 - 2q} & \text{if } q < 0 \\ \left(1 + \frac{1}{q}\right)|t| & \text{if } q > 0 \text{ and } t^2 \geq 2q. \end{cases}$$
$$(4.1.23)$$

It immediately follows from (4.1.5) that the operator B sends $L^{\infty,\infty}$ to $L^{1,\infty}$, the singularities at $\pm\sqrt{2q}$ (if $q > 0$) being locally integrable. Next, $AP = \mathcal{F}(BQ)\mathcal{F}^{-1}$, and

$$(BQw)(t) = \mathrm{char}(t^2 \geq 2q) \sum_{\varepsilon=\pm 1} \varepsilon g(t + \varepsilon\sqrt{t^2 - 2q}) \, w(\varepsilon\sqrt{t^2 - 2q}) : \qquad (4.1.24)$$

it is immediate that BQ is an endomorphism of $L^{\infty,\infty}$. Finally,

$$(QBw)(t) = \mathrm{char}(t^2 \geq 2q) \sum_{\varepsilon=\pm 1} t\, g(t + \varepsilon\sqrt{t^2 - 2q})\, \frac{w(\varepsilon\sqrt{t^2 - 2q})}{\sqrt{t^2 - 2q}}. \qquad (4.1.25)$$

If $q > 0$, we set $s = \varepsilon\sqrt{t^2 - 2q}$, so that $\frac{|t|\,dt}{\sqrt{t^2-2q}} = ds$, which yields

$$\int_{-\infty}^{\infty} (1 + |t|)^M\, |(QBw)(t)|\, dt$$

$$\leq \sum_{\varepsilon'} \int_{-\infty}^{\infty} |g(s + \varepsilon'\sqrt{s^2 + 2q})|\, (1 + \sqrt{s^2 + 2q})^M\, |w(s)|\, ds, \qquad (4.1.26)$$

an inequality which implies that QB preserves the space $L^{1,\infty}$. In the case when $q < 0$, looking at (4.1.5), one observes that the denominator $\sqrt{t^2 - 2q}$ cannot approach zero: the problem, this time, is near the point $t = 0$, where the map $\sqrt{t^2 - 2q}$ cannot serve as a regular coordinate in, say, the C^1-sense: however, the product $t\, w(\varepsilon\sqrt{t^2 - 2q})$ is locally integrable near 0 if w is locally integrable near $\pm\sqrt{-2q}$, so we are done. $\qquad\square$

Remark 4.1.2. Trying to substitute $\mathcal{F}L^{2,\infty}$, with an obvious notation (interpolation) for $\mathcal{F}L^{\infty,\infty}$ and $\mathcal{F}L^{1,\infty}$ just fails: one could replace each space by some space $\mathcal{F}L^{p,\infty}$ with $p > 2$ in the first case, $p < 2$ in the second.

4.2 Improving the estimates

When applying the results of the section that precedes to automorphic distributions, rather than to the individual terms $h_{\nu,k}$ of their Fourier series decompositions, we shall need to consider series of symbols $h_{\nu,k}$ with $k \in \mathbb{Z}$, with coefficients making up sequences bounded by powers of $1 + |k|$. To do so, it is necessary to sharpen the estimates obtained so far.

In Lemma 4.1.2, we have already come across the fact that, in some cases, only the image of $h_{\nu_1,q_1} \# h_{\nu_2,q_2}$ under $2i\pi\mathcal{E}$ could be defined as a tempered distribution. We shall now show that substituting for the operator $2i\pi\mathcal{E}$ a certain polynomial P_ι of Pochhammer's style in $2i\pi\mathcal{E}$ improves the estimates, the more so as $\iota = 0, 1, \ldots$ increases. The main application of the next pair of lemmas is Theorem 4.2.3: a totally different proof of a slightly different version of it will be given later.

Lemma 4.2.1. *Let $H = H(\varepsilon_2, t)$ be a function of the real variable t, depending on a parameter $\varepsilon_2 = \pm 1$ and on a (fixed) pair (q_1, q_2) of nonzero numbers: we assume that one does not have simultaneously $q_1 > 0$ and $q_2 < 0$. In the case when $q_1 + q_2 < 0$, we assume that H is C^∞ on the line, and that both $H(\varepsilon_2, t)$ and $\frac{dH}{dt}(\varepsilon_2, t)$ are bounded by some power of $1 + |t|$; when $q_1 + q_2 \geq 0$, we assume that the restriction of $H(\varepsilon_2; t)$ to each of the intervals $] -\infty, -\sqrt{2q_1 + 2q_2}]$ and*

$[\sqrt{2q_1 + 2q_2},\ +\infty[$ *is C^∞ up to the boundary, and keeps the same bound as before as $|t| \to \infty$. Define the operator B_H by the equation*

$$(B_H w)(t) = \sum_{\varepsilon_2} \operatorname{char}(t^2 \geq 2q_1 + 2q_2)\, H(\varepsilon_2, t)\, \frac{w(\varepsilon_2\sqrt{t^2 - 2q_1 - 2q_2})}{\sqrt{t^2 - 2q_1 - 2q_2}}, \quad w \in \mathcal{S}(\mathbb{R}).$$

(4.2.1)

Then, assuming that $w \in \mathcal{S}(\mathbb{R})$ and, in the case when $q_1 + q_2 \geq 0$, that $w(0) = 0$, one has

$$(\operatorname{mad}(P \wedge Q)\, B_H)\, w = \frac{1}{2i\pi}\, B_{H_1} w$$

with

$$H_1(\varepsilon_2, t) = \varepsilon_2 \sqrt{t^2 - 2q_1 - 2q_2}\, \frac{dH}{dt}(\varepsilon_2, t).$$

(4.2.2)

One can dispense with the condition $w(0) = 0$ in the case when H is given by (4.2.3) below, and g_1 and g_2 are even functions. When either of the two conditions making (4.2.2) valid is ensured, this equation can be iterated as many times as needed.

Proof. First, observe that, assuming that $q_1 q_2 \neq 0$ and excluding the case when $q_1 > 0$ and $q_2 < 0$, one cannot have simultaneously $q_1 > 0$ and $q_1 > q_1 + q_2$, so that the operator $\mathcal{F}A_1 A_2 \mathcal{F}^{-1}$ in (4.1.10) is indeed of the kind B_H, as the factor $\operatorname{char}(t^2 \geq 2q_1)$ can be dispensed with. It suffices to set

$$H(\varepsilon_2, t) = \sum_{\varepsilon = \pm 1} H(\varepsilon_1, \varepsilon_2, t)$$

$$= \sum_{\varepsilon_1 = \pm 1} \frac{g_1(t + \varepsilon_1\sqrt{t^2 - 2q_1})}{\sqrt{t^2 - 2q_1}}\, g_2(\varepsilon_1\sqrt{t^2 - 2q_1} + \varepsilon_2\sqrt{t^2 - 2q_1 - 2q_2}):$$

(4.2.3)

the properties of $H(\varepsilon_2; t)$ assumed in the present lemma are indeed true in view of Lemma 4.1.4, in the proof of which it was shown in particular that the arguments of g_1 and g_2 cannot approach 0 in an uncontrollable way.

As seen there, the operator $\operatorname{mad}(P \wedge Q)\, B_H$ acts from $\mathcal{S}(\mathbb{R})$ to $\mathcal{S}'(\mathbb{R})$ (actually, much better than that). Let us prove (4.2.2), first under the additional assumption that $q_1 + q_2 \leq 0$, under which no characteristic function needs be introduced in (4.2.1). One has

$$2i\pi\, [(\operatorname{mad}(P \wedge Q)\, B_H)\, w](t) = \frac{d}{dt}\, [(B_H(Qw))(t)] - t\, (B_H w')(t) = (B_{\frac{dH}{dt}}(Qw))(t)$$

$$+ \sum_{\varepsilon_1, \varepsilon_2} H(\varepsilon_2, t)\, \left[\frac{d}{dt}\left(\varepsilon_2 w(\varepsilon_2\sqrt{t^2 - 2q_1 - 2q_2}) \right) - t\, \frac{w'(\varepsilon_2\sqrt{t^2 - 2q_1 - 2q_2})}{\sqrt{t^2 - 2q_1 - 2q_2}} \right],$$

(4.2.4)

and the bracket is zero. In the case when $q_1 + q_2 > 0$, the factor $\mathrm{char}(t^2 \geq 2q_1 + 2q_2)$ must be reintroduced: this will not result in the addition of any extra term in the case when, with $s = \varepsilon_2 \sqrt{t^2 - 2q_1 - 2q_2}$, $w(s)$ vanishes at $s = 0$.

But, if one drops this condition, one will find the additional term

$$\sum_{\varepsilon_2} \varepsilon_2 \left[H(\varepsilon_2, \sqrt{2q_1 + 2q_2}) - H(\varepsilon_2, -\sqrt{2q_1 + 2q_2}) \right] w(0). \tag{4.2.5}$$

In the case when H is given by (4.2.3), and the functions g_1 and g_2 are even, this additional term is again zero, as can be seen by changing the summation index ε_1 to $-\varepsilon_1$ in one of the two terms, which does not change the factor $g_2(\varepsilon_1 \sqrt{2q_2})$. Clearly, equation (4.2.2) can always be iterated when valid since $H_1(\varepsilon_2, \pm\sqrt{2q_1 + 2q_2}) = 0$. $\qquad\square$

Looking at (4.1.10), one observes that, while it is easy to cope with arbitrary powers of $\sqrt{t^2 - 2q_1 - 2q_2}$ when using functions u in $\mathcal{S}(\mathbb{R})$, and with powers of $1 + |t|$ by looking at the operator $A_1 A_2$ as having its values in $\mathcal{S}'(\mathbb{R})$, there is a genuine difficulty when trying to save powers of $|q_1|$. This is the reason for the following lemma, the proof of which is remarkably inelegant: it would have simplified computations a lot if we had limited it to the (sufficient) cases when $\iota = 0, 1, 2$, but knowing that we are almost certainly dealing with a general trick — working in a direction opposite to that associated to integrations by parts — is, in our opinion, interesting.

Lemma 4.2.2. *Consider the case when, with the notation of Lemma 4.2.1, the function $H(\varepsilon_2, t) = \sum_{\varepsilon_1 = \pm 1} H(\varepsilon_1, \varepsilon_2, t)$ is given by (4.2.3), with*

$$g_1(\xi) = \left| \frac{\xi}{2} \right|^{-\nu_1}, \quad g_2(\xi) = \left| \frac{\xi}{2} \right|^{-\nu_2}. \tag{4.2.6}$$

Still excluding the case when $q_1 > 0$ and $q_2 < 0$, denote, for $t^2 > 2q_1 + 2q_2 > 0$, as

$$\theta = \varepsilon_2 \sqrt{t^2 - 2q_1 - 2q_2} \, \frac{d}{dt} \tag{4.2.7}$$

the operator that occurs in (4.2.2). Introduce the polynomials

$$P_\iota(\theta) = (\theta - \iota)(\theta - \iota + 1) \ldots (\theta + \iota) \quad \text{for } \iota = 0, 1, \ldots \tag{4.2.8}$$

in the operator θ and set

$$[P_\iota(\theta) H(\varepsilon_1, \varepsilon_2, \cdot)](t) = F_\iota(t) H(\varepsilon_1, \varepsilon_2, t). \tag{4.2.9}$$

Then, for $t^2 > 2q_1 + 2q_2 > 0$ (which implies $t^2 > 2q_1$), $F_\iota(t)$ can be written in a unique way as a polynomial in the indeterminate $\varepsilon_1(t^2 - 2q_1)^{-\frac{1}{2}}$, the coefficients of which are polynomials in the variables t, $\varepsilon_2 \sqrt{t^2 - 2q_1 - 2q_2}$, ν_1, ν_2, but do not depend on any other parameter in a direct way. For $\iota = 0, 1, 2, 3, 4$, the term of lowest degree in the indeterminate $\varepsilon_1(t^2 - 2q_1)^{-\frac{1}{2}}$ of $F_\iota(t)$ has degree at least $\iota + 1$.

Proof. Let us simplify the notation, setting

$$H(\varepsilon_1, \varepsilon_2; t) = \varepsilon_1 \left|\frac{t+r}{2}\right|^{-\nu_1} \left|\frac{r+s}{2}\right|^{-\nu_2} r^{-1}, \tag{4.2.10}$$

with

$$r = \varepsilon_1 \sqrt{t^2 - 2q_1}, \quad s = \varepsilon_2 \sqrt{t^2 - 2q_1 - 2q_2}. \tag{4.2.11}$$

We have now $\theta = s\frac{d}{dt}$, while $\frac{dr}{dt} = \frac{t}{r}$, $\frac{ds}{dt} = \frac{t}{s}$, so that

$$\theta r = \frac{st}{r}, \quad \theta(t+r) = \frac{s(t+r)}{r}, \quad \theta(r+s) = \frac{t(r+s)}{r}, \tag{4.2.12}$$

and

$$F_0(t) = \frac{(\theta \, H(\varepsilon_1, \varepsilon_2, t)}{H(\varepsilon_1, \varepsilon_2, t)} = -(\nu_1 s + \nu_2 t)r^{-1} - st\, r^{-2}. \tag{4.2.13}$$

To simplify the computation further, set

$$\alpha = \nu_1 s + \nu_2 t, \quad \beta = st, \quad \gamma = \nu_1 t + \nu_2 s, \quad \delta = s^2 + t^2, \quad \text{and } \rho = r^{-1}, \tag{4.2.14}$$

so that $F_0 = -\alpha\rho - \beta\rho^2$ and

$$\theta\alpha = \gamma, \quad \theta\gamma = \alpha, \quad \theta\beta = \delta, \quad \theta\delta = 4\beta, \quad \theta\rho = -\beta\rho^3. \tag{4.2.15}$$

For $\iota \geq 1$, one has

$$F_\iota = (\theta^2 - \iota^2) F_{\iota-1} + 2 F_0 \cdot (\theta F_{\iota-1}) + (\theta F_0) \cdot F_{\iota-1} + F_0^2 F_{\iota-1}. \tag{4.2.16}$$

The general properties of the functions F_ι are an immediate consequence of the last relations. Neglecting terms of degree ≥ 5 in r^{-1}, one has

$$\theta F_0 \sim -\gamma\rho - \delta\rho^2 + \alpha\beta\rho^3 + 2\beta^2\rho^4,$$
$$\theta^2 F_0 \sim -\alpha\rho - 4\beta\rho^2 + (\alpha\delta + 2\beta\gamma)\rho^3 + 6\beta\delta\rho^4 \tag{4.2.17}$$

and

$$F_1 = \theta^2 F_0 - F_0 + 3F_0 \cdot \theta F_0 + F_0^3$$
$$\sim (3\alpha\gamma - 3\beta)\rho^2 + (4\alpha\delta + 5\beta\gamma - \alpha^3)\rho^3 + (9\beta\delta - 6\alpha^2\beta)\rho^4. \tag{4.2.18}$$

Next,

$$\theta \, F_1 \sim (3\alpha^2 + 3\gamma^2 - 3\delta)\rho^2 + (9\gamma\delta + 21\alpha\beta - 3\alpha^2\gamma)\rho^3$$
$$+ (9\delta^2 + 42\beta^2 - 18\alpha\beta\gamma - 6\alpha^2\delta)\rho^4 \tag{4.2.19}$$

and

$$\theta^2 F_1 \sim (12\alpha\gamma - 12\beta)\rho^2 + (30\alpha\delta + 57\beta\gamma - 6\alpha\gamma^2 - 3\alpha^3)\rho^3$$
$$+ (162\,\beta\delta - 24\,\beta\gamma^2 - 30\alpha\gamma\delta - 48\,\alpha^2\beta)\rho^4. \tag{4.2.20}$$

Then,

$$F_2 = \theta^2 F_1 - 4F_1 + 2F_0 . \theta F_1 + \theta F_0 . F_1 + F_0^2 F_1$$
$$\sim (20\alpha\delta + 40\beta\gamma - 15\alpha\gamma^2 - 5\alpha^3)\rho^3 + (135\,\beta\delta - 35\,\beta\gamma^2 - 75\,\alpha^2\beta - 55\,\alpha\gamma\delta + 10\,\alpha^3\gamma)\rho^4. \tag{4.2.21}$$

Next,

$$\theta F_2 \sim (60\gamma\delta + 120\alpha\beta - 15\gamma^3 - 45\alpha^2\gamma)\rho^3$$
$$+ (135\,\delta^2 + 540\,\beta^2 - 90\,\gamma^2\delta - 120\,\alpha^2\delta - 440\,\alpha\beta\gamma + 30\,\alpha^2\gamma^2 + 10\,\alpha^4)\rho^4 \tag{4.2.22}$$

and

$$\theta^2 F_2 \sim (180\alpha\delta + 360\beta\gamma - 135\alpha\gamma^2 - 45\alpha^3)\rho^3$$
$$+ (2160\,\beta\delta - 800\,\beta\gamma^2 - 960\,\alpha^2\beta - 880\,\alpha\gamma\delta + 60\,\alpha\gamma^3 + 100\,\alpha^3\gamma)\rho^4. \tag{4.2.23}$$

Hence,

$$F_3 \sim \theta^2 F_2 - 9F_2 - 2\alpha\rho\,\theta F_2 - \gamma\rho\,F_2$$
$$\sim [945\,\beta\delta - 525\,(\beta\gamma^2 + \alpha^2\beta + \alpha\gamma\delta) + 105\,(\alpha^3\gamma + \alpha\gamma^3)]\,\rho^4. \tag{4.2.24}$$

Next,

$$\theta\,F_3 \sim [945\,\delta^2 + 3780\,\beta^2 - 1050\,(\gamma^2\delta + \alpha^2\delta) - 4200\,\alpha\beta\gamma + 630\,\alpha^2\gamma^2 + 105\,(\alpha^4 + \gamma^4)]\,\rho^4 \tag{4.2.25}$$

and

$$\theta^2 F_3 \sim [15120\,\beta\delta - 8400\,(\beta\gamma^2 + \alpha^2\beta + \alpha\gamma\delta) + 1680\,(\alpha\gamma^3 + \alpha^3\gamma)]\,\rho^4 \tag{4.2.26}$$

Then

$$F_4 \sim \theta^2 F_3 - 16\,F_3 \sim 0. \tag{4.2.27}$$

□

Remark 4.2.1. We have not found a proof, along these lines (really a problem in non-commutative algebra, according to (4.2.15)-(4.2.16)), valid for every ι. Having pushed the computation so far would be more than sufficient in view of our forthcoming study of the sharp composition of two Hecke distributions, or in view of the same question involving two Eisenstein distributions $\mathfrak{E}_{\nu_1}$ and $\mathfrak{E}_{\nu_2}$ with $|\mathrm{Re}\,(\nu_1 \pm \nu_2)| < 2$. The lemma that precedes can be appreciated in the following way: with $h = h^1 \# h^2$ (not always a distribution, but its image under $\mathcal{E}$ is, even if $q_1 + q_2 = 0$), it tells precisely in which way estimates of the function H such that $\mathcal{F}\mathrm{Op}(h)\mathcal{F}^{-1} = B_H$ improve, if $\sqrt{t^2 - 2q_1}$ is to be considered as large, when $2i\pi\mathcal{E}h$ is replaced by the image of h under the operator $P_\iota(2i\pi\mathcal{E})$ (observe that $P_\iota(\theta)$ is divisible by θ) with larger values of ι. Here, integral numbers are to be substituted to q_1, q_2 in view of applications to modular distribution theory, so that it may be assumed that $|q_1| \geq 1, |q_2| \geq 1$. The difficulty is gaining on the exponents of q_1 or q_2, while gaining on the exponent of $q_1 + q_2$ is easy (cf. (4.2.29) below), but insufficient when $q_1 q_2 < 0$.

Basing our estimates, in the sequel, on the explicit decomposition of $h^1 \# h^2$ (with $A_k = \mathrm{Op}(h^k)$) given by Theorem 4.5.1 into homogeneous components, we shall avoid later the need for the present lemma. It is extremely likely that generalizing Lemma 4.2.2 to all values of ι (which we have not been able to do by algebraic manipulations) would be an easy matter, after the explicit decompositions in Section 4.5 have been obtained. However, we have not found it useful to do so: it would be contrary to the spirit of this lemma, which is to find an *a priori* reason why applying Pochhammer's style polynomials in $\mathrm{mad}(P \wedge Q)$ should be beneficial to the situation, at the same time providing a much needed verification. The condition $\iota \leq 4$ in Theorem 4.2.3 below will be dispensed with later, in Proposition 4.5.4, under the initial assumption that $|\mathrm{Re}\,(\nu_1 \pm \nu_2)| < 1$. Note that this condition $\iota \leq 4$ is only necessary in the last part of the proof below, dealing with the "difficult" case.

Theorem 4.2.3. *Let* $A_k = \mathrm{Op}\,(h_{\nu_k,q_k})$. *Recall that*

$$P_\iota(2i\pi\mathcal{E}) = (2i\pi\mathcal{E} - \iota)_{2\iota+1} = (2i\pi\mathcal{E} - \iota)(2i\pi\mathcal{E} - \iota + 1)\ldots(2i\pi\mathcal{E} + \iota), \quad (4.2.28)$$

so that $P_\iota(2i\pi\mathcal{E})\,(h_{\nu_1,q_1} \# h_{\nu_2,q_2})$ *is the symbol of* $P_\iota(2i\pi\,\mathrm{mad}(P \wedge Q))\,(A_1 A_2)$. *Given* $\iota = 0, 1, 2, 3, 4$, *a real number* $a < \iota + 1$, *and an arbitrary number* N, *the symbol of* $P_\iota(2i\pi\mathrm{mad}(P \wedge Q))\,(A_1 A_2)$ *is a* $O\big(|q_1|^{\frac{-1-\mathrm{Re}\,(\nu_1+\nu_2)-a}{2}}\,(1 + |q_1 + q_2|)^{-N}\big)$ *in the space* $\mathcal{S}'(\mathbb{R}^2)$.

Proof. Let us start with the observation that since

$$(q_1 + q_2)\,e^{2i\pi(q_1+q_2)\frac{x}{\xi}} = \frac{1}{2i\pi}\xi\frac{\partial}{\partial x}\left(e^{2i\pi(q_1+q_2)\frac{x}{\xi}}\right), \quad (4.2.29)$$

arbitrary powers of $1 + |q_1 + q_2|$ can be gained when estimating symbols of such a kind in the space $\mathcal{S}'(\mathbb{R}^2)$, and it amounts to the same to gain powers of $|q_1|$ or of $|q_2|$: only the sum of exponents matters. Let us first dispose of some easy cases. Recall from (4.1.10) that, if $B_k = \mathcal{F}\,A_k\,\mathcal{F}^{-1}$, one has

$$(B_1 B_2 w)(t) = \mathrm{char}(t^2 \geq 2q_1)\,\mathrm{char}(t^2 \geq 2q_1 + 2q_2)\sum_{\varepsilon_1}\frac{g_1(t + \varepsilon_1\sqrt{t^2 - 2q_1})}{\sqrt{t^2 - 2q_1}}$$

$$\cdot\sum_{\varepsilon_2}\frac{g_2(\varepsilon_1\sqrt{t^2 - 2q_1} + \varepsilon_2\sqrt{t^2 - 2q_1 - 2q_2})}{\sqrt{t^2 - 2q_1 - 2q_2}}\,w(\varepsilon_2\sqrt{t^2 - 2q_1 - 2q_2}), \quad (4.2.30)$$

with $g_k(\xi) = \left|\frac{\xi}{2}\right|^{-\nu_k}$.

If both q_1 and q_2 are negative, there is no singularity to cope with, and the absolute value $\sqrt{t^2 - 2q_1 - 2q_2}$ of the argument of w is $\geq \sqrt{|q_1| + |q_2| + \frac{t^2}{2}}$: then, for every N, the operator $B_1 B_2$ is a $O\left((|q_1| + |q_2|)^{-N}\right)$ as an endomorphism of the space $L^{\infty,\infty}$ (cf. Lemma 4.1.5) of functions on the line which remain essentially bounded after having been multiplied by arbitrary powers of $1 + |t|$. This was of course the easiest case. Consider next the case when $q_1 > 0$ and $q_2 < 0$: then, $q_1 + q_2 < q_1$, and the product of characteristic functions in (4.1.10) reduces to the factor $\operatorname{char}(t^2 \geq 2q_1)$. To cope with the (integrable) singularity of the function $(B_1 B_2 w)(t)$, with $B_1 B_2 = \mathcal{F} A_1 A_2 \mathcal{F}^{-1}$, at $t = \pm\sqrt{2q_1}$, we give an L^1-type of estimate for this function, under assumptions of the L^∞-type regarding w. More precisely, assuming that $w \in L^{\infty,\infty}$ and setting $r = \varepsilon_1 \sqrt{t^2 - 2q_1}$, so that $dt = \dfrac{|r|\, dr}{\sqrt{r^2 + 2q_1}}$, we obtain from (4.1.10) that

$$\int_{-\infty}^{\infty} |(B_1 B_2 w)(t)|\, dt = \sum_{\varepsilon_2} \int_{-\infty}^{\infty}$$

$$\frac{g_1(r + \sqrt{r^2 + 2q_1})}{\sqrt{r^2 + 2q_1}}\, \frac{g_2(r + \varepsilon_2\sqrt{r^2 - 2q_2})}{\sqrt{r^2 - 2q_2}}\, w(\varepsilon_2\sqrt{r^2 - 2q_2})\, dr. \quad (4.2.31)$$

Since $\sqrt{r^2 - 2q_2} \geq \max(\sqrt{r^2 + 2}, \sqrt{|2q_2|})$, this integral is, for arbitrary N, a $O\left(|q_1|^{\frac{\operatorname{Re} \nu_1 - 1}{2}} |q_2|^{-N}\right)$ when w remains in a bounded set of the space of functions rapidly decreasing at infinity.

We have covered the cases when $q_2 < 0$, and we thus assume that $q_2 > 0$ from now on, so that we can dispense with the factor $\operatorname{char}(t^2 \geq 2q_1)$ in (4.2.30). The case when $q_1 > 0$ too is just as easy as the last one: now, there are integrable singularities at $t = \pm\sqrt{2q_1 + 2q_2}$, to be taken care of by the same method as before, obtaining estimates of an L^1-type for $B_1 B_2 w$ with assumptions of an $L^{\infty,\infty}$-type about w. What matters here is gaining powers of q_1^{-1} (or q_2^{-1}) with large exponents: since $t^2 \geq 2q_1$ when $\operatorname{char}(t^2 \geq 2q_1 + 2q_2) \neq 0$, it suffices to bound by arbitrary powers of q_2^{-1} an integral such as $\int_{-\infty}^{\infty} |(B_1 B_2 w)(t)|\, (1 + |t|)^{-N} dt$, which is done as before, setting this time $s = \varepsilon_2 \sqrt{t^2 - 2q_1 - 2q_2}$. As a last easy case, we consider the one in which $q_1 < 0$ and $\frac{q_2}{q_1} \geq -\frac{1}{2}$ (this is somewhat arbitrary: $\frac{q_2}{q_1} > -\delta$ with $\delta < 1$ would do just as well), in other words $q_2 < \frac{|q_1|}{2}$: then, $|q_1| \leq 2|q_1 + q_2|$: again, since arbitrary powers of $(1 + |q_1 + q_2|)^{-1}$ can be "gained", we are done.

The only difficult case is that in which $q_1 < 0$ and $\frac{q_2}{q_1} < -\frac{1}{2}$, or $q_2 > \frac{|q_1|}{2}$. In that case, if $H_{\varepsilon_2}(t) = \sum_{\varepsilon_1 = \pm 1} H(\varepsilon_1, \varepsilon_2, t)$ is the function $H(\varepsilon_2, t)$ introduced in Lemma 4.2.2, one has $(B_1 B_2 w)(t) = \sum_{\varepsilon_2 = \pm 1} (B_{H_{\varepsilon_2}} w)(t)$ with the notation of Lemma 4.2.1. It follows from this lemma that, for $\iota = 0, 1, \ldots$, one has

$$P_\iota(\operatorname{mad}(P \wedge Q))(B_1 B_2) = (2i\pi)^{-\iota} \sum_{\varepsilon_2 = \pm 1} B_{H_{\varepsilon_2, \iota}}, \quad (4.2.32)$$

with $H_{\varepsilon_2,\iota} = P_\iota\left(\varepsilon_2\sqrt{t^2 - 2q_1 - 2q_2}\frac{d}{dt}\right)H_{\varepsilon_2}$. Lemma 4.2.2 finally makes it possible to save an extra factor $\left(\sqrt{t^2 - 2q_1}\right)^{-\iota-1}$, provided that $\iota \leq 4$. $\square$

4.3   A regularization argument

Considering two symbols of the kind $h^j(x,\xi) = f_j(\xi)e^{2i\pi q_j\frac{x}{\xi}}$, we wish to obtain the decomposition of $h^1 \# h^2$ into homogeneous components by an application of Theorem 3.3.1. However, such symbols never lie in $\mathcal{S}(\mathbb{R}^2)$, so that Theorem 3.3.1 cannot be applied directly. But, under the assumption that $f \in \mathcal{S}(\mathbb{R})$ is flat to infinite order at 0, the product of h by a function $\phi\left(\frac{\theta x}{\xi}\right)$, with $\phi \in \mathcal{S}(\mathbb{R})$ and $\theta > 0$, will lie in $\mathcal{S}(\mathbb{R}^2)$. Let us first characterize the Mellin transforms of such functions f (assumed to be even, a permanent assumption on symbols).

Lemma 4.3.1. *Given $A > 0$, and an even function f on $\mathbb{R}^\times = \mathbb{R}\backslash\{0\}$, the following two conditions are equivalent:*

(i) *f is C^∞ on $\mathbb{R}^\times$; given $b < A$ and $N = 0, 1, \ldots$, there exists $C > 0$ such that*

$$\left|\left(\xi\frac{d}{d\xi}\right)^N f(\xi)\right| \leq C|\xi|^{-1}(|\xi| + |\xi|^{-1})^{-b}; \tag{4.3.1}$$

(ii) *there exists a function $\psi = \psi(\nu)$ holomorphic in the strip $\{\nu: |\mathrm{Re}\ \nu| < A\}$, rapidly decreasing as a function of $\mathrm{Im}\ \nu$ in a way uniform with respect to $\mathrm{Re}\ \nu$ under the condition $|\mathrm{Re}\ \nu| \leq b$ for some $b < A$, such that, for every $a \in]-A, A[$, the identity*

$$f(\xi) = \frac{1}{i}\int_{\mathrm{Re}\ \nu=a} \psi(\nu)\,|\xi|^{-\nu-1}d\nu, \quad \xi \neq 0 \tag{4.3.2}$$

holds. Of necessity, such a function ψ is unique.

The space of functions satisfying these conditions for some given $A > 0$ will be denoted as $\mathcal{S}_A(\mathbb{R}^\times)$. In particular, the space $\mathcal{S}_\infty(\mathbb{R}^\times) = \cap_{A>0}\mathcal{S}_A(\mathbb{R}^\times)$ coincides with the space of functions in $\mathcal{S}_{even}(\mathbb{R})$, flat to infinite order at 0.

Proof. Given f, one defines $\psi(\nu) = \frac{1}{2\pi}\int_0^\infty t^\nu f(t)\,dt$, and one applies the Fourier inversion formula, interpreting the definition of ψ as the fact that $\psi(a + i\lambda)$ is the inverse Fourier transform, evaluated at λ, of the function $\tau \mapsto e^{2\pi(a+1)\tau}f(e^{2\pi\tau})$ (there is no need for the exponential change of variable if, unlike this author, you are as familiar with the inverse Mellin transformation as with the inverse Fourier transformation). Also, one writes

$$\nu^N\psi(\nu) = (-1)^N \cdot \frac{1}{2\pi}\int_0^\infty t^\nu\left(t\frac{d}{dt} + 1\right)^N f(t)\,dt. \tag{4.3.3}$$

The remaining details are even easier. $\square$

We shall use the approximation of a symbol $h(x, \xi) = f(\xi)e^{2i\pi q\frac{x}{\xi}}$ by the symbol h^θ obtained as the result of multiplying $h(x, \xi)$ by $\phi\left(\frac{\theta x}{\xi}\right)$, with $\phi \in \mathcal{S}(\mathbb{R})$ and $\theta > 0$ going to zero: we shall assume that $f \in \mathcal{S}_\infty(\mathbb{R})$, so that $h^\theta \in \mathcal{S}_{\text{even}}(\mathbb{R}^2)$.

Lemma 4.3.2. *With $f \in \mathcal{S}_\infty(\mathbb{R})$, and $q \neq 0$, define the operator A_q by the equation $A_q = \mathrm{Op}\left((x, \xi) \mapsto f(\xi)e^{2i\pi q\frac{x}{\xi}}\right)$, so that, with $g(\xi) = \frac{|\xi|}{2}f(\frac{\xi}{2})$, one has, as proved in Lemma 4.1.1*

$$\left(\mathcal{F}(A_q u)\right)(t) = \mathrm{char}(t^2 \geq 2q) \sum_\varepsilon g(t + \varepsilon\sqrt{t^2 - 2q}) \frac{(\mathcal{F}u)(\varepsilon\sqrt{t^2 - 2q})}{\sqrt{t^2 - 2q}}. \quad (4.3.4)$$

Define, for some function $\phi \in \mathcal{S}_{\text{even}}(\mathbb{R})$, satisfying the property that $\widehat{\phi}$ has its support contained in the open interval $] - |q|, |q|[$, the operator A_q^θ as the operator with symbol $h^\theta(x, \xi) = \phi\left(\frac{\theta x}{\xi}\right) f(\xi)e^{2i\pi q\frac{x}{\xi}}$. One has the identity

$$A_q^\theta = \theta^{-1} \int_{-\infty}^\infty \widehat{\phi}\left(\frac{r - q}{\theta}\right) A_r \, dr. \quad (4.3.5)$$

If $\phi(0) = 1$, the operator A_q^θ converges, as $\theta \to 0$, to A_q in the space of weakly continuous linear operators from $\mathcal{S}(\mathbb{R})$ to $\mathcal{S}'(\mathbb{R})$, i.e., $(v|A_q^\theta u) \to (v|A_q u)$ for every pair u, v of functions in $\mathcal{S}(\mathbb{R})$.

Proof. The expression $\left(\mathcal{F}(A_q^\theta u)\right)(t)$ can be obtained by the integral analogous to (4.1.7), only inserting the extra factor $\phi\left(-\frac{2\theta\xi}{t+x}\right)$ under the integral (do not forget the role of the matrix $\left(\begin{smallmatrix} 0 & -1 \\ 1 & 0 \end{smallmatrix}\right)$). Since

$$\int_{-\infty}^\infty \phi\left(-\frac{2\theta\xi}{t+x}\right) \exp\left(2i\pi\xi\left(t - x - \frac{2q}{t+x}\right)\right) d\xi = \left|\frac{t+x}{2\theta}\right| \widehat{\phi}\left(\frac{t^2 - x^2 - 2q}{2\theta}\right),$$
$$\quad (4.3.6)$$

one has

$$\left(\mathcal{F}(A_q^\theta u)\right)(t) = \int_{-\infty}^\infty f(\frac{t+x}{2}) \left|\frac{t+x}{2\theta}\right| \widehat{\phi}\left(\frac{t^2 - x^2 - 2q}{2\theta}\right) (\mathcal{F}u)(x) \, dx. \quad (4.3.7)$$

Making the change of variable $x \mapsto r = \frac{t^2 - x^2}{2}$, so that $t^2 - 2r \geq 0$ and $x = \pm\sqrt{t^2 - 2r}$ and $\left|\frac{dx}{dq}\right| = \frac{1}{\sqrt{t^2 - 2q}}$, one obtains

$$\left(\mathcal{F}(A_q^\theta u)\right)(t) = \sum_{\varepsilon - \pm 1} g(t + \varepsilon\sqrt{t^2 - 2r}) \int_{-\infty}^\infty \theta^{-1}\widehat{\phi}\left(\frac{r - q}{\theta}\right) \frac{(\mathcal{F}u)(\varepsilon\sqrt{t^2 - 2r})}{\sqrt{t^2 - 2r}} \, dr$$

$$= \int_{-\infty}^\infty \theta^{-1}\widehat{\phi}\left(\frac{r - q}{\theta}\right) (\mathcal{F}(A_r u))(t) \, dr. \quad (4.3.8)$$

The second part follows from (4.3.5), together with the fact that, in the space of operators under discussion, the operator A_r depends continuously on r. $\qquad\square$

Lemma 4.3.3. *Let $f_j \in \mathcal{S}_\infty(\mathbb{R})$ $(j = 1, 2)$, and let $q_1, q_2 \in \mathbb{R}$ be such that $q_1 q_2 (q_1 + q_2) \neq 0$. Let $h^j(x, \xi) = f_j(\xi) e^{2i\pi q_j \frac{x}{\xi}}$. With ϕ as defined in Lemma 4.3.2, and $|\theta|$ small enough, set $(h^j)^\theta(x, \xi) = h^j(x, \xi)\, \phi\left(\frac{\theta x}{\xi}\right)$. The symbol $(h^1)^{\theta_1} \# (h^2)^{\theta_2}$ converges to $h^1 \# h^2$ in $\mathcal{S}'(\mathbb{R}^2)$ as $\theta_1, \theta_2 \to 0$.*

Proof. Set $h^j_r(x, \xi) = f_j(\xi) e^{2i\pi r \frac{x}{\xi}}$ so that, in particular, $h^j = h^j_{q_j}$. Set $B^{\theta_j}_j = \mathcal{F}\mathrm{Op}((h^j)^{\theta_j})\mathcal{F}^{-1}$ and $B_{j,r} = \mathcal{F}\mathrm{Op}(h^j_r)\mathcal{F}^{-1}$. One must show that the product $B^{\theta_1}_1 B^{\theta_2}_2$ converges to $B_1 B_2 := B_{1,q_1} B_{2,q_2}$ in the space of weakly continuous operators from $\mathcal{S}(\mathbb{R})$ to $\mathcal{S}'(\mathbb{R})$ as $\theta_1, \theta_2 \to 0$. Fix $\delta > 0$ such that the condition $r_1 r_2 (r_1 + r_2) \neq 0$ holds if $|r_1 - q_1| + |r_2 - q_2| < \delta$: then, if θ_j is small enough for the support of $\widehat{\phi}$ to be contained in the interval $] - \frac{\delta}{\theta_j}, \frac{\delta}{\theta_j}[$, one can apply (4.3.5) to each of the two operators $B^{\theta_j}_j$, obtaining as a result

$$B^{\theta_1}_1 B^{\theta_2}_2 = (\theta_1 \theta_2)^{-1} \int_{\mathbb{R}^2} \widehat{\phi}\left(\frac{r_1 - q_1}{\theta_1}\right) \widehat{\phi}\left(\frac{r_2 - q_2}{\theta_2}\right) B_{1,r_1} B_{2,r_2}\, dr_1\, dr_2. \quad (4.3.9)$$

What remains to be proved is that, for (r_1, r_2) sufficiently close to (q_1, q_2), the product $B_{1,r_1} B_{2,r_2}$ remains in a weakly bounded subset of the space of weakly continuous linear operators from $\mathcal{S}(\mathbb{R})$ to $\mathcal{S}'(\mathbb{R})$ and that, in the same space, this product converges to $B_{1,q_1} B_{2,q_2}$ as $r_1 \to q_1, r_2 \to q_2$. From (4.1.10), one obtains

$$(B_{1,r_1} B_{2,r_2} w)(t) = \mathrm{char}(t^2 \geq 2q_1)\, \mathrm{char}(t^2 \geq 2q_1 + 2q_2) \sum_{\varepsilon_1} \frac{g_1(t + \varepsilon_1\sqrt{t^2 - 2r_1})}{\sqrt{t^2 - 2r_1}}$$

$$\cdot \sum_{\varepsilon_2} \frac{g_2(\varepsilon_1\sqrt{t^2 - 2r_1} + \varepsilon_2\sqrt{t^2 - 2r_1 - 2r_2})}{\sqrt{t^2 - 2r_1 - 2r_2}}\, w(\varepsilon_2\sqrt{t^2 - 2r_1 - 2r_2}). \quad (4.3.10)$$

The points q_1 and $q_1 + q_2$ are by assumption distinct and distinct from 0, so that the five points $\pm\sqrt{2q_1}, \pm\sqrt{2(q_1 + q_2)}$ and 0 are pairwise distinct. Let V be the union of 4 closed intervals centered at the first 4 points, pairwise disjoint and not containing 0: for (r_1, r_2) close enough to (q_1, q_2), each of the points $\pm\sqrt{2r_1}, \pm\sqrt{2(r_1 + r_2)}$ remains in a fixed interval taken from the 4 ones making up V. Then, the singularities involved in (4.3.10) lie, as r_1, r_2 vary, in 4 disjoint closed intervals and do not prevent integrability. Under the assumption that w remains in a bounded subset of $L^\infty(\mathbb{R})$, one sees that the integral of $|B_{1,r_1} B_{2,r_2} w|$ over V will be as small as desired provided that V is small enough. On the other hand, it is clear that a uniform bound on some product $(1 + |t|)^M \left|\frac{dw}{dt}\right|$ will ensure that the integral of $|B_{1,r_1} B_{2,r_2} w - B_{1,q_1} B_{2,q_2} w|$ (or even of the product of this function by $(1 + |t|)^{M'}$ with M' as large as desired) over $\mathbb{R}\backslash V$ goes to zero as $r_1 \to q_1, r_2 \to q_2$. $\qquad \square$

4.4   Computing an elementary integral

We need now to compute an elementary integral: though special cases of the calculations which follow have certainly been made hundreds of times, it would be hard to find a reference giving exactly what we need.

Lemma 4.4.1. *With $j = 0$ or 1, consider, for* Re α, Re β_1, Re β_2 *negative,* Re $(\alpha + \beta_1 + \beta_2 + 2) > 0$ *and $c \in \mathbb{R}$, $c \neq 0, -1$ the (convergent) integral*

$$R^{(j)}(\alpha, \beta_1, \beta_2; c) = \int_{-\infty}^{\infty} |t|_j^{-\alpha-1}|1-t|_j^{-\beta_1-1}|c+t|_j^{-\beta_2-1}dt. \tag{4.4.1}$$

When $c > -1$, $c \neq 0$, this integral can be made explicit as

$$R^{(j)}(\alpha, \beta_1, \beta_2; c) = \Delta_j(-\beta_2, \alpha+\beta_2+1) \, |c|^{-\alpha-\beta_2-1} \, {}_2F_1(\beta_1+1, -\alpha; -\alpha-\beta_2; -c)$$
$$+ \Delta_j(-\beta_1, -\alpha-\beta_2-1) \, {}_2F_1(\beta_2+1, \alpha+\beta_1+\beta_2+2; \alpha+\beta_2+2; -c). \tag{4.4.2}$$

where we have set

$$\Delta_j(x,y) = \frac{\Gamma(x)\Gamma(y)}{\Gamma(x+y)} + (-1)^j \frac{\Gamma(x)\Gamma(1-x-y)}{\Gamma(1-y)} + \frac{\Gamma(1-x-y)\Gamma(y)}{\Gamma(1-x)}$$

$$= \Delta_j(1-x-y, y) = \frac{B_j(x+y)}{B_j(x)B_0(y)}, \tag{4.4.3}$$

a function singular only where $x+y = 1+j+2n$, or $x = -j-2n$, or $y = -2n$, with $n = 0, 1, \ldots$. When $c < 0$, $c \neq -1$, one has

$$R^{(j)}(\alpha, \beta_1, \beta_2; c) = (-1)^j |c|^{-\alpha-\beta_2-1}$$

$$\cdot \Bigg[\Delta_j(\alpha+\beta_1+\beta_2+2, -\beta_1-\beta_2-1) \, {}_2F_1(\beta_1+1, -\alpha; \beta_1+\beta_2+2; c+1)$$

$$+ \Delta_j(-\beta_1, \beta_1+\beta_2+1) \, |1+c|^{-\beta_1-\beta_2-1} \, {}_2F_1(-\alpha-\beta_1-\beta_2-1, -\beta_2; -\beta_1-\beta_2; c+1) \Bigg]. \tag{4.4.4}$$

If Re $x > 0$, Re $y > 0$ *and* Re $(x+y) < 1$, *one has*

$$\Delta_j(x,y) = \int_{-\infty}^{\infty} |t|_j^{-x-y}|1+t|_j^{x-1} \, dt. \tag{4.4.5}$$

Proof. One has the identities

$$R^{(j)}(\alpha, \beta_1, \beta_2; c) = |1+c|_j^{-\alpha-\beta_1-\beta_2-2} \, R^{(j)}\left(\beta_2, \beta_1, \alpha; -\frac{c}{1+c}\right)$$

$$= (-1)^j \, |c|^{-\alpha-\beta_2-1}|1+c|^{-\beta_1-\beta_2-1} \, R^{(j)}(\alpha, \beta_1, -(\alpha+\beta_1+\beta_2+2); -c-1): \tag{4.4.6}$$

the first one is obtained by means of the change of variable $t \mapsto (1+c)t - c$, the second one by means of the change of variable $t \mapsto \frac{ct}{c+1-t}$.

Next, recall the equations (under the preceding assumptions, and with $\gamma > 0$)

$$\int_0^1 t^{-\alpha-1}(1-t)^{-\beta_1-1}(1+\gamma t)^{-\beta_2-1}dt$$

$$= \frac{\Gamma(-\alpha)\Gamma(-\beta_1)}{\Gamma(-\alpha-\beta_1)}\, {}_2F_1(\beta_2+1,-\alpha;-\alpha-\beta_1;-\gamma),$$

$$\int_0^\infty t^{-\beta_2-1}(1+t)^{-\alpha-1}(1+\gamma t)^{-\beta_1-1}dt$$

$$= \frac{\Gamma(-\beta_2)\Gamma(\alpha+\beta_1+\beta_2+2)}{\Gamma(\alpha+\beta_1+2)}\, {}_2F_1(\beta_1+1,-\beta_2;\alpha+\beta_1+2;1-\gamma). \qquad (4.4.7)$$

The first equation can be found in considerably many places, for instance [22, p. 54]. The second follows, after one has performed the change of variable $t = \frac{s}{1-s}$.

Then, assuming $c > 0$, we split the integral (4.4.1) defining $R^{(j)}(\alpha,\beta_1,\beta_2;c)$ into 4 parts, using the points $-c,0$ and 1 to separate the interval. One has

$$\int_{-\infty}^{-c} \ldots = \int_0^\infty t^{-\beta_2-1}(c+t)^{-\alpha-1}(1+c+t)^{-\beta_1-1}dt,$$

$$\int_{-c}^0 \ldots = (-1)^j c^{-\alpha-\beta_2-1}\int_0^1 t^{-\alpha-1}(1-t)^{-\beta_2-1}(1+ct)^{-\beta_1-1}dt,$$

$$\int_1^\infty \ldots = (-1)^j \int_0^\infty t^{-\beta_1-1}(1+t)^{-\alpha-1}(c+1+t)^{-\beta_2-1}dt. \qquad (4.4.8)$$

In each case, including of course that of the original integral $\int_0^1 \ldots$, to be inserted without transformation of the integrand between the last two equations, one can compute the integral by an application of one of the two equations (4.4.7). One obtains, still assuming $c > 0$,

$$R^{(j)}(\alpha,\beta_1,\beta_2;c)$$

$$= c^{-\alpha-\beta_2-1}(1+c)^{-\beta_1-1}\frac{\Gamma(-\beta_2)\Gamma(\alpha+\beta_1+\beta_2+2)}{\Gamma(\alpha+\beta_1+2)}$$

$$\cdot {}_2F_1(\beta_1+1,-\beta_2;\alpha+\beta_1+2;\frac{1}{1+c})$$

$$+ (-1)^j c^{-\alpha-\beta_2-1}\frac{\Gamma(-\alpha)\Gamma(-\beta_2)}{\Gamma(-\alpha-\beta_2)}\, {}_2F_1(\beta_1+1,-\alpha;-\alpha-\beta_2;-c)$$

$$+ c^{-\beta_2-1}\frac{\Gamma(-\alpha)\Gamma(-\beta_1)}{\Gamma(-\alpha-\beta_1)}\, {}_2F_1(\beta_2+1,-\alpha;-\alpha-\beta_1;-\frac{1}{c})$$

$$+ (-1)^j(1+c)^{-\beta_2-1}\frac{\Gamma(-\beta_1)\Gamma(\alpha+\beta_1+\beta_2+2)}{\Gamma(\alpha+\beta_2+2)}$$

$$\cdot {}_2F_1(\beta_2+1,-\beta_1;\alpha+\beta_2+2;\frac{c}{c+1}). \qquad (4.4.9)$$

We take benefit, now, from the set of so-called linear transformations of the hypergeometric function to move all arguments to the value $-c$ by means of the successive transformations $z \mapsto 1 - \frac{1}{z}$ or $\frac{1}{z}$ or $\frac{z}{z-1}$, obtaining [22, p. 48] and, in the first and last cases, [22, p. 47]

$$
{}_2F_1(\beta_1 + 1, -\beta_2; \alpha + \beta_1 + 2; \frac{1}{1+c})
$$
$$
= \frac{\Gamma(\alpha + \beta_1 + 2)\Gamma(\alpha + \beta_2 + 1)}{\Gamma(\alpha + 1)\Gamma(\alpha + \beta_1 + \beta_2 + 2)} (1+c)^{\beta_1+1} \, {}_2F_1(\beta_1 + 1, -\alpha; -\alpha - \beta_2; -c)
$$
$$
+ \frac{\Gamma(\alpha + \beta_1 + 2)\Gamma(-\alpha - \beta_2 - 1)}{\Gamma(\beta_1 + 1)\Gamma(-\beta_2)} c^{\alpha+\beta_2+1}(1+c)^{-\beta_2}
$$
$$
\cdot \, {}_2F_1(\alpha + 1, -\beta_1; \alpha + \beta_2 + 2; -c) \tag{4.4.10}
$$

or

$$
{}_2F_1(\beta_1 + 1, -\beta_2; \alpha + \beta_1 + 2; \frac{1}{1+c})
$$
$$
= \frac{\Gamma(\alpha + \beta_1 + 2)\Gamma(\alpha + \beta_2 + 1)}{\Gamma(\alpha + 1)\Gamma(\alpha + \beta_1 + \beta_2 + 2)} (1+c)^{\beta_1+1} \, {}_2F_1(\beta_1 + 1, -\alpha; -\alpha - \beta_2; -c)
$$
$$
+ \frac{\Gamma(\alpha + \beta_1 + 2)\Gamma(-\alpha - \beta_2 - 1)}{\Gamma(\beta_1 + 1)\Gamma(-\beta_2)} c^{\alpha+\beta_2+1}(1+c)^{\beta_1+1}
$$
$$
\cdot \, {}_2F_1(\beta_2 + 1, \alpha + \beta_1 + \beta_2 + 2; \alpha + \beta_2 + 2; -c), \tag{4.4.11}
$$

next

$$
{}_2F_1(\beta_2 + 1, -\alpha; -\alpha - \beta_1; -\frac{1}{c})
$$
$$
= \frac{\Gamma(-\alpha - \beta_1)\Gamma(\alpha + \beta_2 + 1)}{\Gamma(\beta_2 + 1)\Gamma(-\beta_1)} c^{-\alpha} \, {}_2F_1(\beta_1 + 1, -\alpha; -\alpha - \beta_2; -c)
$$
$$
+ \frac{\Gamma(-\alpha - \beta_1)\Gamma(-\alpha - \beta_2 - 1)}{\Gamma(-\alpha)\Gamma(-\alpha - \beta_1 - \beta_2 - 1)} c^{\beta_2+1}
$$
$$
\cdot \, {}_2F_1(\beta_2 + 1, \alpha + \beta_1 + \beta_2 + 2; \alpha + \beta_2 + 2; -c), \tag{4.4.12}
$$

finally

$$
{}_2F_1(\beta_2 + 1, -\beta_1; \alpha + \beta_2 + 2; \frac{c}{c+1})
$$
$$
= (1+c)^{\beta_2+1} \, {}_2F_1(\beta_2 + 1, \alpha + \beta_1 + \beta_2 + 2; \alpha + \beta_2 + 2; -c). \tag{4.4.13}
$$

All terms on the right-hand side of (4.4.9) become linear combinations, the coefficients of which are products of powers of c and $1 + c$, of the two expressions

$$
G = {}_2F_1(\beta_1 + 1, -\alpha; -\alpha - \beta_2; -c), \quad H = {}_2F_1(\beta_2 + 1, \alpha + \beta_1 + \beta_2 + 2; \alpha + \beta_2 + 2; -c). \tag{4.4.14}
$$

Simplifying the coefficients, we obtain (4.4.2) in the case when $c > 0$.

In the case when $-1 < c < 0$, the first equation (4.4.6) yields

$$R^{(j)}(\alpha, \beta_1, \beta_2; c) = (1 + c)^{-\alpha - \beta_1 - \beta_2 - 2}$$

$$\cdot \left[\Delta_j(-\alpha, \alpha + \beta_2 + 1) \left(-\frac{c}{1+c}\right)^{-\alpha - \beta_2 - 1} {}_2F_1\left(\beta_1 + 1, -\beta_2; -\alpha - \beta_2; \frac{c}{1+c}\right) \right.$$

$$\left. + \Delta_j(-\beta_1, -\alpha - \beta_2 - 1) \, {}_2F_1\left(\alpha + 1, \alpha + \beta_1 + \beta_2 + 2; \alpha + \beta_2 + 2; \frac{c}{1+c}\right) \right].$$

$$(4.4.15)$$

On one hand, one has $\Delta_j(x, y) = \Delta_j(1 - x - y, y)$, so that

$$\Delta_j(-\alpha, \alpha + \beta_2 + 1) = \Delta_j(-\beta_2, \alpha + \beta_2 + 1). \tag{4.4.16}$$

On the other hand, one has the pair of equations (using the transformation $z \mapsto \frac{z}{z-1}$ of the argument [22, p. 47])

$$_2F_1(\beta_1+1, -\beta_2; -\alpha-\beta_2; \frac{c}{1+c}) = (1+c)^{\beta_1+1} \, {}_2F_1(\beta_1+1, -\alpha; -\alpha-\beta_2; -c) \tag{4.4.17}$$

and

$$_2F_1\left(\alpha + 1, \alpha + \beta_1 + \beta_2 + 2; \alpha + \beta_2 + 2; \frac{c}{1+c}\right)$$

$$= (1 + c)^{\alpha + \beta_1 + \beta_2 + 2} \, {}_2F_1(\beta_2 + 1, \alpha + \beta_1 + \beta_2 + 2; \alpha + \beta_2 + 2; -c). \tag{4.4.18}$$

This transforms (4.4.15) to (4.4.2). In the case when $c < 0$, one obtains (4.4.4) by a combination of the second identity (4.4.6) and of (4.4.2).

Finally, let us derive now the second relation (4.4.3) from the first, say in the case when $j = 1$. The expression $\Delta_1(x, y)$ can be transformed, with the help of the formula of complements and of the duplication formula, to

$$\frac{\Gamma(x)\Gamma(y)}{\Gamma(x+y)} \left[1 - \frac{\sin \pi y}{\sin \pi(x+y)} + \frac{\sin \pi x}{\sin \pi(x+y)} \right]$$

$$= \frac{\Gamma(x)\Gamma(y)}{\Gamma(x+y)} \frac{\sin \pi \frac{x+y}{2} + \sin \pi \frac{x-y}{2}}{\sin \pi \frac{x+y}{2}} = \frac{2\,\Gamma(x)\Gamma(y)}{\Gamma(x+y)} \frac{\sin \frac{\pi x}{2} \cos \frac{\pi y}{2}}{\sin \pi \frac{x+y}{2}}$$

$$= \pi^{\frac{1}{2}} \frac{\Gamma(\frac{1+x}{2})}{\Gamma(\frac{2-x}{2})} \frac{\Gamma(\frac{y}{2})}{\Gamma(\frac{1-y}{2})} \frac{\Gamma(\frac{2-x-y}{2})}{\Gamma(\frac{1+x+y}{2})} = \frac{B_1(x+y)}{B_1(x) B_0(y)}. \tag{4.4.19}$$

The equation (4.4.5) is immediate, just cutting the integral into 3 parts.   □

Remarks 4.3.1. (i) each of the two functions $\Delta_j(\dots)$ on the right-hand side of (4.4.2) has a simple pole, as a function of $\alpha + \beta_2 + 1$, when this argument is zero: the residues are ± 2, so that, using also the fact that the two hypergeometric functions on the right-hand side of (4.4.2) coincide when $\alpha + \beta_2 + 1 = 0$, one sees that $R^{(j)}(\dots)$ has no singularity there, as obvious from its integral definition (4.4.1). This will be important in the beginning of the proof of Theorem 4.5.1.

(ii) for $c \neq 0, -1$, the function $R^{(j)}(\alpha, \beta_1, \beta_2; c)$ extends as a meromorphic function of α, β_1, β_2 in the whole of $\mathbb{C}^3$, with explicit poles and polar parts. On the other hand, one should not believe that the values 0 and -1 of c appear as poles in some extension of the function under examination to some domain in $\mathbb{C}$: ramification occurs. As $c \to 0$ from either side, $R^{(j)}(\dots)$ can be written as the sum of a C^∞ function of c and of the product of such a function by $|c|^{-\alpha-\beta_2-1}$; as $c \to -1$ from either side, $R^{(j)}(\dots)$ can be written as the sum of a C^∞ function of $c+1$ and of the product of such a function by $|c+1|^{-\beta_1-\beta_2-1}$. Indeed, let us use a C^∞ partition of unity $1 = \phi_0 + \phi_1 + \phi_\infty$ on the real line, where ϕ_0 is supported near 0 and equal to 1 in some smaller neighbourhood, and ϕ_1 does the same job around $t = 1$. The function $R^{(j)}(\dots)$ differs by some function of c regular near 0 from the integral

$$\int_{-\infty}^{\infty} \phi_0(t)\,(1-t)^{-\beta_1-1}\,|t|_j^{-\alpha-1}|c+t|_j^{-\beta_2-1}\,dt$$

$$= |c|^{-\alpha-\beta_2-1} \int_{-\infty}^{\infty} \phi_0(ct)\,(1-ct)^{-\beta_1-1}|t|_j^{-\alpha-1}|1+t|_j^{-\beta_2-1}. \quad (4.4.20)$$

If $\mathrm{Re}\,\alpha < 0$, $\mathrm{Re}\,\beta_2 < 0$ and $\mathrm{Re}\,(\alpha + \beta_2 + 1) > 0$, it follows with the help of (4.4.5) that the main term of $R^{(j)}(\alpha, \beta_1, \beta_2; c)$ near $c = 0$ is $|c|^{-\alpha-\beta_2-1}\Delta_j(-\beta_2, \alpha+\beta_2+1)$, which agrees with (4.4.2).

The analysis of $R^{(j)}(\dots)$ for c close to -1 is obtained in the same way, replacing the function ϕ_0 by ϕ_1. This provides a much needed verification of the formulas in the last lemma.

4.5 The sharp product of joint eigenfunctions of $\mathcal{E}$, $\xi\frac{\partial}{\partial x}$

The theorems of this section are concerned with computing explicitly the sharp products of any two individual symbols h_{ν_r, q_r} (4.0.1), under the assumption that $|\mathrm{Re}\,(\nu_1 \pm \nu_2)| < 1$. It follows then from Lemma 4.1.2 that, if $q_1 q_2(q_1 + q_2) \neq 0$ (an assumption to be removed in latter parts of the section), this sharp product is well-defined as a tempered distribution: also, it depends analytically on (ν_1, ν_2). To prove the identities which are the object of the next theorem, one can thus address instead the similar problem, in which each of the two individual symbols has been replaced by an integral superposition of such (with respect to ν_r lying on a line $\mathrm{Re}\,\nu_r = a_r$, using analytic continuation at the end) making up a function $h(x, \xi) = f(\xi)e^{2i\pi q\frac{x}{\xi}}$ with $f \in \mathcal{S}_\infty(\mathbb{R})$. This makes it possible to use the approximation process in Lemma 4.3.2, as is necessary since we have only stated and proved the composition formulas (3.3.2) to (3.3.5) in the case of two symbols in the space $\mathcal{S}(\mathbb{R}^2)$.

Theorem 4.5.1. *Let ν_1, ν_2 satisfy the conditions $|\mathrm{Re}\,(\nu_1 \pm \nu_2)| < 1$ and, with $r = 1, 2$, let $h^r = h_{\nu_r, q_r}$, $A_r = \mathrm{Op}(h^r)$. Assume that $q_1 q_2 (q_1 + q_2) \neq 0$ and set $h = h^1 \# h^2$. If $\frac{q_2}{q_1} < 0$, h admits a decomposition into homogeneous components $h = \int_{-\infty}^{\infty} h_{i\lambda} d\lambda$, given by the equation*

$$
h_{i\lambda}(x, \xi) = \frac{2^{\frac{-1+\nu_1+\nu_2-i\lambda}{2}}}{4\pi} \sum_{j=0,1} (-1)^j |\xi|^{-1-i\lambda} e^{2i\pi \frac{(q_1+q_2)x}{\xi}}
$$

$$
\cdot \left[\Delta_j \left(\frac{1+\nu_1-\nu_2+i\lambda}{2}, -i\lambda \right) |q_1|_j^{\frac{-1-\nu_1-\nu_2-i\lambda}{2}} |q_1+q_2|^{i\lambda} \right.
$$

$$
\cdot\, {}_2F_1 \left(\frac{1+\nu_1+\nu_2+i\lambda}{2}, \frac{1-\nu_1+\nu_2+i\lambda}{2}; 1+i\lambda; \frac{q_1+q_2}{q_1} \right)
$$

$$
+ \Delta_j \left(\frac{1-\nu_1-\nu_2-i\lambda}{2}, i\lambda \right) |q_1|_j^{\frac{-1-\nu_1-\nu_2+i\lambda}{2}}
$$

$$
\left. \cdot\, {}_2F_1 \left(\frac{1-\nu_1+\nu_2-i\lambda}{2}, \frac{1+\nu_1+\nu_2-i\lambda}{2}; 1-i\lambda; \frac{q_1+q_2}{q_1} \right) \right].
$$

$$(4.5.1)$$

If $\frac{q_2}{q_1} > -1$, one has

$$
h_{i\lambda}(x, \xi) = 2^{\frac{-1+\nu_1+\nu_2-i\lambda}{2}} \cdot \frac{1}{4\pi} |\xi|^{-i\lambda-1} e^{2i\pi(q_1+q_2)\frac{x}{\xi}} |q_1+q_2|^{i\lambda} \sum_{j=0,1} |q_1|_j^{\frac{-1-\nu_1-\nu_2-i\lambda}{2}}
$$

$$
\cdot \left[\Delta_j \left(\frac{1+\nu_1+\nu_2-i\lambda}{2}, -\nu_2 \right) \right.
$$

$$
\cdot\, {}_2F_1 \left(\frac{1+\nu_1+\nu_2+i\lambda}{2}, \frac{1-\nu_1+\nu_2+i\lambda}{2}; 1+\nu_2; -\frac{q_2}{q_1} \right)
$$

$$
+ \left| \frac{q_1}{q_2} \right|^{\nu_2} \Delta_j \left(\frac{1-\nu_1-\nu_2-i\lambda}{2}, \nu_2 \right)
$$

$$
\left. \cdot\, {}_2F_1 \left(\frac{1-\nu_1-\nu_2+i\lambda}{2}, \frac{1+\nu_1-\nu_2+i\lambda}{2}; 1-\nu_2; -\frac{q_2}{q_1} \right) \right]. \quad (4.5.2)
$$

Proof. As observed in the beginning of Section 4.1, the assumption $q_1 + q_2 \neq 0$ implies that the distribution $|\xi|^{-1-i\lambda} e^{2i\pi \frac{(q_1+q_2)x}{\xi}}$ is an analytic function of λ, even at $\lambda = 0$. On the other hand, it has been observed, in Remark 4.3.1 (i) following the proof of Lemma 4.4.1, that each of the two functions Δ_j present on the right-hand side of (4.5.1) is singular at $\lambda = 0$. However, we have also stressed that the linear combination of these two functions, when the two hypergeometric functions present here are taken as coefficients, is not: besides, the extra factor $|q_1|^{-\frac{i\lambda}{2}} |q_1 + q_2|^{i\lambda}$ which shows up in the first term after one has applied (4.5.22) agrees, when $\lambda = 0$, with the corresponding factor $|q_1|^{\frac{i\lambda}{2}}$ from the second term. To sum this point up, the right-hand side of (4.5.1) has no singularity at $\lambda = 0$, as was to be expected under the assumption that $q_1 q_2(q_1 + q_2) \neq 0$.

To compute $(h^1)^{\theta_1} \# (h^2)^{\theta_2}$, we wish to apply the composition formula provided by the equations (3.3.2) to (3.3.5). This cannot be done directly: however, we may do so if we consider, in place of the pair of functions h_{ν_r, q_r}, a pair of functions $h^r(x, \xi) = f_r(\xi)e^{2i\pi q_r \frac{x}{\xi}}$ with $f_r \in \mathcal{S}_\infty(\mathbb{R})$. Indeed, using Lemma 4.3.3, one sees that $h = h^1 \# h^2$ is the limit as $\theta_1, \theta_2 \to 0$ of the symbol $(h^1)^{\theta_1} \# (h^2)^{\theta_2}$, with

$$(h^1)^{\theta_1}(x, \xi) = f_1(\xi)\,\phi\left(\frac{\theta_1 x}{\xi}\right)e^{2i\pi q_1 \frac{x}{\xi}}, \quad (h^2)^{\theta_2}(x, \xi) = f_2(\xi)\,\phi\left(\frac{\theta_2 x}{\xi}\right)e^{2i\pi q_2 \frac{x}{\xi}} :$$

(4.5.3)

the two symbols $(h^r)^{\theta_r}$ lie in $\mathcal{S}_{\text{even}}(\mathbb{R}^2)$ and their sharp composition can be obtained with the help of equations (3.3.2) and (3.3.5). One can write (4.3.2), here recalled,

$$f_r(\xi) = \frac{1}{i}\int_{\text{Re }\nu_r = a_r} \psi_r(\nu_r)\,|\xi|^{-\nu_r - 1}d\nu_r, \quad \xi \neq 0,$$

(4.5.4)

with $a_r < 0$, or even $a_r = 0$ provided one makes a slight deformation of contour around $\nu_r = 0$. Up to a last integral, with respect to the measure $\psi_1(\nu_1)\psi_2(\nu_2)\,d\nu_1\,d\nu_2$, taken on a product of lines Re $\nu_r = a_r$ (or, in the other direction, taking a density relative to this measure and using the fact that $|\xi|^{-1-\nu_r}$ depends on ν_r in an analytic way), finding a formula for $\left(f_1(\xi)e^{2i\pi q_1 \frac{x}{\xi}}\right) \# \left(f_2(\xi)e^{2i\pi q_2 \frac{x}{\xi}}\right)$ when $f_r \in \mathcal{S}_\infty(\mathbb{R})$ or when $f_r = |\xi|^{-1-\nu_r}$ amounts to the same. We thus go back to the situation when $f_r = h_{\nu_r, q_r}$: then, $\left[(h^1)^{\theta_1}\right]^{\flat}_{\nu_1}(s_1) = \phi(\theta_1 s_1)e^{2i\pi q_1 s_1}$ and $\left[(h^2)^{\theta_2}\right]^{\flat}_{\nu_2}(s_2) = \phi(\theta_2 s_2)e^{2i\pi q_2 s_2}$. According to the recipe between (3.3.2) and (3.3.5), we must compute the limit as $\theta_1, \theta_2 \to 0$ of the integral

$$I_j(\nu_1, \nu_2; i\lambda; s)$$
$$= \int_{\mathbb{R}^2} \phi(\theta_1 s_1)\,\phi(\theta_1 s_2)\,e^{2i\pi(q_1 s_1 + q_2 s_2)}\,|s_1 - s_2|_j^\alpha\,|s - s_1|_j^{\beta_1}\,|s_2 - s|_j^{\beta_2}\,ds_1\,ds_2,$$

(4.5.5)

with

$$\alpha = \frac{-1 + \nu_1 + \nu_2 + i\lambda}{2}, \quad \beta_1 = \frac{-1 + \nu_1 - \nu_2 - i\lambda}{2}, \quad \beta_2 = \frac{-1 - \nu_1 + \nu_2 - i\lambda}{2}.$$

(4.5.6)

One may recall at this point that the index $j = 0, 1$ singles out the commutative and anticommutative parts of the symbol h, defined as

$$h^1 \mathop{\#}_{j} h^2 = \frac{1}{2}\left[h^1 \# h^2 + (-1)^j(h^2 \# h^1)\right].$$

(4.5.7)

We first compute

$$\widetilde{I}(s_2) = \int_{-\infty}^{\infty} \phi(\theta_1 s_1)\, e^{2i\pi q_1 s_1}\, |s_1 - s_2|_j^{\alpha}\, |s - s_1|_j^{\beta_1}\, ds_1, \qquad (4.5.8)$$

obtaining with the help of (1.1.6) the equation

$$\widetilde{I}(s_2) = (-1)^j B_j(-\alpha) B_j(-\beta_1) \int_{-\infty}^{\infty} \theta_1^{-1}\, \left(\mathcal{F}^{-1}\phi\right) \left(\frac{q_1 - r_1}{\theta_1}\right)\, dr_1$$

$$\cdot \int_{-\infty}^{\infty} e^{2i\pi s_2(r_1 - t)} |r_1 - t|_j^{-\alpha-1}\, e^{2i\pi st}\, |t|_j^{-\beta_1-1}\, dt. \quad (4.5.9)$$

Then,

$$I_j(\nu_1, \nu_2; i\lambda;\, s) = (-1)^j B_j(-\alpha) B_j(-\beta_1) \int_{-\infty}^{\infty} \theta_1^{-1}\, \left(\mathcal{F}^{-1}\phi\right) \left(\frac{q_1 - r_1}{\theta_1}\right)\, dr_1$$

$$\cdot \int_{\mathbb{R}^2} \phi(\theta_2 s_2)\, e^{2i\pi s_2(r_1 + q_2 - t)}\, e^{2i\pi st}\, |s_2 - s|_j^{\beta_2}\, |r_1 - t|_j^{-\alpha-1}\, |t|_j^{-\beta_1-1}\, ds_2\, dt. \quad (4.5.10)$$

Now, one has

$$\int_{-\infty}^{\infty} \phi(\theta_2 s_2)\, e^{2i\pi s_2(r_1 + q_2 - t)}\, |s_2 - s|_j^{\beta_2}\, ds_2$$

$$\rightarrow e^{2i\pi s(r_1 + q_2 - t)} (-1)^j B_j(-\beta_2)\, |r_1 + q_2 - t|_j^{-\beta_2-1}, \quad \theta_2 \to 0. \quad (4.5.11)$$

As $\theta_2 \to 0$, the integral $I_j(\nu_1, \nu_2; \nu;\, s)$ thus goes to

$$B_j(-\alpha) B_j(-\beta_1) B_j(-\beta_2) \int_{-\infty}^{\infty} \delta_1^{-1}\, \left(\mathcal{F}^{-1}\phi\right) \left(\frac{q_1 - r_1}{\delta_1}\right)\, dr_1$$

$$\cdot e^{2i\pi s(r_1 + q_2)} \int_{-\infty}^{\infty} |r_1 + q_2 - t|_j^{-\beta_2-1}\, |r_1 - t|_j^{-\alpha-1}\, |t|_j^{-\beta_1-1}\, dt, \quad (4.5.12)$$

an expression the limit of which, as $\theta_1 \to 0$, is

$$B_j(-\alpha) B_j(-\beta_1) B_j(-\beta_2)\, e^{2i\pi s(q_1 + q_2)} \int_{-\infty}^{\infty} |q_1 + q_2 - t|_j^{-\beta_2-1}\, |q_1 - t|_j^{-\alpha-1}\, |t|_j^{-\beta_1-1}\, dt. \qquad (4.5.13)$$

The integral is the same, if $q_1(q_1 + q_2) \neq 0$, as

$$(-1)^j |q_1|_j^{-\alpha-\beta_1-\beta_2-2}\, R^{(j)} \left(\beta_1, \alpha, \beta_2;\, -\frac{q_1 + q_2}{q_1}\right). \qquad (4.5.14)$$

According to this lemma, the limit as $\theta_1, \theta_2 \to 0$ of the integral $I_j(\nu_1, \nu_2; i\lambda;\, s)$ is

$$(-1)^j B_j(-\alpha) B_j(-\beta_1) B_j(-\beta_2)\, R^{(j)} \left(\beta_1, \alpha, \beta_2;\, -\frac{q_1 + q_2}{q_1}\right)$$

$$\cdot |q_1|_j^{-\alpha-\beta_1-\beta_2-2}\, e^{2i\pi(q_1 + q_2)s}. \qquad (4.5.15)$$

Using the recipe provided by the equations (3.3.2) to (3.3.5), and the equation $h_{i\lambda}(x,\xi) = |\xi|^{-1-i\lambda} h_{i\lambda}^{\flat}(\frac{x}{\xi})$, we obtain

$$h_{i\lambda}(x,\xi) = \frac{1}{4\pi} \sum_{j=0,1} C^{(j)}_{\nu_1,\nu_2;i\lambda} (-1)^j B_j(-\alpha) B_j(-\beta_1) B_j(-\beta_2)$$

$$\cdot R^{(j)} \left(\beta_1, \alpha, \beta_2; -\frac{q_1+q_2}{q_1} \right) |q_1|_j^{-\alpha-\beta_1-\beta_2-2} |\xi|^{-1-i\lambda} \exp\left(2i\pi\frac{(q_1+q_2)x}{\xi} \right).$$

$$(4.5.16)$$

Using (3.3.4), we note that the functional equation (1.1.2) of the function B_j can be applied 3 times, which leads to the equation

$$h_{i\lambda}(x,\xi) = 2^{\frac{-1+\nu_1+\nu_2-i\lambda}{2}} \cdot \frac{1}{4\pi} \sum_{j=0,1} (-1)^j R^{(j)} \left(\beta_1, \alpha, \beta_2; -\frac{q_1+q_2}{q_1} \right)$$

$$\cdot |q_1|_j^{-\alpha-\beta_1-\beta_2-2} |\xi|^{-1-i\lambda} \exp\left(2i\pi\frac{(q_1+q_2)x}{\xi} \right) \quad (4.5.17)$$

or, using (4.5.6),

$$h_{i\lambda}(x,\xi) = 2^{\frac{-1+\nu_1+\nu_2-i\lambda}{2}} \cdot \frac{1}{4\pi} \sum_{j=0,1} (-1)^j |\xi|^{-1-i\lambda} e^{2i\pi\frac{(q_1+q_2)x}{\xi}} |q_1|_j^{\frac{-1-\nu_1-\nu_2+i\lambda}{2}}$$

$$\cdot R^{(j)} \left(\frac{-1+\nu_1-\nu_2-i\lambda}{2}, \frac{-1+\nu_1+\nu_2+i\lambda}{2}, \frac{-1-\nu_1+\nu_2-i\lambda}{2}; -\frac{q_1+q_2}{q_1} \right).$$

$$(4.5.18)$$

Up to this point, it has not been necessary to consider separately the (intersecting) cases when $\frac{q_2}{q_1} < 0$ or $\frac{q_2}{q_1} > -1$. In order to obtain (4.5.1), it suffices then to apply (4.4.2), while (4.5.2) is a consequence of (4.4.4). We must not fail to observe, however, that Lemma 4.4.1 is applicable, which demands verifying the inequalities stated in the beginning of this lemma: they are all guaranteed by the assumption $|\mathrm{Re}\,(\nu_1 \pm \nu_2)| < 1$. $\square$

Remark 4.5.1. In view of the amount of computation that led to Theorem 4.5.1, some verification is no luxury: let us verify that the two terms of (4.5.1) are compatible. To do so, we shall verify that using this formula would lead to the identity (in which $\frac{q_2}{q_1} < 0$)

$$h_{\nu_1,q_1} \# \mathcal{G} h_{\nu_2,q_2} = \mathcal{G} \left(h_{\nu_1,q_1} \# h_{\nu_2,q_2} \right), \qquad (4.5.19)$$

which is to be expected in view of the interpretation (cf. what precedes immediately (3.1.15)) of $\mathcal{G}$ as an operation on symbols. One has for $q \neq 0$ the identity $\mathcal{G} h_{\nu,q} = 2^{\nu} |q|^{-\nu} h_{-\nu,q}$, which makes it possible to express the two sides of (4.5.19), starting from (4.5.1). An extra factor $|q_2|^{-\nu}$ will show up in the left-hand side: we get rid of it by means of the identity [22, p. 47]

$$(1 - z)^{a+b-c}\,{}_2F_1(a, b;\, c;\, z) = {}_2F_1(c - a, c - b;\, c;\, z), \quad z < 1, \tag{4.5.20}$$

in which we take $z = \frac{q_1+q_2}{q_1}$, so that $1 - z = \left|\frac{q_2}{q_1}\right|$, the other parameters a, b, c being those involved in the hypergeometric functions which show up in (4.5.1), so that in both cases $a + b - c = \nu_2$. The rest of the verification is straightforward, using also the identity $\Delta_j(x, y) = \Delta_j(1 - x - y, y)$.

Theorem 4.5.2. *Keeping the notation of Theorem 4.5.1, assume that $|\mathrm{Re}\,(\nu_1 \pm \nu_2)| < 1$ and $q_1 \neq 0$, $q_1 + q_2 = 0$, and recall from Lemma 4.1.2 that, while the operator $A_1 A_2$ does not have a symbol in $\mathcal{S}'(\mathbb{R}^2)$, its image under $\mathrm{mad}(P \wedge Q)$ does: to prevent confusion, we shall still denote this symbol as $\mathcal{E}h$, but h is no longer a (tempered) distribution, only a continuous linear form on the image of $\mathcal{S}(\mathbb{R}^2)$ under $\mathcal{E}$ (a quasi-distribution as defined in Remark 4.1.1(ii)). One has the identity $(2i\pi\mathcal{E})h = p.v. \int_{-\infty}^{\infty} ((2i\pi\mathcal{E})h)_{i\lambda}\, d\lambda$, where the sign p.v. indicates that the integral (again, a weak integral in $\mathcal{S}'(\mathbb{R}^2)$) has to be taken in Cauchy's principal sense near $\lambda = 0$, and*

$$((2i\pi\mathcal{E})h)_{i\lambda}(x, \xi) = \frac{1}{4\pi}\, 2^{\frac{-1+\nu_1+\nu_2-i\lambda}{2}} \sum_{j=0,1} (-1)^j\,(-i\lambda)$$

$$\cdot \left[|q_1|_j^{\frac{-1-\nu_1-\nu_2-i\lambda}{2}} \Delta_j\left(\frac{1+\nu_1-\nu_2+i\lambda}{2}, -i\lambda\right) \frac{\zeta(1-i\lambda)}{\zeta(i\lambda)}\, |x|^{-i\lambda}\delta(\xi) \right.$$

$$\left. + |q_1|_j^{\frac{-1-\nu_1-\nu_2+i\lambda}{2}} \Delta_j\left(\frac{1-\nu_1-\nu_2-i\lambda}{2}, i\lambda\right) |\xi|^{-i\lambda-1} \right]. \tag{4.5.21}$$

Proof. From Lemma 4.1.2, $\mathrm{mad}(P \wedge Q)(A_1 A_2)$, with $A_n = \mathrm{Op}\,(h_{\nu_n, q_n})$, is a continuous functions of the pair q_1, q_2 with values in the space of linear operators from $\mathcal{S}(\mathbb{R})$ to $\mathcal{S}'(\mathbb{R})$, provided with the weak topology. Still assuming $q_1 + q_2 \neq 0$, observe that the right-hand side of the equation (a special case of (4.1.2))

$$\mathcal{F}^{\mathrm{symp}}\left(|\xi|^{i\lambda-1} e^{2i\pi(q_1+q_2)\frac{x}{\xi}}\right) = |q_1 + q_2|^{i\lambda}\, |\xi|^{-i\lambda-1} e^{2i\pi(q_1+q_2)\frac{x}{\xi}}, \tag{4.5.22}$$

occurs as a factor in the first term on the right-hand side of (4.5.1). On the other hand, replacing, as desired, h by $2i\pi\mathcal{E}\,h$ amounts to multiplying $h_{i\lambda}$ by $-i\lambda$. This leads, continuing $h_{i\lambda}$ as h_ν for $\mathrm{Re}\,\nu$ close to 0, to

$$(2i\pi\mathcal{E}h)_\nu(x, \xi)$$

$$= 2^{\frac{-1+\nu_1+\nu_2-\nu}{2}} \cdot \frac{1}{4\pi} \sum_{j=0,1} (-1)^j\, |q_1|_j^{\frac{-1-\nu_1-\nu_2-\nu}{2}} \mathcal{F}^{\mathrm{symp}}\left(|\xi|^{\nu-1}\, e^{2i\pi\frac{(q_1+q_2)x}{\xi}}\right)$$

$$\cdot (-\nu)\Delta_j\left(\frac{1+\nu_1-\nu_2+\nu}{2}, -\nu\right) \tag{./.}$$

$$\cdot {}_2F_1\left(\frac{1+\nu_1+\nu_2+\nu}{2}, \frac{1-\nu_1+\nu_2+\nu}{2}; 1+\nu; \frac{q_1+q_2}{q_1}\right)$$

$$+2^{\frac{-1+\nu_1+\nu_2-\nu}{2}} \cdot \frac{1}{4\pi} \sum_{j=0,1} (-1)^j |q_1|_j^{\frac{-1-\nu_1-\nu_2+\nu}{2}} |\xi|^{-1-\nu} e^{2i\pi\frac{(q_1+q_2)x}{\xi}}$$

$$\cdot (-\nu)\Delta_j\left(\frac{1-\nu_1-\nu_2-\nu}{2}, \nu\right)$$

$$\cdot {}_2F_1\left(\frac{1-\nu_1+\nu_2-\nu}{2}, \frac{1+\nu_1+\nu_2-\nu}{2}; 1-\nu; \frac{q_1+q_2}{q_1}\right). \qquad (4.5.23)$$

For $q_1 + q_2 \neq 0$, the distribution $|\xi|^{-1\pm i\lambda} e^{2i\pi\frac{(q_1+q_2)x}{\xi}}$ is defined for all values of λ, even $\lambda = 0$. This is no longer the case when $q_1 + q_2 = 0$, and the extra factor $-i\lambda$ does not completely solve the difficulty because the coefficients $\Delta_j(\frac{1\pm\nu_1-\nu_2\pm\nu}{2}, \mp\nu)$ have also simple poles at $\nu = 0$. However, it is $2i\pi\mathcal{E}h$, i.e., the integral $\frac{1}{i}\int_{\text{Re }\nu=0}(2i\pi\mathcal{E}h)_\nu d\nu$, we are really interested in. To obtain the limit of this integral as $q_1 + q_2 \to 0$, we must make two distinct deformations of contour, turning around $\nu = 0$ on the right in the first term and on the left in the second term. In the limit, we shall obtain the sum of two terms: first, the integral $\int_{-\infty}^{\infty}(2i\pi\mathcal{E}h)_{i\lambda}(x, \xi)\, d\lambda$, taken in Cauchy's principal sense; next, a half-difference of residues (times 2π).

Now, the residue of $|\xi|^{-\nu-1}$ at $\nu = 0$ is $-2\,\delta(\xi)$ (as observed at the end of the proof of Theorem 1.1.7), while, as is easily verified, one has

$$\mathcal{F}^{\text{symp}}\left(|\xi|^{\nu-1}\right) = B_0(1-\nu)\,|x|^{-\nu}\,\delta(\xi) = \frac{\zeta(1-\nu)}{\zeta(\nu)}\,|x|^{-\nu}\,\delta(\xi), \quad \nu \text{ near } 0. \quad (4.5.24)$$

Using also the last expression of $\Delta_j(x, y)$ in (4.4.3) and the fact (1.1.1) that, as $\nu \to 0$, $B_0(1-\nu) \sim \frac{2}{\nu}$ and $B_0(\nu) \sim \frac{\nu}{2}$, one verifies, remarking finally that when $q_1 + q_2 = 0$, both hypergeometric functions which show up reduce to 1, that the residues at $\nu = 0$ of the two terms on the right-hand side of (./.) agree in this case. Finally, the limit as $q_1 + q_2 \to 0$ of $2i\pi\mathcal{E}\,h$ reduces to the integral on the real line, taken in Cauchy's principal sense, indicated. $\qquad \square$

In the next theorem, we consider the case when $q_2 = 0$ but $q_1 \neq 0$. The case when $q_1 = 0$ but $q_2 \neq 0$ is not only totally similar, but can also be reduced to the preceding one, by taking the operator $(A_1 A_2)^* = A_2^* A_1^*$ in place of $A_1 A_2$: replacing an operator by its adjoint is the same as replacing its symbol by the complex conjugate thereof.

Theorem 4.5.3. *Let* $h^1 = h_{\nu_1,q_1}$ *with* $q_1 \neq 0$, *let* $h^2(x, \xi) = h_{\nu_2,0}(x, \xi) = |\xi|^{-1-\nu_2}$ *with* $\text{Re }\nu_2 < 1$, $\nu_2 \neq 0$ *and* $|\text{Re}\,(\nu_1 \pm \nu_2)| < 1$, *and set* $A_r = \text{Op}(h^r)$. *The symbol* h *of the operator* $A_1 A_2$ *is characterized by the equation*

$$h_{i\lambda}(x,\xi) = 2^{\frac{-1+\nu_1+\nu_2-i\lambda}{2}} \cdot \frac{1}{4\pi} |\xi|^{-i\lambda-1} e^{2i\pi q_1 \frac{x}{\xi}} \sum_{j=0,1} |q_1|_j^{\frac{-1-\nu_1-\nu_2+i\lambda}{2}}$$

$$\cdot \Delta_j \left(\frac{1+\nu_1+\nu_2-i\lambda}{2}, -\nu_2 \right). \tag{4.5.25}$$

Proof. Let us first assume that Re $\nu_2 < 0$ so that, with $h^2 = h_{\nu_2,q_2}$, the conditions of Lemma 4.1.1 are satisfied when $q_2 = 0$. Again, with $A_r = \mathrm{Op}(h^r)$, we set (as in the proof of Lemma 4.3.3) $B_r = \mathcal{F} A_r \mathcal{F}^{-1}$. Recalling (4.1.10) and (4.1.11), also that $g_r(\xi) = |\frac{\xi}{2}|^{-\nu_r}$, one has in the case when Re $\nu_2 < 0$ and $q_2 \geq 0$,

$$(B_1 B_2 w)(t) = \mathrm{char}(t^2 \geq 2q_1 + 2q_2) \frac{1}{\sqrt{t^2-2q_1}} \left| \frac{t+\varepsilon_1\sqrt{t^2-2q_1}}{2} \right|^{-\nu_1} F_{q_2}(t), \tag{4.5.26}$$

with (exchanging ε_1 and ε_2 in (4.1.10) when $q_2 \neq 0$)

$$F_{q_2}(t) = \begin{cases} \sum_{\varepsilon_1,\varepsilon_2} \frac{1}{\sqrt{t^2-2q_1-2q_2}} \left| \frac{\varepsilon_2\sqrt{t^2-2q_1}+\varepsilon_1\sqrt{t^2-2q_1-2q_2}}{2} \right|^{-\nu_2} w(\varepsilon_1\sqrt{t^2-2q_1}) \\ \sum_{\varepsilon_1} (\sqrt{t^2-2q_1})^{-\nu_2-1} w(\varepsilon_1\sqrt{t^2-2q_1}) \end{cases}$$

$$\tag{4.5.27}$$

according to whether $q_2 > 0$ or $q_2 = 0$. Clearly, to obtain the limit of the first expression as $q_2 \to 0$, one must take only the term with $\varepsilon_2 = \varepsilon_1$ since Re $\nu_2 < 0$, and one sees that, indeed, it converges pointwise towards the second expression. Nothing else needs be said if $q_1 < 0$ while, if $q_1 > 0$, we look for an estimate of L^1-style of $B_1 B_2 w$, for $w \in \mathcal{S}(\mathbb{R})$, just as done in the proof of Lemma 4.1.5. Setting $s = \varepsilon_1\sqrt{t^2-2q_1-2q_2}$ in the integral of $|B_1 B_2 w|$ on the line, one transforms $\frac{dt}{\sqrt{t^2-2q_1}\sqrt{t^2-2q_1-2q_2}}$ to $\frac{ds}{\sqrt{s^2+2q_2}}$, where the non-integrability near 0 in the limit is taken care of by the factor $|s + \varepsilon_2\sqrt{s^2+2q_2}|^{-\mathrm{Re}\,\nu_2}$ since Re $\nu_2 < 0$. Emphasizing the dependence of A_2 on q_2, denoting it as $A_2(q_2)$, we have just proved that, in the case when Re $\nu_2 < 0$, the operator $A_1 A_2(q_2)$ converges to $A_1 A_2 = A_1 A_2(0)$ in the space of weakly continuous linear operators from $\mathcal{S}(\mathbb{R})$ to $\mathcal{S}'(\mathbb{R})$ as $q_2 > 0$ goes to zero. Starting from (4.5.2) and going to the limit, we obtain (4.5.25).

For later applications, it will improve things to allow the weaker condition Re $\nu_2 < 1$ (a consequence of the assumptions of the theorem under proof), just assuming that $\nu_2 \neq 0$ so that $(x,\xi) \mapsto |\xi|^{-\nu_2-1}$ should make sense as a tempered distribution in $\mathbb{R}^2$. Let us even assume Re $\nu_2 < 2$ and $\nu_2 \neq 0, 1$, and let us write $|\xi|^{-\nu_2-1} = (\nu_2(\nu_2-1))^{-1}\frac{d^2}{d\xi^2}(|\xi|^{-\nu_2+1})$. We use here a second-order operator for the sole reason that, for notational simplicity, we have not analyzed the sharp products of possibly "odd" versions of the function $h_{\nu,q}$. Using the general identity $\mathrm{Op}\left(\frac{\partial h}{\partial\xi}\right) = -2i\pi\,[Q,\mathrm{Op}(h)]$ and the commutation relation

$$A_1[Q,[Q,A_2]] = [Q,[Q,A_1 A_2]] - 2[Q,[Q,A_1]A_2] + [Q,[Q,A_1]]A_2, \tag{4.5.28}$$

one sees that from Lemma 4.1.2 that the operator $A_1 A_2$ acts from $\mathcal{S}(\mathbb{R})$ to $\mathcal{S}'(\mathbb{R})$, hence has a symbol in $\mathcal{S}'(\mathbb{R}^2)$, if $q_1 \neq 0$ and Re $\nu_2 < 2$, $\nu_2 \neq 0, 1$. Then, the decomposition (4.5.25) follows in this case as well.   □

The next proposition provides both an improvement on the result of Theorem 4.2.3, under the additional assumption that $|\text{Re}\,(\nu_1 \pm \nu_2)| < 1$, and a totally different proof of it.

Proposition 4.5.4. *Assume that* $|\text{Re}\,(\nu_1 \pm \nu_2)| < 1$, *and that* $q_1 q_2 \neq 0$. *The result of Theorem 4.2.3 extends to all values of* $\iota = 0, 1, \dots$, *with the following additional improvement: when* $\iota \geq 1$, *the differential operator* $P_\iota(2i\pi\mathcal{E})$ *which occurs in the statement there can be replaced by the differential operator obtained by forgetting the two factors* $2i\pi\mathcal{E} \pm 1$ *from the definition (4.2.28) of* $P_\iota(2i\pi\mathcal{E})$.

Proof. Recall that, in the proof of Theorem 4.2.3, the only "difficult" case was that in which $q_1 < 0$ and $q_2 > \frac{|q_1|}{2}$. In that case, we rely now on the explicit formula (./.). So as to take advantage of the factor $|q_1|_j^{\frac{-1-\nu_1-\nu_2-i\varepsilon\lambda}{2}}$ (recall that our problem is to gain powers of $|q_1|^{-1}$), we shall move the line of integration, replacing the line Re $\nu = 0$ (with $\nu = i\lambda$) by the line Re $(\varepsilon\nu) = a$, with $a > 0$ distinct from an even integer: note that we move the line of integration into different directions in relation to the two terms of (./.). We thus need to estimate the continuation of the product of the last two factors of each term of this expression.

From the definition of the function Δ_j and the classical integral representation [22, p. 54] of the hypergeometric function, obtaining under the assumption that $\frac{q_1+q_2}{q_1} < 1$, i.e., $\frac{q_2}{q_1} < 0$, and $|\text{Re}\,(\nu_1 \pm \nu_2)| < 1$ the expression

$$
\Delta_j \left(\frac{1 + \varepsilon\nu_1 - \nu_2 + \varepsilon\nu}{2}, -\varepsilon\nu \right)
$$
$$
\cdot \,_2F_1 \left(\frac{1 + \varepsilon\nu_1 + \nu_2 + \varepsilon\nu}{2}, \frac{1 - \varepsilon\nu_1 + \nu_2 + \varepsilon\nu}{2}; 1 + \varepsilon\nu; \frac{q_1 + q_2}{q_1} \right)
$$
$$
= \left[\frac{\Gamma(-\varepsilon\nu)\Gamma(1+\varepsilon\nu)}{\Gamma(\frac{1+\varepsilon\nu_1-\nu_2-\varepsilon\nu}{2})\Gamma(\frac{1-\varepsilon\nu_1+\nu_2+\varepsilon\nu}{2})} + (-1)^j + \frac{\Gamma(-\varepsilon\nu)\Gamma(1+\varepsilon\nu)}{\Gamma(\frac{1-\varepsilon\nu_1+\nu_2-\varepsilon\nu}{2})\Gamma(\frac{1+\varepsilon\nu_1-\nu_2+\varepsilon\nu}{2})} \right]
$$
$$
\cdot \int_0^1 t^{\frac{-1-\varepsilon\nu_1+\nu_2+\varepsilon\nu}{2}} (1-t)^{\frac{-1+\varepsilon\nu_1-\nu_2+\varepsilon\nu}{2}} \left(1 - \frac{q_1+q_2}{q_1}t \right)^{\frac{-1-\varepsilon\nu_1-\nu_2-\varepsilon\nu}{2}} dt. \quad (4.5.29)
$$

Since $\frac{q_2}{q_1} < -\frac{1}{2}$, one has $\frac{q_1+q_2}{q_1} < \frac{1}{2}$: also, $|\text{Re}\,(\nu_1 \pm \nu_2)| < 1$, so that the last integral is uniformly bounded when Re $(\varepsilon\nu) \geq 0$. When performing the changes of contour, recall that we do not have to worry, so far as the $d\nu$-integrability on some line is concerned, about powers of $1 + |\text{Im}\,\nu|$, since $c^2 - \nu^2$ (with c depending on how far we wish to get away from the pure imaginary line) can always be replaced by the operator $c^2 + 4\pi^2\mathcal{E}^2$ as many times as needed, and we are only claiming weak bounds in $\mathcal{S}'(\mathbb{R}^2)$. What is important is taking care of the poles, apparent in the expression within brackets on the right-hand side of (4.5.29). The poles originate from the singularities of the bracket and reduce to the set of integers. Then, as the operator $P_\iota(2i\pi\mathcal{E})$ is divisible by each of the two products $\pm 2i\pi\mathcal{E}(\pm 2i\pi\mathcal{E} - 1)\ldots(\pm 2i\pi\mathcal{E} - \iota)$, applying this operator kills all the poles (originating from the factor $\Gamma(-\varepsilon\nu)\Gamma(1 + \varepsilon\nu)$) that appear during the deformation of contour, provided that $a < \iota + 1$ and $q_1 + q_2 \neq 0$. A closer look, however, shows that the point ν such that $\varepsilon\nu = 1$ is actually not a pole of the bracket in (4.5.29): this is the origin of the additional improvement in Proposition 4.5.4. In the case when $q_1 + q_2 = 0$, we may rely on (4.5.21): note that the proof of Theorem 4.5.2 showed that, as soon as one sets apart around 0, each in its proper direction, the two contours on which the two terms of the decomposition of (4.5.21) are to be integrated, the residues of these terms at $\lambda = 0$ must be forgotten.

Whether $q_1 + q_2 \neq 0$ or not, the benefit of our having changed the contour of integration comes from the last factor on each term of the right-hand side of (./.) not yet made use of, to wit the power $|q_1|_j^{\frac{-1-\nu_1-\nu_2-\varepsilon\nu}{2}}$: it leads to Proposition 4.5.4.   □

Remark 4.5.2. The factor $2^{\frac{1+\nu_1+\nu_2-i\lambda}{2}}$ which appears on the right-hand side of most identities in this section would disappear if, in place of the function $h_{\nu,q}$, its rescaled version $h_{\nu,q}^{\text{resc}} = 2^{\frac{-1-\nu}{2}} h_{\nu,q}$ had been used in all occurrences.

Finally, we consider the case of two factors, each of which is proportional to one of the first two terms of the expansion (1.1.38) of some Eisenstein distribution. In this case, as will be seen, the sharp product under study will be of the same kind as one of such two terms: in particular, it will be homogeneous (so that no $d\lambda$-integration occurs in the formula).

Theorem 4.5.5. *One has* $|\xi|^{-\nu_1-1} \# |\xi|^{-\nu_2-1} = |\xi|^{-\nu_1-\nu_2-2}$ *and the equations*

$$\left(|x|^{-\nu_1}\delta(\xi)\right) \# \left(|x|^{-\nu_2}\delta(\xi)\right) = \frac{2^{\nu_1+\nu_2}\,\zeta(\nu_1)\zeta(\nu_2)}{\zeta(1-\nu_1)\zeta(1-\nu_2)}\,|\xi|^{-2+\nu_1+\nu_2}, \qquad (4.5.30)$$

and

$$|\xi|^{-\nu_1-1} \# \left(|x|^{-\nu_2}\,\delta(\xi)\right) = 2^{1+\nu_1}\,\frac{\zeta(\nu_2)\zeta(2+\nu_1-\nu_2)}{\zeta(1-\nu_2)\zeta(-1-\nu_1+\nu_2)}\,|x|^{1+\nu_1-\nu_2}\,\delta(\xi),$$

$$\left(|x|^{-\nu_1}\delta(\xi)\right) \# |\xi|^{-\nu_2-1} = 2^{1+\nu_2}\,\frac{\zeta(\nu_1)\zeta(2-\nu_1+\nu_2)}{\zeta(1-\nu_1)\zeta(-1+\nu_1-\nu_2)}\,|x|^{1-\nu_1+\nu_2}\,\delta(\xi).$$

$$\qquad (4.5.31)$$

The assumptions relative to ν_1, ν_2 *in each case are those, detailed just after* (1.1.5), *which give the power functions involved a meaning: besides, one must avoid the poles* $\mu = 1, 3, \ldots$ *of the factors* $\frac{\zeta(\mu)}{\zeta(1-\mu)} = B_0(\mu)$ *present.*

Proof. The first equation is obvious, since the operator with symbol $h(x, \xi) = f(\xi)$ is the operator A such that $(\mathcal{F}Au)(t) = f(t)(\mathcal{F}u)(t)$. Next, we use (1.1.6), here rewritten as

$$\mathcal{G}\left(|x|^{-\nu}\delta(\xi)\right) = 2^{\nu} \frac{\zeta(\nu)}{\zeta(1-\nu)} |\xi|^{\nu-1}, \tag{4.5.32}$$

and the fact that taking the $\mathcal{G}$-transform of a symbol amounts to multiplying the operator associated to it under the Weyl calculus by the operator $u \mapsto \check{u}$ on the right, or if preferred on the left if the symbol is globally even. It follows that

$$|\xi|^{-\nu_1-1} \# \left(|x|^{-\nu_2}\delta(\xi)\right) = 2^{\nu_2} \frac{\zeta(\nu_2)}{\zeta(1-\nu_2)} |\xi|^{-\nu_1-1} \# \mathcal{G}\left(|\xi|^{\nu_2-1}\right)$$

$$= 2^{\nu_2} \frac{\zeta(\nu_2)}{\zeta(1-\nu_2)} \mathcal{G}\left[|\xi|^{-\nu_1-1} \# |\xi|^{\nu_2-1}\right]$$

$$= 2^{1+\nu_1} \frac{\zeta(\nu_2)\zeta(2+\nu_1-\nu_2)}{\zeta(1-\nu_2)\zeta(-1-\nu_1+\nu_2)} |x|^{1+\nu_1-\nu_2}\delta(\xi).$$

$$\tag{4.5.33}$$

The other equations are obtained in the same way. $\qquad\square$

4.6 Transferring a sharp product $h_{\nu_1,q_1} \# h_{\nu_2,q_2}$ to the half-plane

With the help of the map $\Theta = (\Theta_0, \Theta_1)$, introduced in (2.1.5), we transfer to the half-plane the results, regarding the sharp composition of two functions of type $h_{\nu,q}$, obtained in the last chapter. Recall that Theorems 4.5.1 and 4.5.3 made the sharp composition of any two such functions explicit as an integral superposition of functions $h_{i\lambda,q_1+q_2}$: the parameter $j = 0$ or 1 which shows up in the formulas accounts for the commutative and anticommutative parts of $h_{\nu_1,q_1} \# h_{\nu_2,q_2}$, as defined in (4.5.7). Except for some simple extra factor, the second equation (4.5.2), valid in the case when $\frac{q_2}{q_1} > -1$, is very similar to the one (4.5.1), valid when $\frac{q_2}{q_1} < 0$, after one has exchanged $i\lambda$ and ν_2 and traded $\frac{q_1+q_2}{q_1}$ for $-\frac{q_2}{q_1}$.

The operator Θ_0 transforms distributions in the plane into functions in the hyperbolic half-plane: its adjoint Θ_0^* moves in the other direction, and combining the two as in (3.2.32) makes it possible to replace spectral decompositions in Π relative to Δ by decompositions into homogeneous components in $\mathbb{R}^2$. Note that the factor $\Gamma(\frac{1+i\lambda}{2})\Gamma(\frac{1-i\lambda}{2})$ on the right-hand side of (3.2.32) decreases exponentially at infinity: this may work to our advantage, or create the need for new estimates, according to the direction in which the correspondence between functions or distributions in $\mathbb{R}^2$ or Π is used. On the other hand, as recalled several times, the spectral theory in the plane (relative to $2i\pi\mathcal{E}$) is slightly more precise than the one in the half-plane (relative to Δ). When moving information from the plane to the half-plane, nothing is lost: when moving it in the other direction, it is necessary to use the pair (Θ_0, Θ_1) in place of the sole operator Θ_0.

The two operations, on functions in the half-plane, which are going to play a role similar to the operations $\#_{j}$, with $j = 0$ or 1, on distributions in $\mathbb{R}^2$, are the operations $\times_{j}$ defined in (3.3.12), to wit the pointwise product and half the Poisson bracket, as defined in (3.3.11). Of course, the operations $\#_{j}$ do not transfer globally to the operations $\times_{j}$, just as the sharp operation in the plane does not transfer to the pointwise multiplication in the half-plane. However, in Theorem 3.3.1 and Corollary 3.3.2, we have seen that such a transfer is possible if "localized" with the help of the two corresponding spectral theories available on $\mathbb{R}^2$ and on Π: i.e., explicit λ-dependent factors will appear. But this was proved only if dealing with the sharp product of two (globally even) functions in $\mathcal{S}(\mathbb{R}^2)$: extending these results to the case of two factors of the kind $h_{\nu, q}$ will constitute the main matter of the present section.

Lemma 4.6.1. *Let $q \in \mathbb{R}^\times$, and let $h(x, \xi) = f(\xi)e^{2i\pi q\frac{x}{\xi}}$ satisfy the assumptions of Lemma 4.1.1, here recalled: f is a continuous even function on $\mathbb{R}\backslash\{0\}$, and $|f(\xi)|$ is bounded, for some pair (C, N), by $C\left(|\xi| + |\xi|^{-1}\right)^N$. Then, one has*

$$(\Theta_0 h)(x + iy) = (2y)^{\frac{1}{2}} e^{2i\pi qx} \int_{-\infty}^{\infty} f(\xi) \exp\left(-\pi y\left(2\xi^2 + \frac{q^2}{2\xi^2}\right)\right) d\xi. \quad (4.6.1)$$

In particular, if one sets

$$\mathcal{W}_{\pm\nu, q}(x + iy) = y^{\frac{1}{2}} K_{\frac{\nu}{2}}(2\pi |q| y) e^{2i\pi qx}, \quad (4.6.2)$$

one has, for $q \neq 0$ and an arbitrary $\nu \in \mathbb{C}$,

$$\Theta_0 h_{\nu, q} = 2^{\frac{3+\nu}{2}} |q|^{-\frac{\nu}{2}} \mathcal{W}_{\nu, q}, \quad \Theta_0 \left(\mathcal{F}^{\text{symp}} h_{-\nu, q}\right) = 2^{\frac{3+\nu}{2}} |q|^{\frac{\nu}{2}} \mathcal{W}_{\nu, q}, \quad (4.6.3)$$

and

$$(\Theta_0 h_{\nu, 0})(x + iy) = \pi^{\frac{\nu}{2}} \Gamma(-\frac{\nu}{2}) y^{\frac{\nu}{2}}, \quad \nu \neq 0, 2, \dots. \quad (4.6.4)$$

Proof. Starting from the equation

$$(\Theta_0 h)(z) = 2 \int_{\mathbb{R}^2} f(\xi) e^{2i\pi\frac{qt}{\xi}} \exp\left(-2\pi\frac{|t - z\xi|^2}{y}\right) dt \, d\xi, \quad (4.6.5)$$

setting $z = x + iy$ and integrating first with respect to dt, one obtains (4.6.1). The first equation (4.6.3) follows from (4.6.1) and a usual integral definition of Bessel functions [22, p. 85]; since, on one hand, $\Theta_0 = \Theta_0 \mathcal{G}$ with $\mathcal{G} = 2^{2i\pi\mathcal{E}} \mathcal{F}^{\text{symp}}$, on the other hand $\mathcal{W}_{-\nu, q} = \mathcal{W}_{\nu, q}$, the second equation (4.6.3) then follows from (4.1.2). $\square$

Lemma 4.6.2. *One has, for $q \in \mathbb{R}^\times$ and $y > 0$,*

$$\int_{-\infty}^{\infty} e^{2i\pi qx} \left[\frac{(s-x)^2 + y^2}{y} \right]^{\frac{-1-i\lambda}{2}} dx = \frac{2\,\pi^{\frac{1+i\lambda}{2}}}{\Gamma(\frac{1+i\lambda}{2})} \, y^{\frac{1}{2}} K_{\frac{i\lambda}{2}}(2\pi\,|q|\,y)\,e^{2i\pi qs}. \quad (4.6.6)$$

Proof. The integral transforms to

$$e^{2i\pi qs} y^{\frac{1+i\lambda}{2}} \int_{-\infty}^{\infty} (x^2 + y^2)^{\frac{-1-i\lambda}{2}} e^{2i\pi qx} dx,$$

after which one can use [22, p. 401]. □

Lemma 4.6.3. *Consider on Π a function of the kind $g(x+iy) = e^{2i\pi qx} F(y)$, where $q \in \mathbb{R}^\times$ and F is a continuous function on $]0,\infty[$, bounded by some power of y at infinity and by $Cy^{\frac{1}{2}+\varepsilon}$ for some $\varepsilon > 0$ near $y = 0$. The function $g_{\frac{1+\lambda^2}{4}}$, as defined by (3.2.8), is given as*

$$g_{\frac{1+\lambda^2}{4}}(x+iy) = \frac{\pi^{-2}}{|\Gamma(\frac{1+i\lambda}{2})|^2} \, y^{\frac{1}{2}} K_{\frac{i\lambda}{2}}(2\pi\,|q|\,y)\,e^{2i\pi qx} \int_0^\infty y'^{-\frac{3}{2}} F(y')\,K_{\frac{i\lambda}{2}}(2\pi\,|q|\,y')\,dy'.$$

$$(4.6.7)$$

Proof. Relying on (3.2.30), we compute first, with the help of Lemma 4.6.2, the function

$$(\Theta_0^* g)_{i\lambda}^{\flat}(s) = (2\pi)^{\frac{-3-i\lambda}{2}} \Gamma(\frac{1-i\lambda}{2}) \int_0^\infty F(y) \frac{dy}{y^2} \int_{-\infty}^{\infty} e^{2i\pi qx} \left[\frac{(s-x)^2 + y^2}{y} \right]^{\frac{-1-i\lambda}{2}} dx$$

$$= 2^{\frac{-1-i\lambda}{2}} \pi^{-1} e^{2i\pi qs} \int_0^\infty y^{-\frac{3}{2}} F(y)\,K_{\frac{i\lambda}{2}}(2\pi\,|q|\,y)\,dy. \quad (4.6.8)$$

Next, applying (3.2.30) and, again, Lemma 4.6.2 (with λ replaced by $-\lambda$), we obtain Lemma 4.6.3. □

The factor $|\Gamma(\frac{1+i\lambda}{2})|^{-2}$ on the right-hand side of (4.6.7) is troublesome, in view of the behaviour, recalled in (1.1.8), of the Gamma function at infinity on vertical lines. However, the situation is saved, thanks to the two factors $K_{\frac{i\lambda}{2}}$ and of the following estimate.

Lemma 4.6.4. *For every $\delta \in]0,1[$, one has for some $C > 0$ the estimate*

$$|K_{\frac{i\lambda}{2}}(2y)| \leq C\,|\Gamma(\frac{3+i\lambda}{2})|\,y^{\delta-1}, \quad \lambda \in \mathbb{R}, \ y > 0. \quad (4.6.9)$$

Proof. The result of Lemma 4.6.2 can be rewritten, after a change of variable, as

$$K_{\frac{i\lambda}{2}}(2y) = \pi^{-\frac{1}{2}} y^{-\frac{i\lambda}{2}} \Gamma(\frac{1+i\lambda}{2}) \int_0^\infty (\cosh s)^{-i\lambda} \cos(2y \sinh s)\,ds : \quad (4.6.10)$$

then, after an integration by parts, one can write the integral on the right-hand side as

$$(1 + i\lambda) \int_0^\infty (\cosh s)^{-i\lambda - 2} \sinh s \frac{\sin(2y \sinh s)}{2y} \, ds. \tag{4.6.11}$$

Writing for $0 < \delta < 1$,

$$\left| \frac{\sin(2y \sinh s)}{2y} \right| \leq \min(\sinh s, (2y)^{-1}) \leq (\sinh s)^\delta (2y)^{\delta - 1}, \tag{4.6.12}$$

one obtains (4.6.9). $\square$

Lemma 4.6.5. *Given ν_1, ν_2 such that $|\mathrm{Re}\,(\nu_1 \pm \nu_2)| < 1$ and $q_1, q_2 \in \mathbb{R}$ such that $q_1 q_2 (q_1 + q_2) \neq 0$, one has the estimates $\left(\Theta_0 h_{\nu_1, q_1} \underset{j}{\times} \Theta_0 h_{\nu_2, q_2} \right)_{\frac{1+\lambda^2}{4}} (x + iy) = O(\lambda^2)$ as $|\lambda| \to \infty$, where the implied constant depends only on bounds for $(\mathrm{Im}\,z)^{\pm 1}$ if q_1, q_2 and $q_1 + q_2$ are bounded and bounded away from zero.*

Proof. According to Lemma 4.6.1, one has if $\nu_r \neq 0$

$$\left| (\Theta_0 h_{\nu_r, q_r}) (x + iy) \right| = \begin{cases} O\left(y^{\frac{1 - |\mathrm{Re}\,\nu_r|}{2}} \right), & y \to 0, \\ O\left(y^{-N} \right), & y \to \infty, \end{cases} \tag{4.6.13}$$

where N is arbitrary: when $\nu_r = 0$, an (unimportant) extra factor $\log y$ is needed on the right-hand side. Then, from Lemma 4.6.1 again, $\left(\Theta_0 h_{\nu_1, q_1} \underset{0}{\times} \Theta_0 h_{\nu_2, q_2} \right) (x + iy)$ can be written as $e^{2i\pi(q_1 + q_2)x} F(y)$, where $F(y)$ is rapidly decreasing at infinity and bounded, as $y \to 0$, by $C\, y^{\frac{2 - |\mathrm{Re}\,\nu_1| - |\mathrm{Re}\,\nu_2|}{2}}$, with a possible extra factor $|\log y|$ or $(\log y)^2$. As $|\mathrm{Re}\,(\nu_1 \pm \nu_2)| < 1$, Lemma 4.6.3 applies, and (4.6.7) gives an integral expression of $\left(\Theta_0 h_{\nu_1, q_1} \underset{j}{\times} \Theta_0 h_{\nu_2, q_2} \right)_{\frac{1+\lambda^2}{4}} (x + iy)$. Noting that the dy'-integral will still be convergent, given what we know about $F(y')$, if δ is close enough to 1, we obtain the part of Lemma 4.6.5 dealing with the pointwise product as a consequence of the estimate (4.6.9).

To treat the "$j = 1$"-part, we shall show that using the Poisson bracket of the two functions $\Theta_0 h_{\nu_r, q_r}$ involved in place of their pointwise product actually improves the estimates: starting from

$$\left\{ y^{\frac{1}{2}} K_{\frac{\nu_1}{2}} (2\pi |q_1| y) e^{2i\pi q_1 x}, \, y^{\frac{1}{2}} K_{\frac{\nu_2}{2}} (2\pi |q_2| y) e^{2i\pi q_2 x} \right\}$$

$$= 2i\pi y^2 \left[-q_2 \frac{d}{dy} \left(y^{\frac{1}{2}} K_{\frac{\nu_1}{2}} (2\pi |q_1| y) \right) y^{\frac{1}{2}} K_{\frac{\nu_2}{2}} (2\pi |q_2| y) \right.$$

$$\left. + q_1 y^{\frac{1}{2}} K_{\frac{\nu_1}{2}} (2\pi |q_1| y) \frac{d}{dy} \left(y'^{\frac{1}{2}} K_{\frac{\nu_2}{2}} (2\pi |q_2| y) \right) \right] e^{2i\pi(q_1 + q_2)x}, \tag{4.6.14}$$

and using the identity [22, p. 67]

$$\frac{d}{dy}\left(y^{\frac{1}{2}} K_{\frac{\nu}{2}}(2\pi |q| y)\right)$$

$$= \pi |q| y^{\frac{1}{2}}\left[-\left(1+\frac{1}{\nu}\right) K_{\frac{\nu-2}{2}}(2\pi |q| y) - \left(1-\frac{1}{\nu}\right) K_{\frac{\nu+2}{2}}(2\pi |q| y)\right], \quad (4.6.15)$$

it suffices to observe that the extra factor y^2 present on the right-hand side of (4.6.14) more than compensates the effect, so far as integrability near $y' = 0$ of the integral involved in (4.6.7) is concerned, of the shift by ± 1 of the parameter $\frac{\nu_1}{2}$ or $\frac{\nu_2}{2}$.

Let us remark that the result of Lemma 4.6.5 remains true if one replaces there h_{ν_r,q_r} by a function $h^r(x,\xi) = f_r(\xi)e^{2i\pi q \frac{x}{\xi}}$ with $|f_r(\xi)| \le C |\xi|^{-1-a_r}$, provided that $|a_1| + |a_2| < 1$: indeed, this follows from the inequality $|(\Theta_0 h^r)(z)| \le (\Theta_0 h_{a_r,0})(z)$, a consequence of (4.6.1).   □

Theorem 4.6.6. *Let* $h^r = h_{\nu_r,q_r}$, *with* $|\mathrm{Re}\,(\nu_1 \pm \nu_2)| < 1$ *and* $q_1 q_2 (q_1 + q_2) \neq 0$. *With* $j = 0$ *or* 1, *one has, in the weak sense in* $\mathcal{S}'(\mathbb{R}^2)$,

$$h^1 \underset{j}{\#} h^2 = \frac{\pi}{2}(-i)^j \int_{-\infty}^{\infty} \left(\prod_{\eta_1,\eta_2=\pm1} \Gamma\left(\frac{1+\eta_1\nu_1+\eta_2\nu_2+\eta_1\eta_2 i\lambda+2j}{4}\right)\right)^{-1}$$

$$\cdot \left[\Theta_0^*\left(\Theta_0 h^1 \underset{j}{\times} \Theta_0 h^2\right)\right]_{i\lambda} d\lambda \quad (4.6.16)$$

and, in $C^\infty(\Pi)$,

$$\Theta_0\left(h^1 \underset{j}{\#} h^2\right) = \pi^2(-i)^j$$

$$\cdot \int_{-\infty}^{\infty} \frac{\Gamma(\frac{1+i\lambda}{2})\Gamma(\frac{1-i\lambda}{2})}{\prod_{\eta_1,\eta_2=\pm1}\Gamma\left(\frac{1+\eta_1\nu_1+\eta_2\nu_2+i\eta_1\eta_2\lambda+2j}{4}\right)} \left[\Theta_0 h^1 \underset{j}{\times} \Theta_0 h^2\right]_{\frac{1+\lambda^2}{4}} d\lambda. \quad (4.6.17)$$

Proof. Let us prove first the equation

$$\Theta_0\left(h^1 \# h^2\right) = \pi^2(4\Delta)^3 \sum_{j=0,1}(-i)^j$$

$$\cdot \int_{-\infty}^{\infty} \frac{\Gamma(\frac{1+i\lambda}{2})\Gamma(\frac{1-i\lambda}{2})}{\prod_{\eta_1,\eta_2=\pm1}\Gamma\left(\frac{1+\eta_1\nu_1+\eta_2\nu_2+i\eta_1\eta_2\lambda+2j}{4}\right)} \left[\Theta_0 h^1 \underset{j}{\times} \Theta_0 h^2\right]_{\frac{1+\lambda^2}{4}} \frac{d\lambda}{(1+\lambda^2)^3}.$$

$$(4.6.18)$$

Replacing h^1 and h^2 by functions in $\mathcal{S}_{\text{even}}(\mathbb{R}^2)$, decomposing them according to (3.2.1) as integral superpositions of functions homogeneous of degrees $-1-i\lambda_1$ and

$-1 - i\lambda_2$ (where $i\lambda_r$ is to be substituted for ν_r), one obtains a weak (i.e., integral relative to λ_1, λ_2) version of (4.6.18) as a consequence of Corollary 3.3.2: indeed, with $f = \Theta_0 h^1 \times_j \Theta_0 h^2$, $f_{\frac{1+\lambda^2}{4}}$ is a generalized eigenfunction of Δ for the eigenvalue $\frac{1+\lambda^2}{4}$. Next, this equation, still taken in the weak integral sense, extends to the case when $h^r(x, \xi) = f_r(\xi)e^{2i\pi q_r \frac{x}{\xi}}$ for some pair of functions $f_r \in \mathcal{S}_\infty(\mathbb{R}^\times)$. To see this, it suffices to use the approximation process in Lemma 4.3.2: as $\theta_1, \theta_2 \to 0$, the operator with symbol $h^1 \# h^2$ is the weak limit (in the space of operators from $\mathcal{S}(\mathbb{R})$ to $\mathcal{S}'(\mathbb{R})$) of the operator with symbol $(h^1)^{\theta_1} \# (h^2)^{\theta_2}$. This takes care of the limiting process, so far as the left-hand side of (4.6.17) is concerned. The appropriate limit is also reached on the right-hand side in view of Lebesgue's convergence theorem and (4.3.5) since, as a consequence of the remark at the end of the proof of Lemma 4.6.5 (we recall that the estimates there are uniform with respect to q_1, q_2 as long as q_1, q_2 and $q_1 + q_2$ are bounded and bounded away from 0) and of (1.1.8)

$$\left| \frac{\Gamma(\frac{1+i\lambda}{2})\Gamma(\frac{1-i\lambda}{2})}{\prod_{\eta_1, \eta_2 = \pm 1} \Gamma\left(\frac{1+\eta_1\nu_1+\eta_2\nu_2+i\eta_1\eta_2\lambda+2j}{4}\right)} \right| \leq C\left(1 + |\lambda|\right)^{1-2j}, \qquad (4.6.19)$$

the factor $(1+\lambda^2)^{-3}$ suffices to ensure convergence. Since the densities ψ_r in (4.5.4) are analytic functions of ν_r, the equation (4.6.17), valid when each symbol h^r is an integral superposition, with respect to $\nu_r \in i\mathbb{R}$, of functions h_{ν_r, q_r}, is also valid for every pair of individual such functions.

Again, we can make from (4.6.18) an identity involving $\Theta_1(h^1 \# h^2)$ in place of the Θ_0-transform, just inserting an extra factor $-i\lambda$ on the right-hand side. As already mentioned immediately after (2.1.14), the symbol $h^1 \# h^2$ we are interested in is characterized by its (Θ_0, Θ_1)-transform. This leads to an identity which is almost the identity (4.6.16) to be proved. Indeed, since $\pi^2 \mathcal{E}^2$ transfers under Θ_0 or Θ_1 to $\Delta - \frac{1}{4}$, what we obtain in place of (4.6.16) is an identity with the same left-hand side, while the following two modifications have occurred on the right-hand side: an extra factor $(1 + \lambda^2)^{-3}$ shows up under the integral sign, and the operator $(1 + 4\pi^2\mathcal{E}^2)^3$ is applied to the result. These two modifications of course cancel each other. This concludes the proof of Theorem 4.6.6: the role of the extra factor $(1 + \lambda^2)^{-3}$, in (4.6.18), was only temporary. $\qquad\square$

Theorem 4.6.7. *The result of Theorem 4.6.6 remains true if the assumptions relative to q_1, q_2 are replaced by the conditions $q_1 \neq 0$, $q_2 = 0$. Assuming now that $q_1 \neq 0$, $q_1 + q_2 = 0$ and setting $A_r = \mathrm{Op}(h_{\nu_r, q_r})$, the symbol of the operator $(2i\pi \, \mathrm{mad}(P \wedge Q))(A_1 A_2)$ is given as the integral, to be understood in Cauchy's principal sense around $\lambda = 0$ and weakly convergent in $\mathcal{S}'(\mathbb{R}^2)$,*

$$\frac{\pi}{2} \sum_{j=0,1} (-i)^j \int_{-\infty}^{\infty} (-i\lambda) \, d\lambda \left(\prod_{\eta_1,\eta_2=\pm 1} \Gamma\left(\frac{1 + \eta_1\nu_1 + \eta_2\nu_2 + i\,\eta_1\eta_2\lambda + 2j}{4} \right) \right)^{-1}$$

$$\cdot \left[\Theta_0^* \left(\Theta_0 h^1 \underset{j}{\times} \Theta_0 h^2 \right) \right]_{i\lambda}. \quad (4.6.20)$$

Proof. Recall from Lemma 4.1.2 that in the case when $q_1 + q_2 = 0$, one cannot, in general, define $h_{\nu_1,q_1} \# h_{\nu_2,q_2}$ as a distribution: only the image of this object under $\mathcal{E}$ is: it is the symbol of the operator $\mathrm{mad}(P \wedge Q)(A_1 A_2)$. The second part of Theorem 4.6.7 is then a consequence of Theorem 4.5.2, with the same details as those given for proving Theorem 4.6.6: the insertion of the factor $-i\lambda$ is equivalent to the application of the operator $2i\pi\mathcal{E}$. The reason why this does not suffice to make the integral (4.6.20) under examination a convergent one, only one convergent in Cauchy's sense, was made clear in the proof of Theorem 4.5.2: if wishing to obtain a truly convergent integral, it would suffice to apply $A_1 A_2$ the square of the operator $2i\pi\,\mathrm{mad}(P \wedge Q)$, which would result in replacing the factor $-i\lambda$ by its square too.

Considering now the case when $q_1 \neq 0$ and $q_2 = 0$, let us give, for a change, a direct proof of the fact that the symbol $h = h_{\nu_1,q_1} \# h_{\nu_2,0}$ is given as

$$h = \frac{\pi}{2} \sum_{j=0,1} (-i)^j \int_{-\infty}^{\infty} \left(\prod_{\eta_1,\eta_2=\pm 1} \Gamma\left(\frac{1 + \eta_1\nu_1 + \eta_2\nu_2 + i\,\eta_1\eta_2\lambda + 2j}{4} \right) \right)^{-1}$$

$$\cdot \left[\Theta_0^* \left(\Theta_0 h^1 \underset{j}{\times} \Theta_0 h^2 \right) \right]_{i\lambda} d\lambda, \quad (4.6.21)$$

the same equation as in the case when $q_1 q_2 (q_1 + q_2) \neq 0$. The function $(\Theta_0 h_{\nu_1,q_1})$ $(x + iy)$ is still given by (4.6.3). On the other hand,

$$(\Theta_0 h_{\nu_2,0})(x + iy) = 2 \int_{\mathbb{R}^2} |\beta|^{-1-\nu_2} \exp\left(-2\pi \frac{(b - x\beta)^2 + y^2\beta^2}{y} \right) db \, d\beta$$

$$= 2^{\frac{1+\nu_2}{2}} \pi^{\frac{\nu_2}{2}} \Gamma(-\frac{\nu_2}{2}) y^{\frac{1+\nu_2}{2}}. \quad (4.6.22)$$

Using (3.2.11) and (3.2.20), we obtain

$$\left[\Theta_0^* \left(\Theta_0 h_{\nu_1,q_1} \underset{0}{\times} \Theta_0 h_{\nu_2,0} \right) \right]_{i\lambda}^b (s) = 2^{\frac{1+\nu_1+\nu_2-i\lambda}{2}} \pi^{\frac{-3+\nu_2-i\lambda}{2}} \Gamma(-\frac{\nu_2}{2}) \Gamma(\frac{1+i\lambda}{2})$$

$$\cdot |q_1|^{-\frac{\nu_1}{2}} \int_0^{\infty} y^{\frac{-2+\nu_2}{2}} K_{\frac{\nu_1}{2}}(2\pi \, |q_1| \, y) \, dy \int_{-\infty}^{\infty} \left[\frac{(s-x)^2 + y^2}{y} \right]^{\frac{-1-i\lambda}{2}} e^{2i\pi q_1 x} \, dx,$$

$$(4.6.23)$$

the same as

$$2^{\frac{3+\nu_1+\nu_2-i\lambda}{2}} \pi^{\frac{-2+\nu_2}{2}} \Gamma(-\frac{\nu_2}{2}) |q_1|^{\frac{-\nu_1+i\lambda}{2}} e^{2i\pi q_1 s}$$

$$\cdot \int_0^\infty y^{\frac{-1+\nu_2}{2}} K_{\frac{\nu_1}{2}}(2\pi |q_1| y) K_{\frac{i\lambda}{2}}(2\pi |q_1| y)\, dy$$

$$= 2^{\frac{-3+\nu_1+\nu_2-i\lambda}{2}} \pi^{-\frac{3}{2}} \frac{\Gamma(-\frac{\nu_2}{2})}{\Gamma(\frac{1+\nu_2}{2})} |q_1|^{\frac{-1-\nu_1-\nu_2+i\lambda}{2}} e^{2i\pi q_1 s}$$

$$\cdot \prod_{\eta_1,\eta_2=\pm 1} \Gamma\left(\frac{1+\eta_1\nu_1+\nu_2+\eta_2 i\lambda}{4}\right), \tag{4.6.24}$$

after one has used the integral [22, p. 101]. Multiplying this, as invited to do by (4.6.21), by

$$\frac{\pi}{2}\left(\prod_{\eta_1,\eta_2=\pm 1} \Gamma\left(\frac{1+\eta_1\nu_1+\eta_2\nu_2+i\,\eta_1\eta_2\lambda+2j}{4}\right)\right)^{-1}, \tag{4.6.25}$$

one obtains the expression

$$2^{\frac{-5+\nu_1+\nu_2-i\lambda}{2}} \pi^{-\frac{1}{2}} \frac{\Gamma(-\frac{\nu_2}{2})}{\Gamma(\frac{1+\nu_2}{2})} |q_1|^{\frac{-1-\nu_1-\nu_2+i\lambda}{2}} e^{2i\pi q_1 s} \frac{\Gamma(\frac{1+\nu_1+\nu_2-i\lambda}{4})\Gamma(\frac{1-\nu_1+\nu_2+i\lambda}{4})}{\Gamma(\frac{1+\nu_1-\nu_2-i\lambda}{4})\Gamma(\frac{1-\nu_1-\nu_2+i\lambda}{4})}, \tag{4.6.26}$$

which we must compare to the expression

$$h^b_{i\lambda}(s) = 2^{\frac{-5+\nu_1+\nu_2-i\lambda}{2}} \pi^{-1} e^{2i\pi q_1 s} |q_1|^{\frac{-1-\nu_1-\nu_2+i\lambda}{2}} \Delta_0\left(\frac{1+\nu_1+\nu_2-i\lambda}{2}, -\nu_2\right) \tag{4.6.27}$$

arising from the "$j=0$"-part of (4.5.25). Using (4.4.3), one obtains the identity of the two expressions, which proves the "$j=0$"-part (the commutative one) of the first part of Theorem 4.6.7.

Detailing the "$j=1$"-part can be dispensed with: it is entirely similar, since the Poisson bracket of $\Theta_0 h^1$ and $\Theta_0 h^2$ reduces in this case to

$$\{\Theta_0 h_{\nu_1,q_1}, \Theta_0 h_{\nu_2,0}\} = y^2 \frac{\partial}{\partial x}(\Theta_0 h^1) \times \frac{\partial}{\partial y}(\Theta_0 h^2)$$

$$= 2^{\frac{2+\nu_1+\nu_2}{2}} \pi^{\frac{\nu_2}{2}} (1+\nu_2)\Gamma(-\nu_2)(2i\pi q_1) e^{2i\pi q_1 x} y^{\frac{3+\nu_2}{2}} K_{\frac{\nu_1}{2}}(2\pi |q_1| y), \tag{4.6.28}$$

an expression identical, up to some changes of parameters, to that of the pointwise product, with improved integrability near $y=0$. $\qquad\square$

Chapter 5

The sharp composition of modular distributions

In this chapter, we compute the sharp product of two modular distributions, a well-defined notion provided that we define it as a quasi-distribution (of a species as close to that of a distribution as is possible).

The proof is still based, ultimately, on the composition formulas (3.3.2)-(3.3.5). However, it would be quite difficult, if at all possible, to apply these directly in the automorphic case, because of the singularities of objects such as $(h_1)^\flat_{i\lambda_1}$ in such a case. For instance [34, p.17], $\mathfrak{E}^\flat_\nu$ is a linear form on the space $C^\infty_{-\nu}$ of C^∞ vectors u of the representation $\pi_{-\nu}$ (it extends the representation $\pi_{i\lambda}$ the definition of which was recalled in (3.2.3), but is not unitary unless Re $\nu = 0$: it is, however, a representation by bounded operators in a weighted L^2-space on the line, depending on Re ν) defined as follows: recalling that, for $u \in C^\infty_{-\nu}$,

$$u(s) = u_\infty |s|^{-1+\nu} + \mathrm{O}(|s|^{-2+\mathrm{Re}\ \nu}), \quad |s| \to \infty, \tag{5.0.1}$$

one has

$$\langle \mathfrak{E}^\flat_\nu, u \rangle = \frac{1}{2} \sum_{|m|+|n|\neq 0} |m|^{\nu-1} u(\frac{n}{m}), \tag{5.0.2}$$

with the convention that

$$|m|^{\nu-1} u(\frac{n}{m}) = u_\infty |n|^{\nu-1} \quad \text{when } m = 0, n \neq 0. \tag{5.0.3}$$

The singularities are difficult to handle in this realization: the "distribution" so obtained is, in some sense, "carried" by the set of rational points of the projective line.

Instead, we shall take advantage of the Fourier series decompositions (1.1.38) and (1.2.4) and apply the results of Chapter 4 providing the spectral decompositions of sharp products of two distributions of type $h_{\nu,k}$. We shall not make use of the Hilbert space $L^2(\Gamma \backslash \mathbb{R}^2)$, briefly alluded to in (3.2.61), which might make it possible, up to a point, to use Hilbert space methods in the automorphic situation. Still, as long as this is possible, we shall decompose automorphic distributions into modular distributions the degrees of homogeneity of which will lie on the line $-1 + i\mathbb{R}$: but this line ought to be thought of as a line of reference rather than a spectral line, and exceptional terms, Eisenstein distributions or automorphic distributions related to derivatives of such with respect to the parameter, lying outside this line, will sometimes have to be added. This chapter is, again, rather technical, and contour deformations will have to be made repeatedly. The (b, M)-trick from Remark 3.2.1 will help in this respect, making integrability estimates at infinity unnecessary: the pole-chasing will keep us busy enough.

Moving back and forth between the plane and the half-plane, as done in Section 4.6, will help. We shall start this chapter with some reminders about the Roelcke-Selberg decomposition (the decomposition of automorphic functions in Π into non-holomorphic modular forms) and establish its analogue (both more general and less precise, since we shall dispense with any L^2-theory) in the automorphic distribution environment.

5.1   The decomposition of automorphic distributions

Recall the Roelcke-Selberg decomposition (2.1.22) ([16, Section 15] or [14, p. 112])

$$f(z) = \Phi^0 + \frac{1}{8\pi} \int_{-\infty}^{\infty} \Phi(i\lambda)\, E_{\frac{1-i\lambda}{2}}(z)\, d\lambda + \sum_{r \geq 1} \sum_{\ell} \Phi^{r,\ell}\, \mathcal{M}_{r,\ell}(z), \qquad (5.1.1)$$

valid for an arbitrary function $f \in L^2(\Gamma \backslash \Pi)$: the notation is not meant to imply that, as a function of λ, $\Phi(i\lambda)$ should in general extend away from the real line. Uniqueness of the coefficients making this identity valid is ensured if one imposes the condition $\zeta^*(-i\lambda)\Phi(i\lambda) = \zeta^*(i\lambda)\Phi(-i\lambda)$. Then, one has $\Phi^0 = \frac{3}{\pi} \int_{\Gamma \backslash \Pi} f(z)\, dm(z)$ since the area of the fundamental domain is $\frac{\pi}{3}$, next $\Phi(i\lambda) = \left(E_{\frac{1-i\lambda}{2}} \,|\, f \right)_{L^2(\Gamma \backslash \Pi)} = \int_{\Gamma \backslash \Pi} E_{\frac{1+i\lambda}{2}}(z)\, f(z)\, dm(z)$ for almost every λ, finally $\Phi^{r,\ell} = (\mathcal{M}_{r,\ell} \,|\, f)_{L^2(\Gamma \backslash \Pi)}$. Here, $\mathcal{M}_{r,\ell}$ is L^2-normalized. Also,

$$\| f \|^2_{L^2(\Gamma \backslash \Pi)} = \frac{\pi}{3} |\Phi^0|^2 + \frac{1}{8\pi} \int_{-\infty}^{\infty} |\Phi(i\lambda)|^2\, d\lambda + \sum_{r \geq 1} \sum_{\ell} |\Phi^{r,\ell}|^2. \qquad (5.1.2)$$

Our program in this section is to obtain an analogous theorem in the automorphic distribution environment. Starting from a distribution

$$\mathfrak{S}(x, \xi) = g_0(x)\, \delta(\xi) + \sum_{k \in \mathbb{Z}} f_k(\xi)\, \exp\left(2i\pi \frac{kx}{\xi} \right), \qquad (5.1.3)$$

assumed to be automorphic, we wish to obtain, under suitable assumptions regarding the coefficients, the first of which is of course that the function g_0, as well as all functions f_k, should be even, a decomposition

$$\mathfrak{S} = C_\infty + C_0\delta + \sum_{j \text{ with } \mu_j \neq \pm 1} C_j \mathfrak{E}_{\mu_j} + \frac{1}{8\pi}\int_{-\infty}^{\infty} \Psi(i\lambda)\,\mathfrak{E}_{i\lambda}\,d\lambda + \sum_{r \neq 0}\sum_{\ell} \Psi^{r,\ell}\mathfrak{N}_{r,\ell}.$$

$$(5.1.4)$$

Note that such a kind of expansion cannot be valid for arbitrary automorphic distributions since the $\frac{d}{d\lambda}$-derivative of the Eisenstein distribution $\mathfrak{E}_{i\lambda}$ is also automorphic (but not modular): this does not happen in the usual theory in the hyperbolic half-plane because one specializes there in a Hilbert space frame and only Eisenstein series $E_{\frac{1-i\lambda}{2}}$, not their $\frac{d}{d\lambda}$-derivatives, occur as generalized eigenfunctions in the spectral decompositions of automorphic functions in the space $L^2(\Gamma\backslash\Pi)$.

A finite number, only, of μ_j's, should be present on the right-hand side, all of which should have a nonzero real part: recall that $\mathfrak{E}_\mu$ is defined only for $\mu \neq \pm 1$. On the other hand, as remarked at the end of the proof of Theorem 1.1.7, the Eisenstein distribution $\mathfrak{E}_{i\lambda}$, in contrast to the individual terms of its Fourier series expansion, is well-defined for $\lambda = 0$: anyway, under the assumptions relative to $\mathfrak{S}$ to be made later, one will have $\Psi(0) = 0$.

We must, at this point, recall Proposition 2.1.1 and Remark 2.1.1 (i), in which the two properly normalized automorphic distributions $\mathfrak{N}_{r,\ell}$ and $\mathfrak{N}_{-r,\ell}$ lying above a given $L^2(\Gamma\backslash\Pi)$-normalized Hecke eigenform $\mathcal{M}_{r,\ell}$ have been defined: they differ by some sign in the exponent of $\left|\frac{k}{t^2}\right|$ in (2.1.19), so that $\mathfrak{N}_{\pm r,\ell}$ is homogeneous of degree $-1 \mp i\lambda_r$. They are related under the transformation $\mathcal{F}^{\text{symp}}$: since both will generally occur in the decomposition of $\mathfrak{S}$, it has been found necessary to let in (5.1.4) r run through $\mathbb{Z}^\times$, not through the set $\{1, 2, \ldots\}$ as in (5.1.1). Another difference between the expected decomposition (5.1.4) and its automorphic function analogue (5.1.1) has the same origin: since $\mathfrak{E}_{i\lambda}$ and $\mathfrak{E}_{-i\lambda}$ are Fourier related rather than proportional (as $E_{\frac{1-i\lambda}{2}}$ and $E_{\frac{1+i\lambda}{2}}$), one may not demand in general that $\Psi(i\lambda)$ and $\Psi(-i\lambda)$ should relate in any definite way, in contrast to $\Phi(i\lambda)$ and $\Phi(-i\lambda)$.

We consider first the case when the terms $g_0(x)\delta(\xi)$ and $f_0(\xi)$ are absent from the expansion (5.1.3): in this case, we wish to obtain an expansion of $\mathfrak{S}$ into a series of Hecke distributions. We have not strived for maximum generality: our point is to obtain a result under conditions general enough to apply to the question of the sharp product of two modular distributions.

Lemma 5.1.1. *For some constant $C > 0$, one has for every $r = 1, 2, \ldots$ the estimate*

$$C^{-1}\lambda_r^{-\frac{1}{2}} \leq \|V^*\mathfrak{N}_{\pm r,\ell}\|_{L^2(\Gamma\backslash\Pi)} \leq C\,\lambda_r^{\frac{1}{2}}.\qquad(5.1.5)$$

On the other hand, the numbers of pairs (r, ℓ) with $r = 1, 2, \ldots$ such that $\lambda_r \leq A$ is, as $A \to \infty$, of the size of A^2.

Proof. Recall (as mentioned between (2.1.21) and (2.1.22)) that, besides the Hecke eigenform $\mathcal{M}_{r,\ell}$, it is customary to introduce a proportional eigenform $\mathcal{N}_{r,\ell}$, characterized by the fact that the first coefficient of its Fourier expansion should be 1: this is especially helpful when considering the effect on eigenforms of the Hecke operators. According to [26], one has for some constant $C > 0$ the inequality

$$C^{-1}\left|\Gamma(\frac{i\lambda_r}{2})\right| \leq \|\mathcal{N}_{r,\ell}\|_{L^2(\Gamma\backslash\Pi)} \leq C\left|\Gamma(1 + \frac{i\lambda_r}{2})\right|. \tag{5.1.6}$$

In view of (3.2.20), recalling also that $\mathfrak{N}^{\text{resc}}_{\pm r,\ell} = 2^{-\frac{1}{2}+i\pi\mathcal{E}}\mathfrak{N}_{\pm r,\ell}$, one has $\Theta_0\mathfrak{N}^{\text{resc}}_{\pm r,\ell} = V^*\pi^{\frac{1}{2}-i\pi\mathcal{E}}\Gamma(\frac{1}{2} + i\pi\mathcal{E})\,\mathfrak{N}_{\pm r,\ell}$: then, from (2.1.25),

$$\mathcal{N}_{r,\ell} = \pi^{\frac{1\pm i\lambda_r}{2}}\Gamma(\frac{1\mp i\lambda_r}{2})\,V^*\mathfrak{N}_{\pm r,\ell} \tag{5.1.7}$$

and (5.1.5) follows. The last assertion, or the more precise version $\#\{(r,\ell)\colon \lambda_r \leq A\} \sim \frac{A^2}{48}$, is a consequence of Selberg's trace formula [16, p.391]. $\qquad\square$

Before stating the next theorem, recall that, in the automorphic distribution environment, defining λ_r, when $r \geq 1$, as the positive number such that $\frac{1+\lambda_r^2}{4}$ should be the eigenvalue attached to the corresponding Hecke eigenform $\mathcal{M}_{r,\ell}$, we also set $\lambda_{-r} = -\lambda_r$: in this way, the Hecke distribution $\mathfrak{N}_{r,\ell}$ is homogeneous of degree $-1 - i\lambda_r$, whatever the sign of r.

Proposition 5.1.2. *Let $\mathfrak{S}$ be an automorphic distribution with a weakly convergent expansion in $\mathcal{S}'(\mathbb{R}^2)$ of the form*

$$\mathfrak{S}(x,\xi) = \sum_{k\neq 0} f_k(\xi)\exp\left(2i\pi\frac{kx}{\xi}\right) \tag{5.1.8}$$

where, for every k, f_k is an even distribution on the line, C^∞ in $\mathbb{R}^\times$, and the collection (f_k) satisfies for some pair $\delta > 0$, $N > 0$ and every $j = 0, 1, \ldots$ the estimate $\left|\left(\xi\frac{d}{d\xi}\right)^j f_k(\xi)\right| \leq C(j)\,(1+|k|)^N\left(|\xi|+|\xi|^{-1}\right)^\delta$, with $C(j)$ depending only on j. For m large enough (in a way depending only on the pair δ, N), the transforms of $\mathfrak{S}$ and $(2i\pi\mathcal{E})\mathfrak{S}$ under the operator W_m introduced in (3.2.43) both lie in $L^2(\Gamma\backslash\Pi)$. The distribution $\mathfrak{S}$ admits an expansion as a series (convergent in $\mathcal{S}'(\mathbb{R}^2)$) of Hecke distributions. In terms of the (Roelcke-Selberg) expansions

$$W_m\left((2i\pi\mathcal{E})^\kappa\mathfrak{S}\right) = \sum_{r\geq 1}\sum_\ell A^{r,\ell}_\kappa \mathcal{M}_{r,\ell}, \qquad \kappa = 0, 1 \tag{5.1.9}$$

of $W_m\mathfrak{S}$ and $W_m((2i\pi\mathcal{E})\mathfrak{S})$, it is given as the series

$$\mathfrak{S} = \sum_{r\in\mathbb{Z}^\times}\sum_\ell \Psi^{r,\ell}\mathfrak{N}_{r,\ell}, \tag{5.1.10}$$

with

$$\Psi^{r,\ell} = (-1)^m 2^{m-\frac{3}{2}} \pi^{\frac{1+i\lambda_r}{2}} \frac{\Gamma(m+\frac{1-i\lambda_r}{2})}{\|\mathcal{N}_{|r|,\ell}\|} \left[A_0^{|r|,\ell} - \frac{1}{i\lambda_r} A_1^{|r|,\ell} \right]. \tag{5.1.11}$$

Proof. We must first observe that, despite the fact that f_k may not be locally summable near 0, the function $f_k(\xi) \exp\left(2i\pi \frac{kx}{\xi}\right)$ is well defined as a tempered distribution: the argument is strictly the same as that, given in the beginning of Section 4.1, which concerned the functions $h_{\nu,k}$ with $k \neq 0$. Now, given $z \in \Pi$, the function $(x,\xi) \mapsto g_m\left(\frac{|x-z\xi|}{(\mathrm{Im}\, z)^{\frac{1}{2}}}\right)$ has a compact support and, while never C^∞, is as smooth as desired provided m is chosen large enough: this makes it possible to test each distribution $f_k(\xi) \exp\left(2i\pi \frac{kx}{\xi}\right)$ on it.

Recall the definition (3.2.43) of the operator W_m from functions in $\mathbb{R}^2$ to functions in Π. One has, denoting as (x',ξ) the current point of $\mathbb{R}^2$ so as to save x for $z = x + iy$,

$$(W_m \mathfrak{S})(z) = \sum_{k \in \mathbb{Z}^\times} \int_{\mathbb{R}^2} f_k(\xi) e^{2i\pi \frac{kx'}{\xi}} g_m\left(\frac{|x'-z\xi|}{y^{\frac{1}{2}}}\right) dx'\, d\xi. \tag{5.1.12}$$

After a translation $x' \mapsto x' + x\xi$, followed by the change $x' = ty\xi$, one obtains

$$(W_m \mathfrak{S})(x+iy) = y \sum_{k \in \mathbb{Z}^\times} e^{2i\pi kx} \int_{-\infty}^{\infty} |\xi|\, f_k(\xi)\, d\xi \int_{-\infty}^{\infty} e^{2i\pi kyt} g_m\left(y^{\frac{1}{2}}\xi\sqrt{1+t^2}\right) dt. \tag{5.1.13}$$

With the notation in Lemma 3.2.6,

$$(W_m \mathfrak{S})(x+iy) = y \sum_{k \in \mathbb{Z}^\times} e^{2i\pi kx} \int_{-\infty}^{\infty} |\xi|\, f_k(\xi)\, I_m(2\pi ky, y^{\frac{1}{2}}\xi)\, d\xi, \tag{5.1.14}$$

where

$$I_m(2\pi ky, y^{\frac{1}{2}}\xi) = C(m)\, |k|^{-m} y^{\frac{-1-m}{2}}\, |\xi|^{m-1}(1-y\xi^2)_+^{\frac{m}{2}}\, J_m\left(2\pi \frac{|k|\, y^{\frac{1}{2}}}{|\xi|}\sqrt{1-y\xi^2}\right) \tag{5.1.15}$$

for some constant $C(m)$. For $m = 0, 1, \ldots$, the Bessel function on the right-hand side of (5.1.15) is bounded by a constant depending only on m, and the estimate

$$|I_m(2\pi ky, y^{\frac{1}{2}}\xi)| \leq C'(m)\, |k|^{-m} y^{\frac{-1-m}{2}}\, |\xi|^{m-1} \mathrm{char}(|\xi| < y^{-\frac{1}{2}}) \tag{5.1.16}$$

makes it possible to bound the $d\xi$-integral on the right-hand side of (5.1.14) by any power of $|k|^{-1}y^{-1}$ for some choice of m. This implies that the automorphic function $W_m \mathfrak{S}$ is well-defined and lies in $L^2(\Gamma\backslash\Pi)$, provided that m is large enough (in a way depending only on the pair δ, N which defined the assumptions made regarding f_k).

The function $W_m \mathfrak{S}$ admits in the Hilbert space $L^2(\Gamma \backslash \Pi)$ a convergent expansion of the type given in (5.1.9), with $\sum_{r \geq 1} \sum_\ell |A_0^{r,\ell}|^2 < \infty$. Since our assumptions involve bounds not only for the functions f_k, but for the functions $\xi \mapsto \xi f_k'(\xi)$ as well, a similar expansion exists if one replaces $\mathfrak{S}$ by $(2i\pi \mathcal{E})\mathfrak{S}$.

Next, we justify the equation linking the Fourier coefficients of $\mathfrak{S}$ to those of $W_m((2i\pi \mathcal{E})^\kappa \mathfrak{S})$. In view of (3.2.20) and (3.2.44), one has generally

$$\Theta_0 h = (-2)^m W_m (2\pi)^{\frac{1}{2} - i\pi \mathcal{E}} \Gamma\left(m + \frac{1}{2} + i\pi \mathcal{E}\right) h : \tag{5.1.17}$$

combining this with $\mathcal{N}_{|r|,\ell} = \Theta_0 \mathfrak{N}_{r,\ell}^{\mathrm{resc}} = 2^{-1-\frac{i\lambda_r}{2}} \Theta_0 \mathfrak{N}_{r,\ell}$ (2.1.25), one has for $r \in \mathbb{Z}^\times$

$$\mathcal{N}_{|r|,\ell} = (-1)^m 2^{m-\frac{1}{2}} \pi^{\frac{1-i\lambda_r}{2}} \Gamma\left(m + \frac{1 - i\lambda_r}{2}\right) W_m \mathfrak{N}_{r,\ell}. \tag{5.1.18}$$

The coefficients $\Psi^{r,\ell}$ with $r \in \mathbb{Z}^\times$ are determined by the necessary conditions that, with

$$\mathfrak{S} = \sum_{r \geq 1} \sum_\ell \left[\Psi^{r,\ell} \mathfrak{N}_{r,\ell} + \Psi^{-r,\ell} \mathfrak{N}_{-r,\ell} \right], \tag{5.1.19}$$

the pair of equations (5.1.9) must be satisfied. Since, with $r \geq 1$, one has $\mathcal{N}_{r,\ell} = \| \mathcal{N}_{r,\ell} \| \mathcal{M}_{r,\ell}$, an application of (5.1.19) reduces this pair of equations, if one sets for $r \in \mathbb{Z}^\times$

$$\Xi^{r,\ell} = \pi^{\frac{-1-i\lambda_r}{2}} \left(\Gamma\left(m + \frac{1 - i\lambda_r}{2}\right) \right)^{-1} \Psi^{r,\ell}, \tag{5.1.20}$$

to the pair (in which $r \geq 1$)

$$(-i\lambda_r)^\kappa \Xi^{r,\ell} + (i\lambda_r)^\kappa \Xi^{-r,\ell} = (-1)^m 2^{m-\frac{1}{2}} \| \mathcal{N}_{r,\ell} \|^{-1} A_\kappa^{r,\ell} : \tag{5.1.21}$$

this implies (5.1.11).

It is easy to see that, if m is large enough, the series (5.1.10), with coefficients as defined in (5.1.11), or even the multiple series obtained when expanding each Hecke distribution into a Fourier series, converges in $\mathcal{S}'(\mathbb{R}^2)$. Indeed, if one replaces the parameter (r, ℓ) by a unique parameter n, allowing repetition of the eigenvalue while keeping the fact that the absolute value of λ_n must be a non-decreasing function of $|n|$, it follows from Selberg's estimate (the last assertion in Lemma 5.1.1) that providing polynomial bounds in terms of λ_r is just as good as providing polynomial bounds in terms of n. But, when only convergence in $\mathcal{S}'(\mathbb{R}^2)$ is asserted, arbitrary powers of $-i\lambda_r$ can be replaced by the application of corresponding powers of $2i\pi \mathcal{E}$. The coefficient in front of the right-hand side of (5.1.11) is bounded by $C |\lambda_r|^{m+\frac{1}{2}}$ in view of (5.1.6). Finally, in the Fourier series defining each Hecke eigenform $\mathfrak{N}_{r,\ell}$, the coefficients of which are, up to bounded factors, borrowed from those of the associated Hecke eigenform $\mathcal{N}_{|r|,\ell}$ (Proposition 2.1.1), one can replace powers of k by corresponding powers of the operator $(2i\pi)^{-1}\xi\frac{\partial}{\partial x}$ and take benefit of the fact that, in a uniform way, the kth Fourier coefficient of $\mathcal{N}_{|r|,\ell}$ is bounded by some power of k (the exponent could be chosen arbitrarily small if the Ramanujan-Petersson conjecture were proved, but this is not necessary here).

Finally, what remains to be proved is that if $\mathfrak{S}$ is an automorphic distribution such that $W_m\left((2i\pi\mathcal{E})^\kappa\mathfrak{S}\right) = 0$ for $\kappa = 0$ or 1, the distribution $\mathfrak{S}$ reduces to zero. As seen in Proposition 3.1.4, this becomes true if W_m is replaced by Θ_0. Then, using (again) the relation

$$\Theta_0 = V^*(2\pi)^{\frac{1}{2}-i\pi\mathcal{E}}\Gamma(\frac{1}{2}+i\pi\mathcal{E}) = (-2)^m W_m \pi^{\frac{1}{2}-i\pi\mathcal{E}}\Gamma(m+\frac{1}{2}+i\pi\mathcal{E}), \quad (5.1.22)$$

one obtains

$$\Theta_0(2\pi)^{-\frac{1}{2}+i\pi\mathcal{E}}\Gamma(m+\frac{1}{2}-i\pi\mathcal{E}) = (-2)^m W_m\Gamma(m+\frac{1}{2}-i\pi\mathcal{E})\Gamma(m-\frac{1}{2}-i\pi\mathcal{E})$$

$$= (-2)^m\Gamma\left(m+\frac{1}{2}+i\sqrt{\Delta-\frac{1}{4}}\right)\Gamma\left(m+\frac{1}{2}-i\sqrt{\Delta-\frac{1}{4}}\right)W_m \quad (5.1.23)$$

if one denotes as $\Gamma\left(m+\frac{1}{2}+i\sqrt{\Delta-\frac{1}{4}}\right)\Gamma\left(m+\frac{1}{2}-i\sqrt{\Delta-\frac{1}{4}}\right)$ the operator transforming a function f in Π into the function g such that

$$g_{\frac{1+\lambda^2}{4}}$$

$$= \Gamma(m+\frac{1+i\lambda}{2})\Gamma(m+\frac{1-i\lambda}{2})f_{\frac{1+\lambda^2}{4}} = \frac{1}{2\pi}\left(\frac{1+i\lambda}{2}\right)_m\left(\frac{1-i\lambda}{2}\right)_m[\Theta_0(\Theta_0^* f)_{i\lambda}]$$

$$(5.1.24)$$

(3.2.32), or

$$g = \frac{1}{2\pi}\Delta(\Delta+2)\dots(\Delta+m^2+m)\Theta_0\Theta_0^* f. \quad (5.1.25)$$

From the assumption that $W_m\left((2i\pi\mathcal{E})^\kappa\mathfrak{S}\right) = 0$ for $\kappa = 0$ or 1, it thus follows that $(2\pi)^{-\frac{1}{2}+i\pi\mathcal{E}}\Gamma(m+\frac{1}{2}-i\pi\mathcal{E})\mathfrak{S} = 0$, so that $\mathfrak{S} = 0$. This concludes the proof of Proposition 5.1.2. □

Remarks 5.1.1. (i) in the last paragraph of the proof, we transferred the operator $\Gamma(m+\frac{1}{2}-i\pi\mathcal{E})\Gamma(m+\frac{1}{2}-i\pi\mathcal{E})$ to a function (in the spectral-theoretic sense) of the operator Δ: this was possible only because we started from an even function of $2i\pi\mathcal{E}$;

(ii) we used the operator W_m, not the operator Θ_0, in the proof, because the distribution $\mathfrak{S}$ could be proved to depend continuously, in the space of tempered distributions, on the pair $(W_m\mathfrak{S}, W_m((2i\pi\mathcal{E})\mathfrak{S}))$ of functions in $L^2(\Gamma\backslash\Pi)$: the "continuity part" of this statement would be totally false if W_m were replaced by Θ_0. On the other hand, Θ_0 has the advantage (linked to its role (3.1.19) in pseudodifferential analysis) that $\Theta_0\mathfrak{S}$ characterizes exactly the $\mathcal{G}$-invariant part of $\mathfrak{S}$. The function $W_m\mathfrak{S}$ characterizes the part of $\mathfrak{S}$ invariant under the more complicated involution

$$(2\pi)^{-2i\pi\mathcal{E}}\frac{\Gamma(m+\frac{1}{2}+i\pi\mathcal{E})}{\Gamma(m+\frac{1}{2}-i\pi\mathcal{E})}\,\mathcal{G} = \mathcal{G}\,(2\pi)^{2i\pi\mathcal{E}}\frac{\Gamma(m+\frac{1}{2}-i\pi\mathcal{E})}{\Gamma(m+\frac{1}{2}+i\pi\mathcal{E})}, \tag{5.1.26}$$

which does not preserve the whole space $\mathcal{S}'(\mathbb{R}^2)$. Nevertheless, in the case when one can take for δ (the number introduced in Proposition 5.1.2) a number < 2, it will map the class of distributions $f_k(\xi)\exp\left(2i\pi\frac{kx}{\xi}\right)$ which are the terms of the decomposition (5.1.3), to the $\mathcal{G}$-transform of this class. Indeed, one has [22, p.91]

$$(2\pi)^{2i\pi\mathcal{E}}\frac{\Gamma(m+\frac{1}{2}-i\pi\mathcal{E})}{\Gamma(m+\frac{1}{2}+i\pi\mathcal{E})} = \int_0^\infty t^{-i\pi\mathcal{E}}J_{2m}4\pi t)\,dt, \tag{5.1.27}$$

and the image under this operator of the distribution $f_k(\xi)\exp\left(2i\pi\frac{kx}{\xi}\right)$ is the distribution $g_k(\xi)\exp\left(2i\pi\frac{kx}{\xi}\right)$, with

$$g_k(\xi) = \int_0^\infty f_k\left(t^{-\frac{1}{2}}\xi\right)J_{2m}(4\pi t)\,t^{-\frac{1}{2}}dt. \tag{5.1.28}$$

This (generally divergent) integral is to be interpreted as follows. Writing $1 = \phi(t) + (1 - \phi(t))$, where the C^∞ functions $\phi(t)$ is supported in $\{|t| \geq 1\}$, and $\phi(t) = 0$ for $t \geq 2$, the part of the integral in which $1 - \phi(t)$ has been inserted as a factor is convergent as soon as $m \geq 1$. In the other part, we use the asymptotic expansion [22, p.139]

$$t^{-\frac{1}{2}}J_{2m}(4\pi t) \sim \sum_{n\geq 0} t^{-n-1}[a_n\cos(4\pi t) + b_n\sin(4\pi t)], \quad t \to \infty, \tag{5.1.29}$$

and we make just one integration by parts, isolating the trigonometric factor: one ends up with a convergent integral when $\delta < 2$. The function g_k then satisfies the same estimates as the function f_k.

After, under the assumptions of Proposition 5.1.2, $\mathfrak{S}$ has been proved to be an automorphic distribution, nothing prevents one (and this will be useful later) from characterizing the coefficients of its expansion (5.1.10) by those of the expansions of $\Theta_\kappa\mathfrak{S} = \Theta_0\left((2i\pi)^\kappa\mathfrak{S}\right)$ instead of $W_m\left((2i\pi)^\kappa\mathfrak{S}\right)$.

Corollary 5.1.3. *Under the assumptions of Proposition 5.1.2, assume that*

$$\Theta_\kappa\mathfrak{S} = \sum_{r\geq 1}\sum_\ell B_\kappa^{r,\ell}\mathcal{N}_{r,\ell}, \quad \kappa = 0,1. \tag{5.1.30}$$

Then, one has, for $r \in \mathbb{Z}^\times$,

$$\Psi^{r,\ell} = 2^{-2-\frac{i\lambda_r}{2}} \left(B_0^{|r|,\ell} - \frac{1}{i\lambda_r} B_1^{|r|,\ell} \right). \tag{5.1.31}$$

Proof. The calculations are simpler than the corresponding ones in the proof of Proposition 5.1.2, since (5.1.18) is to be replaced by the equation $\Theta_\kappa \mathfrak{N}_{r,\ell} = (-i\lambda_r)^\kappa 2^{1+\frac{i\lambda_r}{2}} \mathcal{N}_{|r|,\ell}$, and we have to solve the pair of equations ($\kappa = 0$ or 1, $r \geq 1$)

$$(-i\lambda_r)^\kappa 2^{1+\frac{i\lambda_r}{2}} \Psi^{r,\ell} + (i\lambda_r)^\kappa 2^{1-\frac{i\lambda_r}{2}} \Psi^{-r,\ell} = B_\kappa^{r,\ell}. \tag{5.1.32}$$

$\square$

In the following theorem, we decompose fairly general automorphic distributions (at least, sufficiently general for our purposes) into modular distributions: this is an analogue of the Roelcke-Selberg theorem, for automorphic distributions this time.

We shall consider an automorphic distribution $\mathfrak{S}$ with the Fourier expansion (5.1.3), where, for $k \neq 0$, the functions f_k are even, C^∞ in $\mathbb{R}^\times$ and satisfy bounds $|f_k(\xi)| \leq C (1 + |k|)^N (|\xi| + |\xi|^{-1})^\delta$ for some fixed triple (C, N, δ) of positive numbers; so far as f_0 and g_0 are concerned, we assume that they are well-defined even distributions on the line, C^∞ outside 0 and polynomially bounded at infinity. We set

$$M_0(\mu) = 2 \int_{-1}^{1} f_0(\xi) |\xi|^\mu d\xi, \quad N_0(\mu) = 2 \int_{-1}^{1} g_0(x) |x|^{\mu-1} dx,$$

$$M_\infty(\mu) = 4 \int_{1}^{\infty} f_0(\xi) \xi^\mu d\xi, \quad N_\infty(\mu) = 4 \int_{1}^{\infty} g_0(x) x^{\mu-1} dx : \tag{5.1.33}$$

as the distribution f_0 is C^∞ outside 0, it can be tested on the function $\xi \mapsto |\xi|^\mu \mathrm{char}(|\xi| \leq 1)$ if Re μ is large. From the assumptions just made, the integrals defining $M_0(\mu)$ and $N_0(\mu)$ are convergent for Re μ large enough, and those defining $M_\infty(\mu)$ and $N_\infty(\mu)$ are convergent for $-$Re μ large enough.

Theorem 5.1.4. *With the notation just introduced, we assume that the functions $M_0(\mu)$ and $N_0(\mu)$ extend as continuous functions in the half-plane Re $\mu \geq 0$, holomorphic in the interior, satisfying the usual condition of "polynomial boundedness" at infinity in vertical strips, with the possible exception of a finite number of simple poles μ_j ($j = 1, 2, \dots$) with Re $\mu_j > 0$ and $\mu_j \neq 1$. We make just the same assumptions about the functions $M_\infty(\mu)$ and $N_\infty(\mu)$, only replacing the preceding half-plane by the half-plane Re $\mu \leq 0$, allowing then a finite number of poles μ_j ($j = -1, -2, \dots$) with Re $\mu_j < 0$: this time, the number -1 is not prevented from being one of the μ_j's. Next we assume that the functions $\frac{M_0(\mu)+M_\infty(\mu)}{\zeta(-\mu)}$ and $\frac{N_0(\mu)+N_\infty(\mu)}{\zeta(1-\mu)}$, both well-defined on the real line, agree there, and we denote as*

$\Psi(i\lambda)$ *their common value at* $\mu = i\lambda$. *Finally, we assume that, for* Re $\mu_j > 0$,

$$\frac{1}{4\,\zeta(-\mu_j)}\,\mathrm{Res}_{\mu=\mu_j}\,M_0(\mu) = \frac{1}{4\,\zeta(1-\mu_j)}\,\mathrm{Res}_{\mu=\mu_j}\,N_0(\mu), \quad \text{to be denoted as } C_j$$

(5.1.34)

(if, say, $\mu_j = 2, 4, \ldots$, *the condition means that* $\mathrm{Res}_{\mu=\mu_j}\,M_0(\mu) = 0$; *if* $\zeta(1-\mu_j) = 0$, *it means that* $\mathrm{Res}_{\mu=\mu_j}\,N_0(\mu) = 0$), *and that, for* Re $\mu_j < 0$,

$$\frac{1}{4\,\zeta(-\mu_j)}\,\mathrm{Res}_{\mu=\mu_j}\,M_\infty(\mu) = \frac{1}{4\,\zeta(1-\mu_j)}\,\mathrm{Res}_{\mu=\mu_j}\,N_\infty(\mu), \quad \text{to be denoted as } -C_j$$

(5.1.35)

(the meaning is similar to the one above if $\zeta(-\mu_j) = 0$ *or* $\zeta(1-\mu_j) = 0$).

Then, the automorphic distribution $\mathfrak{S}$ admits a decomposition of the type (5.1.4) into modular distributions. The density $\Psi(i\lambda)$ has already been made explicit, and one has $C_\infty = \frac{1}{4}\mathrm{Res}_{\mu=-1}M_\infty(\mu)$ and $C_0 = -\frac{1}{2}N_0(1)$.

Proof. Before giving the proof, let us remark that the possible pole $\mu_j = -1$ contributes a constant: there is no Eisenstein distribution $\mathfrak{E}_{-1}$. One has if a and b are large enough and $\xi \neq 0$ or $x \neq 0$ according to the cases,

$$f_0(\xi)\,\mathrm{char}(|\xi| \leq 1) = \frac{1}{8i\pi}\int_{\mathrm{Re}\,\mu=b} M_0(\mu)\,|\xi|^{-1-\mu}d\mu,$$

$$g_0(x)\,\mathrm{char}(|x| \leq 1)\,\delta(\xi) = \frac{1}{8i\pi}\int_{\mathrm{Re}\,\mu=b} N_0(\mu)\,|x|^{-\mu}\delta(\xi)\,d\mu,$$

$$f_0(\xi)\,\mathrm{char}(|\xi| \geq 1) = \frac{1}{8i\pi}\int_{\mathrm{Re}\,\mu=-a} M_\infty(\mu)\,|\xi|^{-1-\mu}d\mu,$$

$$g_0(x)\,\mathrm{char}(|x| \geq 1)\,\delta(\xi) = \frac{1}{8i\pi}\int_{\mathrm{Re}\,\mu=-a} N_\infty(\mu)\,|x|^{-\mu}\delta(\xi)\,d\mu. \quad (5.1.36)$$

Let us move the lines of integration to the cases when $a = b = 0$. One must not forget the (simple) poles, isolating the number -1 (which may, or not, be a pole of M_∞, certainly not of N_∞) and, adding the 4 equations, one obtains an identity which, in the domain where $x \neq 0$, reads

$$f_0(\xi) + g_0(x)\delta(\xi)$$
$$= \frac{1}{4}\mathrm{Res}_{\mu=-1}M_\infty(\mu) + \sum_j C_j\left[\zeta(-\mu_j)\,|\xi|^{-\mu_j-1} + \zeta(1-\mu_j)\,|x|^{-\mu_j}\delta(\xi)\right]$$
$$+ \frac{1}{8\pi}\int_{-\infty}^{\infty}\Psi(i\lambda)\left[\zeta(-i\lambda)|\xi|^{-1-i\lambda} + \zeta(1-i\lambda)|x|^{-i\lambda}\delta(\xi)\right]d\lambda. \quad (5.1.37)$$

However, what we really wish to obtain is not the pointwise decomposition of $f_0(\xi) + g_0(x)\delta(\xi)$ for $x \neq 0$, but the decomposition of this distribution into homogeneous components in $\mathcal{S}'(\mathbb{R}^2)$. Under the given assumptions, one has $\Psi(0) = 0$, and the integral term is meaningful as a distribution. But, looking at the second integral from the list (5.1.36), one must take into account the fact that the distribution $|x|^{-\mu}$ has at $\mu = 1$ a pole with residue $-2\delta(x)$, so that one must add to the right-hand side of (5.1.37) the extra term $-\frac{1}{2} N_0(1)\,\delta$, with $\delta = \delta(x)\delta(\xi)$.

If one sets

$$\mathfrak{T}(x,\xi) = \mathfrak{S}(x,\xi) - C_\infty + C_0\delta - \sum_{\mu_j \neq -1} C_j \mathfrak{E}_{\mu_j}(x,\xi) - \frac{1}{8\pi} \int_{-\infty}^{\infty} \Psi(i\lambda)\,\mathfrak{E}_{i\lambda}(x,\xi)\,d\lambda,$$

(5.1.38)

the distribution $\mathfrak{T}$ is automorphic and admits the Fourier expansion

$$\mathfrak{T}(x,\xi) = \sum_{k \neq 0} \widetilde{f_k}(\xi)\,e^{2i\pi \frac{kx}{\xi}},$$

(5.1.39)

with

$$\widetilde{f_k}(\xi) = f_k(\xi) - \sum_{j \text{ with } \mu_j \neq -1} C_j\,\sigma_{\mu_j}(|k|)\,|\xi|^{-1-\mu_j} - \frac{1}{8\pi} \int_{-\infty}^{\infty} \sigma_{i\lambda}(|k|)\,\Psi(i\lambda)\,|\xi|^{-1-i\lambda}d\lambda.$$

(5.1.40)

To ensure summability of the last integral, we use the (b, M)-trick from Remark 3.2.1 and we rewrite it as

$$\left(\xi\frac{d}{d\xi} + 1 - b\right)^M \cdot \frac{1}{8\pi} \int_{-\infty}^{\infty} \sigma_{i\lambda}(|k|)\,\Psi(i\lambda)\,|\xi|^{-1-i\lambda}(b - i\lambda)^{-M}d\lambda$$

for a sufficiently large M, and b such that $b > \delta + 1$ and $b + \operatorname{Re} \mu_j > 0$ for every j, a choice to be justified presently. Letting the operator $\left(\xi\frac{d}{d\xi} + 1 - b\right)^M$ act on $f_k(\xi)$ has the same effect as letting, at the end, the operator $(2i\pi\mathcal{E} - b)^M$ act on the product of $f_k(\xi)$ by $e^{2i\pi \frac{kx}{\xi}}$. Next, we observe that if $|f(\xi)| \leq C\left(|\xi| + |\xi|^{-1}\right)^\delta$, the function (cf. (3.2.4))

$$\left[\left(\xi\frac{d}{d\xi} + 1 - b\right)^{-1} f\right](\xi) = -\int_1^\infty t^{-b} f(t\xi)\,dt,$$

(5.1.41)

well-defined if $b > \delta + 1$, satisfies the same estimate as f. This action of the resolvent of the operator $\xi\frac{d}{d\xi}$ can thus be iterated. Finally, under the constraints indicated about b, one can for every M apply the resolvent $(2i\pi\mathcal{E} - b)^{-M}$ to all terms of the decomposition of the series (5.1.39) obtained from (5.1.40), finding as a result an identity $\mathfrak{T} = (2i\pi\mathcal{E} - b)^M\mathfrak{T}_1$, where the distribution $\mathfrak{T}_1$ is automorphic (it is a continuous superposition of rescaled versions of $\mathfrak{T}$), and given as a series $\sum_{k\neq 0} g_k(\xi)\,e^{2i\pi \frac{kx}{\xi}}$, where the functions g_k satisfy for some pair (N', δ') the estimates demanded from the family (f_k) in Proposition 5.1.2. Hence, $\mathfrak{T}_1$ can be decomposed as a series, convergent in $\mathcal{S}'(\mathbb{R}^2)$, of Hecke distributions: this concludes the proof of Theorem 5.1.4.　□

5.2   On the product or Poisson bracket of two Hecke eigenforms

As a preparation toward the computation of the sharp product of two Hecke eigenforms, we need to obtain the result of testing the product or Poisson bracket of two Hecke eigenforms against an Eisenstein series. The proof (just Rankin's trick) improves, in the case of the Poisson bracket, on the rather complicated one (it addressed a more general question) given in [35, p.177]. The formula for the first question could also be derived from the computation [24] of the decomposition of the product of a Hecke eigenform by an Eisenstein series; Poisson brackets, on the other hand, are just as useful as pointwise products, but have probably not benefitted of the same consideration from arithmeticians.

Given an automorphic function f in the hyperbolic half-plane with the Fourier series expansion

$$f(x + iy) = \sum_{k \in \mathbb{Z}} A_k(y)\, e^{2i\pi kx}, \tag{5.2.1}$$

we first need to recall the way the density Φ from its Roelcke-Selberg expansion (5.1.1) can be recovered. This is based on the so-called Rankin-Selberg trick, or on improvements making it possible to minimize the assumptions about f. The following version is reproduced from [39, p.89], and might be compared to Lemma 5.3.3, which has the same role in connection to the discrete part of the spectral decomposition of f.

Lemma 5.2.1. *Let f be an automorphic function, such that f and Δf are square-integrable in D. Let (2.1.22) be its Roelcke-Selberg expansion, and let (5.2.1) be its Fourier expansion. The function*

$$C_0^-(\mu) = \frac{1}{8\pi} \int_0^1 A_0(y)\, y^{-\frac{3}{2}}\, \frac{(\pi y)^{-\frac{\mu}{2}}}{\Gamma(-\frac{\mu}{2})}\, dy \tag{5.2.2}$$

is holomorphic in the half-plane Re $\mu < -1$ *and extends as a meromorphic function in the half-plane* Re $\mu < 0$ *with an only possible simple pole at $\mu = -1$. The residue of $C_0^-(\mu)$ at this point is $-\frac{\Phi^0}{4\pi}$, so that, in particular, f is orthogonal to constants if and only if the function C_0^- is holomorphic throughout the half-plane* Re $\mu < 0$. *The function*

$$C_0^+(\mu) = -\frac{1}{8\pi} \int_1^\infty A_0(y)\, y^{-\frac{3}{2}}\, \frac{(\pi y)^{-\frac{\mu}{2}}}{\Gamma(-\frac{\mu}{2})}\, dy \tag{5.2.3}$$

is holomorphic in the half-plane Re $\mu > 0$. *Finally, the function $\lambda \mapsto C_0^-(-\varepsilon + i\lambda) - C_0^+(\varepsilon + i\lambda)$ has, as $\varepsilon \to 0$, a limit in the space $L^2_{\mathrm{loc}}(\mathbb{R})$, which coincides with the function*

$$\lambda \mapsto \frac{1}{8\pi}\, \frac{\pi^{-\frac{i\lambda}{2}}}{\Gamma(-\frac{i\lambda}{2})}\, \Phi(-i\lambda). \tag{5.2.4}$$

Let us observe that, even though the factor $\frac{1}{8\pi}\frac{\pi^{-\frac{\mu}{2}}}{\Gamma(-\frac{\mu}{2})}$ is present both in the functions $C_0^{\pm}(\mu)$ and in (5.2.4), one could not completely dispense with it as it kills poles in the half-plane $\operatorname{Re}\mu > 0$. On the other hand, it is the kind of "Archimedian" factor that occurs quite naturally in analysis in the half-plane, not in automorphic distribution theory.

Consider now two Hecke eigenforms $\mathcal{N}_1$ and $\mathcal{N}_2$, eigenfunctions of the automorphic Laplacian for the eigenvalues $\frac{1-\nu_1^2}{4}$ and $\frac{1-\nu_2^2}{4}$ (ν_1 and ν_2 are pure imaginary), with the Fourier expansions (2.1.23), in which we take a_k or b_k for the coefficient b_k relative to the expansion of $\mathcal{N}_1$ (resp. $\mathcal{N}_2$). The "product L-function" or "convolution L-function" relative to the pair $\mathcal{N}_1, \mathcal{N}_2$ is the one defined for $\operatorname{Re} s$ large by the equation

$$L(s, \mathcal{N}_1 \times \mathcal{N}_2) = \zeta(2s) \sum_{k \geq 1} a_k b_k\, k^{-s}. \tag{5.2.5}$$

The definition extends to the case when one of the Hecke eigenforms (say, normalized in Hecke's way), or both, is replaced by an Eisenstein series $\frac{1}{2}E^{*}_{\frac{1-\nu}{2}}$ (again, this normalization gives in (2.1.18) the value 1 to the coefficient a_1 of $y^{\frac{1}{2}}K_{\frac{\nu}{2}}(2\pi y)\,e^{2i\pi x}$): note that the coefficients $\zeta^{*}(1 \pm \nu)$ of the first two terms of the expansion (2.1.18) play no role in the definition of the product L-function in this case. In any of the two cases just considered, however, the product L-function reduces to a pointwise product of L-functions of a single non-holomorphic modular form, as will be seen immediately after (5.2.26).

Define the parities of $\mathcal{N}_1$ and $\mathcal{N}_2$ as the numbers ε_1 and ε_2 equal to 0 or 1 such that $a_{-k} = (-1)^{\varepsilon_1} a_k$ and $b_{-k} = (-1)^{\varepsilon_2} b_k$. We assume in all this section that $\mathcal{N}_1$ and $\mathcal{N}_2$ are normalized by the conditions $a_1 = 1$, $b_1 = 1$, which implies that all their coefficients are real, and $\overline{\mathcal{N}}_2 = (-1)^{\varepsilon_2}\mathcal{N}_2$.

Proposition 5.2.2. *Assume $|\operatorname{Re}\nu| < 1$. If the Hecke eigenforms $\mathcal{N}_1$ and $\mathcal{N}_2$ have the same parity, one has*

$$\int_{\Gamma\backslash\Pi} E_{\frac{1-\nu}{2}}(z)\,\mathcal{N}_1(z)\,\overline{\mathcal{N}}_2(z)\,dm(z)$$
$$= \frac{\pi^{\nu-1}}{4\,\zeta^{*}(\nu)} L\left(\frac{1-\nu}{2}, \mathcal{N}_1 \times \mathcal{N}_2\right) \Gamma\left(\frac{1-\nu+\nu_1+\nu_2}{4}\right)$$
$$\cdot\, \Gamma\left(\frac{1-\nu+\nu_1-\nu_2}{4}\right) \Gamma\left(\frac{1-\nu-\nu_1+\nu_2}{4}\right) \Gamma\left(\frac{1-\nu-\nu_1-\nu_2}{4}\right). \tag{5.2.6}$$

If they have distinct parities, one has

$$\frac{1}{2} \int_{\Gamma \backslash \Pi} E_{\frac{1-\nu}{2}}(z) \, \{\mathcal{N}_1(z), \overline{\mathcal{N}}_2(z)\} \, dm(z)$$

$$= \frac{\pi^{\nu-1}}{4i \, \zeta^*(\nu)} L\left(\frac{1-\nu}{2}, \mathcal{N}_1 \times \mathcal{N}_2\right) \Gamma\left(\frac{3-\nu+\nu_1+\nu_2}{4}\right)$$

$$\cdot \Gamma\left(\frac{3-\nu+\nu_1-\nu_2}{4}\right) \Gamma\left(\frac{3-\nu-\nu_1+\nu_2}{4}\right) \Gamma\left(\frac{3-\nu-\nu_1-\nu_2}{4}\right). \qquad (5.2.7)$$

If the pair of parities of $\mathcal{N}_1$ and $\mathcal{N}_2$ does not agree with the choice of pointwise product or Poisson bracket in the way indicated above, the corresponding integral is zero.

Proof. The density Φ in the spectral decomposition of the product $\mathcal{N}_1(z)\,\overline{\mathcal{N}}_2(z)$ is given by the equation

$$\Phi(-i\lambda) = \int_{\Gamma \backslash \Pi} E_{\frac{1-i\lambda}{2}}(z) \, \mathcal{N}_1(z) \, \overline{\mathcal{N}}_2(z) \, dm(z). \qquad (5.2.8)$$

Since the product, or Poisson bracket, under consideration is rapidly decreasing as Im $z \to \infty$, the recipe given in Lemma 5.2.1 reduces to the equation

$$\Phi(-i\lambda) = \int_0^\infty A_0(y) \, y^{\frac{-3-i\lambda}{2}} \, dy, \qquad (5.2.9)$$

where $A_0(y)$ is the "constant" (i.e., independent of x) term of the Fourier expansion of $\mathcal{N}_1(z)\,\overline{\mathcal{N}}_2(z)$, and the value of the divergent integral is to be understood as the value when $\nu = i\lambda$ of the analytic continuation of the function $\Phi(-\nu) = \int_0^\infty A_0(y) \, y^{\frac{-3-\nu}{2}} \, dy$, initially defined for Re ν sufficiently negative to ensure summability. One obtains from (2.1.23) that

$$A_0(y) = 2y \sum_{k \geq 1} a_k b_k \, K_{\frac{\nu_1}{2}}(2\pi \, |k| \, y) \, K_{\frac{\nu_2}{2}}(2\pi \, |k| \, y). \qquad (5.2.10)$$

The integral (5.2.9) becomes a so-called Weber-Schafheitlin integral, made explicit in [22, p.101], which we shall quote in greater generality for future reference:

$$\int_0^\infty y^{\frac{-1-\nu}{2}} K_{\frac{\nu_1}{2}}(2\pi \, |k_1| \, y) \, K_{\frac{\nu_2}{2}}(2\pi \, |k_2| \, y) \, dy$$

$$= \frac{\pi^{\frac{\nu-1}{2}}}{8} |k_1|^{\frac{\nu-\nu_2-1}{2}} |k_2|^{\frac{\nu_2}{2}} \left(\Gamma(\frac{1-\nu}{2})\right)^{-1} \Gamma\left(\frac{1+\nu_1+\nu_2-\nu}{4}\right)$$

$$\cdot \Gamma\left(\frac{1+\nu_1-\nu_2-\nu}{4}\right) \Gamma\left(\frac{1-\nu_1+\nu_2-\nu}{4}\right) \Gamma\left(\frac{1-\nu_1-\nu_2-\nu}{4}\right)$$

$$\cdot \, {}_2F_1\left(\frac{1+\nu_1+\nu_2-\nu}{4}, \frac{1-\nu_1+\nu_2-\nu}{4}; \frac{1-\nu}{2}; 1-\left(\frac{k_2}{k_1}\right)^2\right),$$

$$\text{Re} \, (1 \pm \nu_1 \pm \nu_2 - \nu) > 0. \qquad (5.2.11)$$

In our case, the hypergeometric function reduces to 1, and one obtains when $-\mathrm{Re}\,\nu$ is large

$$
\Phi(-\nu) = \frac{1}{4}\frac{\pi^{\frac{\nu-1}{2}}}{\Gamma(\frac{1-\nu}{2})} \cdot \sum_{k\geq 1} a_k b_k\, k^{\frac{\nu-1}{2}} \Gamma\left(\frac{1+\nu_1+\nu_2-\nu}{4}\right)
$$

$$
\cdot\, \Gamma\left(\frac{1+\nu_1-\nu_2-\nu}{4}\right)\Gamma\left(\frac{1-\nu_1+\nu_2-\nu}{4}\right)\Gamma\left(\frac{1-\nu_1-\nu_2-\nu}{4}\right):
$$

$$(5.2.12)$$

using (5.2.5), one obtains (5.2.6).

We make now the computations involving the Poisson bracket (3.3.11), still, up to some point, in greater generality than what is currently needed, writing with k_1 and $k_2 \in \mathbb{Z}$

$$
\frac{1}{2}\left\{y^{\frac{1}{2}}K_{\frac{\nu_1}{2}}(2\pi\,|k_1|\,y)\,e^{2i\pi k_1 x},\ y^{\frac{1}{2}}K_{\frac{\nu_2}{2}}(2\pi\,|k_2|\,y)\,e^{2i\pi k_2 x}\right\}
$$

$$
\cdot\, i\pi k\, y^{\frac{5}{2}} e^{2i\pi(k_1+k_2)x}\left[k_2\frac{\partial}{\partial y}\left(y^{\frac{1}{2}}K_{\frac{\nu_1}{2}}\right)K_{\frac{\nu_2}{2}} + k_1 K_{\frac{\nu_1}{2}}\frac{\partial}{\partial y}\left(y^{\frac{1}{2}}K_{\frac{\nu_2}{2}}\right)\right], \quad (5.2.13)
$$

where the argument of $K_{\frac{\nu_1}{2}}$ (and of related functions in what follows) is $2\pi\,|k_1|\,y$ and that of $K_{\frac{\nu_2}{2}}$ is $2\pi\,|k_1|\,y$. Using [22, p.67]

$$
\frac{\partial}{\partial y}\left(y^{\frac{1}{2}}K_{\frac{\nu_1}{2}}\right) = \frac{1}{2}y^{-\frac{1}{2}}K_{\frac{\nu_1}{2}} - \pi\,|k|\,y^{\frac{1}{2}}\left(K_{\frac{\nu_1-2}{2}} + K_{\frac{\nu_1+2}{2}}\right), \qquad (5.2.14)
$$

one has

$$
\frac{1}{2}\left\{y^{\frac{1}{2}}K_{\frac{\nu_1}{2}}(2\pi\,|k_1|\,y)\,e^{2i\pi k_1 x},\ y^{\frac{1}{2}}K_{\frac{\nu_2}{2}}(2\pi\,|k_2|\,y)\,e^{-2i\pi k_2 x}\right\}
$$

$$
= e^{2i\pi(k_1+k_2)x}\times\left\{\frac{i\pi}{2}y^2(k_1+k_2)K_{\frac{\nu_1}{2}}K_{\frac{\nu_2}{2}}\right.
$$

$$
\left. -\, i\pi^2 y^3\left[|k_1|\,k_2\left(K_{\frac{\nu_1-2}{2}} + K_{\frac{\nu_1+2}{2}}\right)K_{\frac{\nu_2}{2}} + k_1\,|k_2|\,K_{\frac{\nu_1}{2}}\left(K_{\frac{\nu_2-2}{2}} + K_{\frac{\nu_2+2}{2}}\right)\right]\right\}.
$$

$$(5.2.15)$$

We specialize now in the case, of current interest to us, when $k_1 = k$, $k_2 = -k$. The new function $\Phi(-\nu)$ (assuming that $\varepsilon_1 + \varepsilon_2 = 1$) is

$$
\Phi(-\nu) = 2i\pi \sum_{k\geq 1} k\, a_k b_k \int_0^\infty y^{\frac{1-\nu}{2}} K_{\frac{\nu_1}{2}} K_{\frac{\nu_2}{2}}\, dy
$$

$$
-\, 2i\pi^2 \sum_{k\geq 1} k^2 a_k b_k \int_0^\infty y^{\frac{3-\nu}{2}}\left[\left(K_{\frac{\nu_1-2}{2}} + K_{\frac{\nu_1+2}{2}}\right)K_{\frac{\nu_2}{2}} + K_{\frac{\nu_1}{2}}\left(K_{\frac{\nu_2-2}{2}} + K_{\frac{\nu_2+2}{2}}\right)\right] dy.
$$

$$(5.2.16)$$

One uses again the Weber-Schafheitlin integral: set

$$\rho_1 = \frac{3 - \nu + \nu_1 + \nu_2}{2}, \quad \rho_2 = \frac{3 - \nu + \nu_1 - \nu_2}{2},$$

$$\rho_3 = \frac{3 - \nu - \nu_1 + \nu_2}{2}, \quad \rho_4 = \frac{3 - \nu - \nu_1 - \nu_2}{2}. \tag{5.2.17}$$

The first line on the right-hand side of (5.2.16) contributes (just replace ν by $\nu - 2$ in (5.2.11))

$$\frac{i}{4} \frac{\pi^{\frac{\nu-1}{2}}}{\Gamma(\frac{3-\nu}{2})} \sum_{k \geq 1} a_k b_k \, k^{\frac{\nu-1}{2}} \prod_{n=1}^{4} \Gamma(\frac{\rho_n}{2}). \tag{5.2.18}$$

The sum on the second line contributes to $\Phi(-\nu)$ the expression

$$\frac{1}{4i} \frac{\pi^{\frac{\nu-1}{2}}}{\Gamma(\frac{5-\nu}{2})} \sum_{k \geq 1} a_k b_k \, k^{\frac{\nu-1}{2}} \left[\Gamma(\frac{\rho_1}{2}) \Gamma(\frac{\rho_2}{2}) \Gamma(\frac{\rho_3}{2} + 1) \Gamma(\frac{\rho_4}{2} + 1) \right.$$

$$+ \Gamma(\frac{\rho_1}{2} + 1) \Gamma(\frac{\rho_2}{2} + 1) \Gamma(\frac{\rho_3}{2}) \Gamma(\frac{\rho_4}{2})$$

$$\left. + \Gamma(\frac{\rho_1}{2}) \Gamma(\frac{\rho_2}{2} + 1) \Gamma(\frac{\rho_3}{2}) \Gamma(\frac{\rho_4}{2} + 1) + \Gamma(\frac{\rho_1}{2} + 1) \Gamma(\frac{\rho_2}{2}) \Gamma(\frac{\rho_3}{2} + 1) \Gamma(\frac{\rho_4}{2}) \right]. \tag{5.2.19}$$

Since

$$\rho_3 \rho_4 + \rho_1 \rho_2 + \rho_2 \rho_4 + \rho_1 \rho_3 = (3 - \nu)^2, \tag{5.2.20}$$

this contribution reduces to

$$\frac{i}{4} \frac{\pi^{\frac{\nu-1}{2}}}{\Gamma(\frac{3-\nu}{2})} \sum_{k \geq 1} a_k b_k \, k^{\frac{\nu-1}{2}} \Gamma\left(\frac{3 - \nu + \nu_1 + \nu_2}{4}\right) \Gamma\left(\frac{3 - \nu + \nu_1 - \nu_2}{4}\right)$$

$$\cdot \Gamma\left(\frac{3 - \nu - \nu_1 + \nu_2}{4}\right) \Gamma\left(\frac{3 - \nu - \nu_1 - \nu_2}{4}\right). \tag{5.2.21}$$

Since $\frac{(3-\nu)^2}{4\Gamma(\frac{5-\nu}{2})} - \frac{1}{\Gamma(\frac{3-\nu}{2})} = \frac{1}{\Gamma(\frac{1-\nu}{2})}$, we obtain (5.2.7).

The last assertion of Proposition 5.2.2 results from the fact that the product of two Hecke eigenforms with distinct parities, or the Poisson bracket of two Hecke eigenforms of the same parity, changes to its negative under the symmetry $z \mapsto -\bar{z}$. $\qquad\square$

One may cover both formulas by the single one to follow:

$$\int_{\Gamma \backslash \Pi} \frac{1}{2} E^*_{\frac{1-\nu}{2}}(z) \left(\mathcal{N}_1 \underset{j}{\times} \overline{\mathcal{N}}_2 \right)(z) \, dm(z)$$

$$= \frac{\pi^{\nu-1}}{8 \, i^j} L\left(\frac{1-\nu}{2}, \mathcal{N}_1 \times \mathcal{N}_2\right) \prod_{\eta_1, \eta_2 = \pm 1} \Gamma\left(\frac{1 + \eta_1 \nu_1 + \eta_2 \nu_2 - \nu + 2j}{4}\right). \tag{5.2.22}$$

Since the left-hand side is invariant under the change $\nu \mapsto -\nu$, it is an entirely straightforward matter, using (1.1.1), to obtain that

$$L\left(\frac{1-\nu}{2}, \mathcal{N}_1 \times \mathcal{N}_2\right) B_j\left(\frac{1-\nu_1-\nu_2+\nu}{2}\right)B_j\left(\frac{1+\nu_1+\nu_2+\nu}{2}\right)$$

is invariant under the change $(\nu_2, \nu) \mapsto (-\nu_2, -\nu)$. (5.2.23)

A fundamental property of L-functions, and product L-functions alike, is that they admit Eulerian products. Going back to Proposition 2.1.1, denote as (b_k) the set of coefficients entering the Fourier series decomposition (2.1.23) of $\mathcal{N}$. Let $\mathfrak{N}$ be any of the two Hecke distributions above $\mathcal{N}$, with the degree of homogeneity $-1 - \nu = -1 - i\lambda$: then, the set of coefficients (β_k) entering the Fourier series decomposition (2.1.24)

$$\mathfrak{N}(x, \xi) = \frac{1}{2} \sum_{k \in \mathbb{Z}^\times} \beta_k |\xi|^{-\nu_1 - 1} \exp\left(2i\pi k \frac{x}{\xi}\right) \tag{5.2.24}$$

of $\mathfrak{N}$ is given (2.1.26) as $\beta_k = |k|^{\frac{i\lambda}{2}} b_k$. The L-function associated to $\mathcal{N}$ admits the Eulerian expansion

$$L(s, \mathcal{N}) = \prod_p \left(1 - b_p\, p^{-s} + p^{-2s}\right)^{-1} = \prod_p \left[(1 - \theta_p p^{-s})(1 - \theta_p^{-1} p^{-s})\right]^{-1} \tag{5.2.25}$$

if θ_p is any of the two roots of the equation $\theta_p^2 - b_p\theta_p + 1 = 0$. Recall from (2.1.30) that a character χ leading to the Hecke distribution $\mathfrak{N}$ according to the construction in Theorem 1.2.2 could be defined by any collection $(\chi(p))$, in which, for every p, $\chi(p) = p^{\frac{i\lambda}{2}}\theta_p$ for any of the two possible choices of θ_p.

In the case of product L-functions ([4, p.73] or [15, p.231], though these references emphasize, rather, the case of modular forms of holomorphic type), one has with an obvious notation

$$L\left(s, \mathcal{N}_1 \times \mathcal{N}_2\right) = \prod_p \prod_{\eta_1, \eta_2 = \pm 1} \left[1 - (\chi_1(p))^{\eta_1}(\chi_2(p))^{\eta_2} p^{-\frac{1}{2}(\eta_1\nu_1 + \eta_2\nu_2) - s}\right]^{-1}.$$

(5.2.26)

If the Eisenstein series $\frac{1}{2}E^*_{\frac{1-\nu_2}{2}}$ is substituted for $\mathfrak{N}_2$, one must take $\chi_2 = 1$ (cf. Remark 1.2.1(v)), obtaining

$$L\left(s, \mathcal{N}_1 \times \frac{1}{2}E^*_{\frac{1-\nu_2}{2}}\right)$$

$$= \prod_{\eta_2 = \pm 1} \prod_p \left[\left(1 - \chi_1(p)\, p^{-\frac{\nu_1}{2} - \eta_2\frac{\nu_2}{2} - s}\right)\left(1 - \chi_1(p)\, p^{\frac{\nu_1}{2} - \eta_2\frac{\nu_2}{2} - s}\right)\right]^{-1}$$

$$= \prod_{\eta_2 = \pm 1} \prod_p \left[\left(1 - \theta_p^{(1)} p^{-\eta_2\frac{\nu_2}{2} - s}\right)\left(1 - \left(\theta_p^{(1)}\right)^{-1} p^{-\eta_2\frac{\nu_2}{2} - s}\right)\right]^{-1}$$

$$= \prod_{\eta_2 = \pm 1} L\left(\frac{\eta_2 \nu_2}{2} + s, \mathcal{N}_1\right). \tag{5.2.27}$$

Finally,

$$L\left(s, \mathcal{N}_1 \times \frac{1}{2} E^*_{\frac{1-\nu_2}{2}}\right) = L(s + \frac{\nu_2}{2}, \mathcal{N}_1) L(s - \frac{\nu_2}{2}, \mathcal{N}_1) \tag{5.2.28}$$

and, in the same way,

$$L\left(s, \frac{1}{2} E^*_{\frac{1-\nu_1}{2}} \times \frac{1}{2} E^*_{\frac{1-\nu_2}{2}}\right) = L\left(s + \frac{\nu_2}{2}, \frac{1}{2} E^*_{\frac{1-\nu_1}{2}}\right) L\left(s - \frac{\nu_2}{2}, \frac{1}{2} E^*_{\frac{1-\nu_1}{2}}\right)$$
$$= \zeta\left(s + \frac{\nu_2 - \nu_1}{2}\right) \zeta\left(s + \frac{-\nu_2 + \nu_1}{2}\right) \zeta\left(s + \frac{\nu_2 + \nu_1}{2}\right) \zeta\left(s - \frac{\nu_2 + \nu_1}{2}\right). \tag{5.2.29}$$

In view of the computation of the discrete part of the spectral decomposition of sharp products of modular distributions, we need to write the modifications of (5.2.22) obtained when testing a pointwise product, or a Poisson bracket, of two non-holomorphic modular forms, one at least of which is an Eisenstein series, against a Hecke eigenform $\mathcal{N}_{r,\ell}$ rather than an Eisenstein series. No Euler product formula seems to be known for the integral on the fundamental domain of the product of three Hecke eigenforms: cf. however [9, 41, 13] for some formulas close to this. When dealing with a product, or a Poisson bracket, of a Hecke eigenform by an Eisenstein distribution, the following identity takes us back to the result (5.2.22) already obtained:

Proposition 5.2.3. *One has the identity (with $r \geq 1$, and $\mathcal{N}_1$ denoting an arbitrary Hecke eigenform)*

$$\int_{\Gamma \backslash \Pi} \overline{\mathcal{N}}_{r,\ell} \left(\mathcal{N}_1 \times_j \frac{1}{2} E^*_{\frac{1-\nu_2}{2}}\right) dm = \int_{\Gamma \backslash \Pi} \frac{1}{2} E^*_{\frac{1-\nu_2}{2}} \left(\overline{\mathcal{N}}_{r,\ell} \times_j \mathcal{N}_1\right) dm. \tag{5.2.30}$$

Alleviating notation, we prove that, more generally,

Lemma 5.2.4. *Given three non-holomorphic modular forms $\mathcal{N}, \mathcal{N}_1, \mathcal{N}_2$, one at least of which is a cusp-form (for convergence), one has*

$$\int_{\Gamma \backslash \Pi} \overline{\mathcal{N}} \{\mathcal{N}_1, \mathcal{N}_2\} dm = \int_{\Gamma \backslash \Pi} \mathcal{N}_2 \{\overline{\mathcal{N}}, \mathcal{N}_1\} dm. \tag{5.2.31}$$

Proof. We make use of the usual fundamental domain $\{z \colon |\operatorname{Re} z| \leq \frac{1}{2}, |z| \geq 1\}$ of Γ in Π. Writing $\partial_x = \frac{\partial}{\partial x}$, $\partial_y = \frac{\partial}{\partial y}$, one makes the left-hand side of (5.2.31) explicit as

$$\int_{\Gamma \backslash \Pi} \overline{\mathcal{N}} \left[-\partial_y \mathcal{N}_1 . \partial_x \mathcal{N}_2 + \partial_x \mathcal{N}_1 . \partial_y \mathcal{N}_2\right] dx\, dy.$$

Now, one has

$$-\overline{N}\,\partial_y\,\mathcal{N}_1\,.\,\partial_x\,\mathcal{N}_2 = \partial_x\left[-\overline{N}\,\partial_y\,\mathcal{N}_1\,.\mathcal{N}_2\right] + \mathcal{N}_2\,\partial_x\left(\overline{N}\,\partial_y\,\mathcal{N}_1\right),$$
$$\overline{N}\,\partial_x\,\mathcal{N}_1\,.\,\partial_y\,\mathcal{N}_2 = \partial_y\left[\overline{N}\,\partial_x\,\mathcal{N}_1\,.\mathcal{N}_2\right] - \mathcal{N}_2\,\partial_y\left(\overline{N}\,\partial_y\,\mathcal{N}_1\right). \qquad (5.2.32)$$

Adding the two equations, one obtains

$$\int_{\Gamma\backslash\Pi}\overline{N}\left\{\mathcal{N}_1,\mathcal{N}_2\right\}dm = \int_{\Gamma\backslash\Pi}\mathcal{N}_2\left\{\overline{N},\mathcal{N}_1\right\}dm$$
$$+ \int_{\Gamma\backslash\Pi}\left[\partial_y\left(\overline{N}\,\partial_x\,\mathcal{N}_1\,.\mathcal{N}_2\right) - \partial_x\left(\overline{N}\,\partial_y\,\mathcal{N}_1\,.\mathcal{N}_2\right)\right]dx\,dy. \qquad (5.2.33)$$

Setting $F = \overline{N}\,\mathcal{N}_2$, still an automorphic function, we are left with proving that

$$A:\ = \int_{\Gamma\backslash\Pi}\left[-\partial_y\left(F\,\partial_x\mathcal{N}_1\right) + \partial_x\left(F\,\partial_y\mathcal{N}_1\right)\right]dx\,dy = 0. \qquad (5.2.34)$$

Denoting now the partial derivatives of $\mathcal{N}_1$ as $\partial_1\mathcal{N}_1$ and $\partial_2\mathcal{N}_1$ rather than $\partial_x\mathcal{N}_1$ and $\partial_y\mathcal{N}_1$, one has, using first the invariance of automorphic functions under the translation $x + iy \mapsto x + 1 + iy$,

$$A = \int_{-\frac{1}{2}}^{\frac{1}{2}} F(x, \sqrt{1-x^2})\,(\partial_1\mathcal{N}_1)\,(x, \sqrt{1-x^2})\,dx$$
$$- \int_{\frac{\sqrt{3}}{2}}^{1}\left[F(\sqrt{1-y^2}, y)\,(\partial_2\mathcal{N}_1)\,(\sqrt{1-y^2}, y)\right.$$
$$\left. - F(-\sqrt{1-y^2}, y)\,(\partial_2\mathcal{N}_1)\,(-\sqrt{1-y^2}, y)\right]dy. \qquad (5.2.35)$$

Making the change of variable $y = \sqrt{1 - x^2}$ in the integral on the second line, one transforms it into

$$\int_{0}^{\frac{1}{2}} \frac{x}{\sqrt{1-x^2}}\left[F(x, \sqrt{1-x^2})\left(\partial_2\mathcal{N}_1\right)(x, \sqrt{1-x^2})\right.$$
$$\left. - F(-x, \sqrt{1-x^2})\,(\partial_2\mathcal{N}_1)\,(-x, \sqrt{1-x^2})\right]dx.$$

Now, since F is automorphic, one has $F(x, \sqrt{1-x^2}) = F(-x, \sqrt{1-x^2})$; one obtains when taking the "total" derivative with respect to x of the similar equation involving $\mathcal{N}_1$ in place of F the identity

$$(\partial_1\mathcal{N}_1)\,(x, \sqrt{1-x^2}) + (\partial_1\mathcal{N}_1)\,(-x, \sqrt{1-x^2})$$
$$= \frac{x}{\sqrt{1-x^2}}\left[(\partial_2\mathcal{N}_1)\,(x, \sqrt{1-x^2}) - (\partial_2\mathcal{N}_1)\,(-x, \sqrt{1-x^2})\right]. \qquad (5.2.36)$$

The two lines of the expression (5.2.35) of A thus cancel each other. $\qquad\square$

Another application of the lemma is the identity

$$\int_{\Gamma\backslash\Pi} \overline{\mathcal{N}}_{r,\ell} \left(\frac{1}{2}E^*_{\frac{1-\nu_1}{2}} \underset{j}{\times} \frac{1}{2}E^*_{\frac{1-\nu_2}{2}}\right) dm = \int_{\Gamma\backslash\Pi} \frac{1}{2}E^*_{\frac{1-\nu_2}{2}} \left(\overline{\mathcal{N}}_{r,\ell} \underset{j}{\times} \frac{1}{2}E^*_{\frac{1-\nu_1}{2}}\right) dm,$$

(5.2.37)

needed in view of the computation of the left-hand side: both sides reduce to zero unless j corresponds to the parity of $\mathcal{N}_{r,\ell}$. Looking back at the proof of Proposition 5.2.2, we observe that replacing one of the two Hecke eigenforms there by an Eisenstein series does not destroy it. For in the pointwise product or the Poisson bracket of a Hecke eigenform by an Eisenstein series, both decomposed into Fourier series, we must always associate a factor $e^{2i\pi kx}$ from a term of the first series with a factor $e^{-2i\pi kx}$ from a term of the second series in order to contribute to the function denoted as Φ in the proof: the two exceptional terms from the Fourier expansion of the Eisenstein series are thus not relevant to this computation. Just as in (5.2.22), we obtain (with j indicating the parity of $\mathcal{N}_{r,\ell}$)

$$\int_{\Gamma\backslash\Pi} \frac{1}{2}E^*_{\frac{1-\nu_2}{2}} \left(\overline{\mathcal{N}}_{r,\ell} \underset{j}{\times} \frac{1}{2}E^*_{\frac{1-\nu_1}{2}}\right) dm$$

$$= \frac{\pi^{\nu_2-1}}{8\,i^j} L\left(\frac{1-\nu_2}{2}, \overline{\mathcal{N}}_{r,\ell} \times \frac{1}{2}E^*_{\frac{1-\nu_1}{2}}\right) \prod_{\eta_1,\eta=\pm1} \Gamma\left(\frac{1+\eta_1\nu_1-\nu_2+\eta\,i\lambda_r+2j}{4}\right).$$

(5.2.38)

Again, $\overline{\mathcal{N}}_{r,\ell} = (-1)^j \mathcal{N}_{r,\ell}$. Using (5.2.22), we obtain finally

$$\int_{\Gamma\backslash\Pi} \overline{\mathcal{N}}_{r,\ell} \left(\frac{1}{2}E^*_{\frac{1-\nu_1}{2}} \underset{j}{\times} \frac{1}{2}E^*_{\frac{1-\nu_2}{2}}\right) dm = \frac{\pi^{\nu_2-1}}{8} i^j L\left(\frac{1+\nu_1-\nu_2}{2}, \mathcal{N}_{r,\ell}\right)$$

$$\cdot L\left(\frac{1-\nu_1-\nu_2}{2}, \mathcal{N}_{r,\ell}\right) \prod_{\eta_1,\eta=\pm1} \Gamma\left(\frac{1+\eta_1\nu_1-\nu_2+\eta\,i\lambda_r+2j}{4}\right). \quad (5.2.39)$$

5.3   The sharp product of two Hecke distributions

Recall from Lemma 4.1.2 that, setting $A_{\nu,k} = \mathrm{Op}\,(h_{\nu,k})$, a composition such as $A_{\nu_1,k_1}A_{\nu_2,k_2}$, while always well-defined as an operator from $\mathcal{S}(\mathbb{R})$ to $\mathcal{S}'(\mathbb{R})$ if $k_1 + k_2 \neq 0$, is not necessarily so when $k_1 + k_2 = 0$. However, if one multiplies the operator $A_{\nu_1,k_1}A_{\nu_2,k_2}$, on the left or on the right, by $P = \frac{1}{2i\pi}\frac{d}{dx}$, the operator obtained always sends $\mathcal{S}(\mathbb{R})$ to $\mathcal{S}'(\mathbb{R})$: in particular, such is the case for the operator

$$2i\pi\,\mathrm{mad}(P \wedge Q)\,(A_{\nu_1,k_1}A_{\nu_2,k_2}) = 2i\pi\,[P\,A_{\nu_1,k_1}A_{\nu_2,k_2}\,Q - Q\,A_{\nu_1,k_1}A_{\nu_2,k_2}\,P].$$

(5.3.1)

Recall from (3.4.2) that we have generalized our definition of the operator $\mathrm{mad}(P \wedge Q)$ acting on operators A, so as to make it possible to perform on A the conjugation by any metaplectic operator.

Whenever two operators $A_1 = \mathrm{Op}(h^1)$, $A_2 = \mathrm{Op}(h^2)$ are such that $\mathrm{mad}(P \wedge Q)(A_1 A_2)$ is a meaningful operator from $\mathcal{S}(\mathbb{R})$ to $\mathcal{S}'(\mathbb{R})$ in the more general sense just alluded to, we shall denote the symbol of $2i\pi \,\mathrm{mad}(P \wedge Q)(A_1 A_2)$ as $\mathcal{M}(h^1, h^2)$. In the case when $A_1 A_2$ sends $\mathcal{S}(\mathbb{R})$ to $\mathcal{S}'(\mathbb{R})$, so that it has a symbol $h^1 \# h^2$, one has (3.1.14)

$$\mathcal{M}(h^1, h^2) = 2i\pi \mathcal{E}\left[h^1 \# h^2\right]. \tag{5.3.2}$$

Only assuming that $\mathcal{M}(h^1, h^2)$ is well-defined in the above sense, so is the symbol $\mathcal{M}(h^1 \circ g^{-1}, h^2 \circ g^{-1})$ for every $g \in SL(2, \mathbb{R})$, and one has

$$\mathcal{M}(h_1 \circ g^{-1}, h_2 \circ g^{-1}) = [\mathcal{M}(h_1, h_2)] \circ g^{-1} : \tag{5.3.3}$$

this is a consequence of (3.4.1) and (3.4.3). It applies in particular in the case when $A_1 = A_{\nu_1, k_1}$ and $A_2 = A_{\nu_2, k_2}$.

Recall that, in Remark 4.1.1(ii), we introduced the notion of quasi-distribution, meaning in the simplest case a continuous linear form on the space image of $\mathcal{S}(\mathbb{R}^2)$ under the operator $2i\pi\mathcal{E}$: this is precisely the situation we are dealing with in relation with the pair $(h^1, h^2) = (h_{\nu_1, k_1}, h_{\nu_2, k_2})$. In the case when $k_1 + k_2 = 0$, one cannot define $h_{\nu_1, k_1} \# h_{\nu_2, k_2}$ as a distribution, but one can still define it as a quasi-distribution: however, prudence demands that we should denote it as $\mathrm{Sharp}(h^1, h^2)$ rather than $h^1 \# h^2$. Whether $k_1 + k_2 \neq 0$ or not, one always has

$$\mathcal{M}(h^1, h^2) = 2i\pi \mathcal{E}\left[\mathrm{Sharp}(h^1, h^2)\right], \tag{5.3.4}$$

but the expression within brackets on the right-hand side is a quasi-distribution only, the image under $2i\pi\mathcal{E}$ of which is indeed, like the left-hand side, a distribution.

Theorem 5.3.1. *Denote as* $\mathrm{Vect}_\nu(\delta)$ *the Banach space of symbols consisting of series*

$$\mathfrak{S}(x, \xi) = \frac{1}{2} \sum_{k \in \mathbb{Z}^\times} \beta_k \, |\xi|^{-\nu-1} \exp\left(2i\pi k \frac{x}{\xi}\right) \tag{5.3.5}$$

such that the coefficients satisfy the estimate $|\beta_k| \leq C \, |k|^{\frac{\delta}{2}}$ *for some* $C > 0$. *Let* Vect_ν *be the dense subspace consisting of symbols defined by the same series, assuming that only finitely many coefficients are nonzero. Assume that*

$$\delta_1 + \delta_2 - \mathrm{Re}\,(\nu_1 + \nu_2) < 1. \tag{5.3.6}$$

Then, the bilinear map $\mathcal{M} : \mathrm{Vect}_{\nu_1} \times \mathrm{Vect}_{\nu_2} \to \mathcal{S}'(\mathbb{R}^2)$ *extends continuously to the space* $\mathrm{Vect}_{\nu_1}(\delta_1) \times \mathrm{Vect}_{\nu_2}(\delta_2)$. *Moreover,* $\mathcal{M}(\mathfrak{S}_1, \mathfrak{S}_2)$ *is an automorphic distribution if both* $\mathfrak{S}_1$ *and* $\mathfrak{S}_2$ *are.*

Proof. Using Proposition 4.5.4, we must show that it is possible to choose $a < 2$ such that

$$\sum_{k_1, k_2 \in \mathbb{Z}^\times} |k_1|^{\frac{\delta_1}{2}} |k_2|^{\frac{\delta_2}{2}} |k_1|^{\frac{-1 - \mathrm{Re}\,(\nu_1 + \nu_2) - a}{2}} (1 + |k_1 + k_2|)^{-N} < \infty \tag{5.3.7}$$

for some choice of N. This is immediate, writing $|k_2|^{\frac{\delta_2}{2}} \le C\,|k_1|^{\frac{\delta_2}{2}}\,(1+|k_1+k_2|)^{\frac{|\delta_2|}{2}}$. The last assertion is a consequence of (5.3.3).   □

Theorem 5.3.1 will apply in the case of two Eisenstein distributions $\mathfrak{S}_k = \mathfrak{E}_{\nu_k}$ provided that $|\mathrm{Re}\,(\nu_1 \pm \nu_2)| < 1$. Indeed, the k_1th Fourier coefficient of the distribution $\mathfrak{E}_{\nu_1}$ is $\sigma_{\nu_1}(|k_1|)$, a $\mathrm{O}\left(|k_1|^{(\mathrm{Re}\,\nu_1)_++\varepsilon}\right)$, with $(\mathrm{Re}\,\nu_1)_+ = \max(0, \mathrm{Re}\,\nu_1)$: the required convergence follows. On the other hand, as quoted in [14, p.128], there is a deep bound $b_k = \mathrm{O}\left(|k|^{\frac{5}{28}+\varepsilon}\right)$, a step on the Ramanujan-Petersson conjecture, for the Fourier coefficients of any Hecke eigenform $\mathcal{N}_{r,\ell}$: any estimate of the same kind with an exponent $< \frac{1}{4}$ would do just as well for our purposes. From the link (2.1.26) between these coefficients and the coefficients β_k of the Hecke distributions $\mathfrak{N}_{\pm r,\ell}$, it follows that Theorem 5.3.1 applies also in the case of the sharp product of two Hecke distributions. This gives $\mathcal{M}\left(\mathfrak{N}_{r_1,\ell_1},\,\mathfrak{N}_{r_2,\ell_2}\right)$ a meaning as a distribution (alternatively, it gives $\mathrm{Sharp}\left(\mathfrak{N}_{r_1,\ell_1},\,\mathfrak{N}_{r_2,\ell_2}\right)$ a meaning as a quasi-distribution). We address now the question of decomposing $\mathcal{M}\left(\mathfrak{N}_{r_1,\ell_1},\,\mathfrak{N}_{r_2,\ell_2}\right)$ into homogeneous components.

Then, we shall observe that the quasi-distribution $\mathrm{Sharp}\left(\mathfrak{N}_{r_1,\ell_1},\,\mathfrak{N}_{r_2,\ell_2}\right)$ coincides with the restriction to $(2i\pi\mathcal{E})\,\mathcal{S}(\mathbb{R}^2)$ of a genuine distribution. This does not mean that it identifies with it, only with the class of this distribution modulo the addition of an automorphic distribution homogeneous of degree -1, to wit of a multiple of $\mathfrak{E}_0$.

Starting from two modular distributions $\mathfrak{S}_1$ and $\mathfrak{S}_2$, our program is to obtain (for some specific choice of the left-hand side, in the sense just indicated) an identity

$$\mathrm{Sharp}\,(\mathfrak{S}_1,\,\mathfrak{S}_2) = \frac{1}{8\pi}\int_{-\infty}^{\infty}\Omega_{i\lambda}(\mathfrak{S}_1,\,\mathfrak{S}_2)\,\mathfrak{E}_{i\lambda}^{\mathrm{resc}}\,\frac{d\lambda}{\zeta(i\lambda)\zeta(-i\lambda)}$$

$$+\frac{1}{2}\sum_{r\in\mathbb{Z}^\times}\sum_{\ell}\Omega_{r,\ell}(\mathfrak{S}_1,\,\mathfrak{S}_2)\,\frac{\Gamma(\frac{i\lambda_r}{2})\Gamma(-\frac{i\lambda_r}{2})}{\|\mathcal{N}_{|r|,\ell}\|^2}\,\mathfrak{N}_{r,\ell}^{\mathrm{resc}}$$

$$+\ \text{exceptional terms}:\tag{5.3.8}$$

the last line stands for the sum of a finite number of Eisenstein distributions, and will reduce to zero unless both $\mathfrak{S}_1$ and $\mathfrak{S}_2$ are Eisenstein distributions. The rescaling operation has been put into action so as to make the sought-after coefficients $\Omega_{i\lambda}(\mathfrak{S}_1, \mathfrak{S}_2)$ and $\Omega_{r,\ell}(\mathfrak{S}_1, \mathfrak{S}_2)$ slightly nicer: they will be even more so if we apply it to the distributions $\mathfrak{S}_1$ and $\mathfrak{S}_2$ as well. A decomposition such as (5.3.8) will turn out, however, to be available in the case of two Eisenstein distributions $\mathfrak{E}_{\nu_1}$ and $\mathfrak{E}_{\nu_2}$ with $|\mathrm{Re}\,(\nu_1 \pm \nu_2)| < 1$ only under the extra assumption that $\nu_1 \pm \nu_2 \ne 0$: a modification will have to be made in the remaining cases.

Remark 5.3.1. (not indispensable for further reading). The distribution $\mathfrak{B}$ with the spectral decomposition

$$\mathfrak{B} = \frac{1}{4\pi} \int_{-\infty}^{\infty} \mathfrak{E}_{i\lambda} \frac{d\lambda}{\zeta(i\lambda)\zeta(-i\lambda)} + \frac{1}{2} \sum_{r\in\mathbb{Z}^\times} \sum_\ell \frac{\Gamma(\frac{i\lambda_r}{2})\Gamma(-\frac{i\lambda_r}{2})}{\|\mathcal{N}_{|r|,\ell}\|^2} \mathfrak{N}_{r,\ell} \qquad (5.3.9)$$

was introduced [35, p.34] under the name of "Bezout distribution". It has a simple direct definition as $\frac{1}{2}\sum_{g\in\Gamma/\Gamma_\infty^\circ} \mathfrak{b} \circ g^{-1}$, where $\Gamma_\infty^\circ = \{(\begin{smallmatrix}1 & b\\ 0 & 1\end{smallmatrix}) : b \in \mathbb{Z}\}$ and $\mathfrak{b}(x,\xi) = 2^{2i\pi x}\delta(\xi-1)$. However, this series does not converge in $\mathcal{S}'(\mathbb{R}^2)$: to make it convergent, one must apply it, termwise, the operator $\pi^2\mathcal{E}^2(\pi^2\mathcal{E}^2+1)\dots(\pi^2\mathcal{E}^2+(\ell-1)^2)$ for some choice of $\ell \geq 1$, obtaining a true distribution $\mathfrak{B}^\ell$ as a result. This is another case when a preliminary application of some Pochhammer polynomial in $i\pi\mathcal{E}$ or $2i\pi\mathcal{E}$ (possibly reducing to $(2i\pi\mathcal{E})^2$) ensures convergence. The distributions $\mathfrak{B}^\ell$ are invariant under $\mathcal{F}^{\text{symp}}$ and are therefore characterized by the Θ_0-transforms of their rescaled versions. These transforms are [35, p.26] special cases of a family of automorphic functions introduced by Selberg [25].

As soon as both distributions $\mathcal{M}(\mathfrak{S}_1, \mathfrak{S}_2)$ and $\mathcal{M}(\mathfrak{S}_2, \mathfrak{S}_1)$ have been defined, it is useful to set, for $j = 0$ or 1,

$$\mathcal{M}_j(\mathfrak{S}_1, \mathfrak{S}_2) = \frac{1}{2}\left[\mathcal{M}(\mathfrak{S}_1, \mathfrak{S}_2) + (-1)^j \mathcal{M}(\mathfrak{S}_2, \mathfrak{S}_1)\right], \quad j = 0,1. \qquad (5.3.10)$$

We extend the notation to the case of the automorphic quasi-distribution $\text{Sharp}(\mathfrak{S}_1, \mathfrak{S}_2)$, setting of course

$$\text{Sharp}_j(\mathfrak{S}_1, \mathfrak{S}_2) = \frac{1}{2}\left[\text{Sharp}(\mathfrak{S}_1, \mathfrak{S}_2) + (-1)^j \text{Sharp}(\mathfrak{S}_2, \mathfrak{S}_1)\right], \quad j = 0,1.$$
$$(5.3.11)$$

In the remainder of this section, we consider two Hecke distributions $\mathfrak{N}_1$ and $\mathfrak{N}_2$: to simplify notation, we denote as $-1 - \nu_1$ (resp. $-1 - \nu_2$) their degrees of homogeneity, in place of $-1 - i\lambda_{r_1}$ (resp. $-1 - i\lambda_{r_2}$): we denote as $\varepsilon_1 = 0$ or 1 and ε_2 their parities under the map $(x,\xi) \mapsto (-x,\xi)$, which are the same as the parities of the associated Hecke eigenforms. Set (5.2.24)

$$\mathfrak{N}_1(x,\xi) = \frac{1}{2}\sum_{k\in\mathbb{Z}^\times} \alpha_k |\xi|^{-\nu_1-1} \exp\left(2i\pi k\frac{x}{\xi}\right) \qquad (5.3.12)$$

and define in the same way the coefficients β_k of $\mathfrak{N}_2$. As it follows from the proof of Theorem 5.3.1, the symbol $\mathcal{M}(\mathfrak{N}_1, \mathfrak{N}_2)$ lies in $\mathcal{S}'(\mathbb{R}^2)$ and is given as the series, convergent in that space,

$$\mathcal{M}(\mathfrak{N}_1, \mathfrak{N}_2) = \frac{1}{4}\sum_{k_1,k_2\in\mathbb{Z}^\times} \alpha_{k_1}\beta_{k_2} \mathcal{M}(h_{\nu_1,k_1}, h_{\nu_2,k_2}). \qquad (5.3.13)$$

From the identity (3.3.10), and the fact that Eisenstein distributions are invariant under the map J that occurs there, a fact not destroyed by the application of a polynomial in $2i\pi\mathcal{E}$, it follows that, given the Hecke distributions $\mathfrak{N}_1$ and $\mathfrak{N}_2$, it is only for $j \equiv \varepsilon_1 + \varepsilon_2 \bmod 2$ that the automorphic distribution

$\mathcal{M}_j\left(\mathfrak{N}_1,\, \mathfrak{N}_2\right)$ will have Eisenstein distributions in its spectral decomposition. In the next proposition, we compute the continuous (Eisenstein) part only of the distribution $\mathcal{M}_j\left(\mathfrak{N}_1,\, \mathfrak{N}_2\right)$.

Proposition 5.3.2. *Let $\mathfrak{N}_1$ and $\mathfrak{N}_2$ be two Hecke distributions, with the degrees of homogeneity $-1 - \nu_1$ and $1 - \nu_2$ and the parities ε_1 and ε_2; let $j = 0$ or 1. Set* (2.1.25) $\mathcal{N}_1 = \Theta_0 \mathfrak{N}_1^{\mathrm{resc}},\, \mathcal{N}_2 = \Theta_0 \mathfrak{N}_2^{\mathrm{resc}}$. *The symbol $\mathfrak{T}_j\colon\, = \mathcal{M}_j\left(\mathfrak{N}_1,\, \mathfrak{N}_2\right)$ admits a decomposition of the kind*

$$\mathfrak{T}_j = \frac{1}{8\pi} \int_{-\infty}^{\infty} \Psi_j(i\lambda)\, \mathfrak{E}_{i\lambda}d\lambda + \sum_{r \neq 0} \sum_{\ell} \Psi_j^{r,\ell} \mathfrak{N}_{r,\ell}. \tag{5.3.14}$$

The function Ψ_j is zero unless $\varepsilon_1 + \varepsilon_2 \equiv j \bmod 2$, in which case it is given by the equation

$$\Psi_j(i\lambda) = \frac{(-1)^{\varepsilon_2}}{\zeta(i\lambda)\zeta(-i\lambda)}\, 2^{\frac{-1+\nu_1+\nu_2-i\lambda}{2}}\, (-i\lambda)$$

$$\cdot B_j\big(\frac{1+\nu_1+\nu_2+i\lambda}{2}\big)\, B_j\big(\frac{1-\nu_1-\nu_2+i\lambda}{2}\big)\, L\big(\frac{1-i\lambda}{2},\, \mathcal{N}_1 \times \mathcal{N}_2\big). \tag{5.3.15}$$

Proof. The distribution $\mathfrak{T}_j$ will not immediately satisfy the assumptions which will make Theorem 5.1.4 applicable: but, as will be seen later (cf. the "(b, M)-trick" below), it will be proved to be the image under some even polynomial in $2i\pi\mathcal{E}$ of a distribution that does. It is automorphic as a consequence of Theorem 5.3.1, and we must first decompose it as a (Fourier) series of the kind (5.1.3). One has

$$\mathfrak{T}_j = \mathcal{M}_j\left(\mathfrak{N}_1,\, \mathfrak{N}_2\right) = \frac{1}{4} \sum_{k_1, k_2 \in \mathbb{Z}^\times} \alpha_{k_1} \beta_{k_2} \mathcal{M}_j\left(h_{\nu_1, k_1},\, h_{\nu_2, k_2}\right) \tag{5.3.16}$$

and one can apply Theorem 4.6.6 in the case when $k_1 + k_2 \neq 0$, Theorem 4.6.7 when $k_1 + k_2 = 0$. Considering the expansion (5.1.3) relative to $\mathfrak{T}_j$, the coefficients that need the most detailed attention are g_0 and f_0.

To start with, one has

$$f_0(\xi) + g_0(x)\, \delta(\xi) = \frac{1}{4} \sum_{k \in \mathbb{Z}^\times} \alpha_k \beta_{-k}\, \mathcal{M}_j\left(h_{\nu_1, k},\, h_{\nu_2, -k}\right). \tag{5.3.17}$$

Indeed, recall, as explained immediately after (4.1.4), that the exponents k_1 and k_2 simply add up when considering the sharp product of two functions of type $h_{\nu, k}$ or, when needed (which is the case here), the image under $2i\pi\mathcal{E}$ of such a possibly undefined sharp product: the term $f_0(\xi) + g_0(x)\, \delta(\xi)$ is obtained when adding up all terms corresponding to pairs k_1, k_2 such that $k_1 + k_2 = 0$. In that case, the applicable formula is (4.5.21), which gives $\mathcal{M}\left(h_{\nu_1, q_1},\, h_{\nu_2, q_2}\right)$ after its right-hand side has been integrated on the real line with respect to $d\lambda$. We must thus set $q_1 = k, q_2 = -k$, obtaining the identities

$$f_0(\xi) = \frac{(-1)^j}{16\pi} \sum_{k \in \mathbb{Z}^\times} \alpha_k \beta_{-k} \int_{-\infty}^{\infty} 2^{\frac{-1+\nu_1+\nu_2-i\lambda}{2}} (-i\lambda)$$

$$\cdot |k|_j^{\frac{-1-\nu_1-\nu_2+i\lambda}{2}} \Delta_j \left(\frac{1 - \nu_1 - \nu_2 - i\lambda}{2}, i\lambda \right) |\xi|^{-1-i\lambda} d\lambda \quad (5.3.18)$$

and

$$g_0(x)\,\delta(\xi) = \frac{(-1)^j}{16\pi} \sum_{k \in \mathbb{Z}^\times} \alpha_k \beta_{-k} \int_{-\infty}^{\infty} 2^{\frac{-1+\nu_1+\nu_2-i\lambda}{2}} (-i\lambda) \frac{\zeta(1 - i\lambda)}{\zeta(i\lambda)}$$

$$\cdot |k|_j^{\frac{-1-\nu_1-\nu_2-i\lambda}{2}} \Delta_j \left(\frac{1 + \nu_1 - \nu_2 + i\lambda}{2}, -i\lambda \right) |x|^{-i\lambda}\delta(\xi)\,d\lambda. \quad (5.3.19)$$

Some examination of these integrals is necessary. First, recall from (4.4.3) and (1.1.1) that the singularities of a factor $\Delta_j(x, y)$ are located at $x = -j - 2n$, $y = -2n$ or $x + y = 1 + j + 2n$ for some $n = 0, 1, \ldots$: they do not concern us here (they may when we change the contour of integration), except for the one at $\lambda = 0$, which may be considered as being taken care of by the factor $-i\lambda$. However, a simple pole at $\lambda = 0$ will remain in each of the two equations, originating either from the distribution $|\xi|^{-1-i\lambda}$ or from the scalar factor $\zeta(1 - i\lambda)$: but, as shown in the proof of Theorem 4.5.2, we obtain a correct formula for $f_0(\xi) + g_0(x)\,\delta(\xi)$ if we forget about these singularities and make in the two $d\lambda$ (or $d\nu$ with $\nu = i\lambda$) integrals small changes of contours around 0, in two different directions.

Next, the integrand in (5.3.18) or (5.3.19) has at most polynomial increase in λ, as a look at the decomposition (4.4.3) of $\Delta_j(x, y)$ as the sum of three terms will confirm: in the present case, only the one in the middle fails to be rapidly decreasing at infinity. Finally, throughout Chapter 4, we have made it clear that all $d\lambda$-integrals of functions of (x, ξ), or distributions, were to be understood in the weak sense in $\mathcal{S}'(\mathbb{R}^2)$. In this sense, the integral (5.3.18) is convergent: it suffices to write for some large pair (b, M) the usual identity $\langle |\xi|^{-i\lambda}, \phi \rangle = (i\lambda - b)^{-M} \langle |\xi|^{-i\lambda}, (2i\pi\mathcal{E} - b)^M \phi \rangle$: just the same trick will work with the terms (originating from an application of Theorem 4.6.6 rather than Theorem 4.6.7) for which $k_1 + k_2 \neq 0$.

As required by (5.1.33), we must test f_0 on a function $\phi = \phi(\xi)$ such as $|\xi|^\mu \text{char}(-1 \leq \xi \leq 1)$. This is quite possible, but the operation, if done under the integral sign, would not lead to a convergent $d\lambda$-integral. The simplest solution, which will apply just as well to the other terms to be considered, consists in applying Theorem 5.1.4 to the distribution $\mathfrak{R}_M = (2i\pi\mathcal{E} - b)^{-M}(-2i\pi\mathcal{E} - b)^{-M}\mathfrak{T}_j$ as defined by way of inserting in all $d\lambda$-integrals appearing from an application of Theorems 4.6.6 and 4.6.7 the extra factor $(-i\lambda - b)^{-M}(i\lambda - b)^{-M}$: this is again the (b, M)-trick from Remark 3.2.1, and an almost identical one was used around (5.1.41). At the end, it would suffice to write $\mathfrak{T}_j = (-2i\pi\mathcal{E} - b)^M (2i\pi\mathcal{E} - b)^M \mathfrak{R}_M$: since all Eisenstein distributions and Hecke distributions are generalized eigenfunctions of the operator $(\pm 2i\pi\mathcal{E} - b)^M$, finding an expansion of the kind (5.1.4) for $\mathfrak{R}_M$ is just as good as finding one for $\mathfrak{T}_j$. Actually, it is not necessary to worry about M, which would disappear at the end anyway, when applying both an operator and its inverse: applying an operator in the variables (x, ξ) to a function of (x, ξ, λ) and specializing λ to some particular value λ' — we shall see below that the computation of Ψ reduces to doing just that — are of course two operations which commute with each other. In order to alleviate notation, we shall simply forget about it, just remembering that the (b, M)-trick will solve all difficulties of integrability linked to the polynomial behaviour, at infinity, of functions of λ. For a clear understanding of the distinction to be made between an operator $2i\pi\mathcal{E} - b$ with b large and an operator such as $2i\pi\mathcal{E}$ or $2i\pi\mathcal{E} \pm 1$ present as a factor in $P_1(2i\pi\mathcal{E})$, one may if needed have another look at the argument developed in Remark 3.2.1.

Our discussion of the convergence of the expressions for $f_0(\xi)$ and $g_0(x)$ is not over: though the convergence of the k-series of integrals is guaranteed by Theorem 5.3.1, we shall benefit, in a moment, from the observation that a preliminary deformation of contour in the $d\lambda$-integral, replacing it by a $d\nu$-integral on a line Re $\nu = -c$ in the case of $f_0(\xi)$, Re $\nu = c$ in the case of $g_0(x)$, makes it possible to save an extra factor $|k|^{-c}$. Not forgetting the link (2.1.26) between a_k and α_k, or b_k and β_k, one sees that taking c reasonably close to 2 (any number $> \frac{12}{7}$ would do, considering the current state of the Ramanujan-Petersson conjecture), will ensure convergence of the k-series: taking $c < 2$ avoids introducing unwanted poles of the Δ_j factors. Under this change, the distribution $|\xi|^{-1-i\lambda}$ only improves locally, while the distribution $|x|^{-i\lambda}$ certainly deteriorates when becoming $|x|^{-\nu}$ with Re $\nu \leq c$, but its only pole, at $\nu = 1$, is taken care of by the factor $\zeta(i\lambda) = \zeta(\nu)$ present in the denominator of (5.3.19).

After we have replaced the $d\lambda$-integrals involved in (5.3.18) and (5.3.19) by $d\nu$-integrals taken respectively on the above-defined lines Re $\nu = -c$ and Re $\nu = c$, we use (5.2.5) and write in the first case (use (2.1.26))

$$\sum_{k\in\mathbb{Z}^\times} \alpha_k \beta_{-k} |k|_j^{\frac{-1-\nu_1-\nu_2+\nu}{2}} = \sum_{k\in\mathbb{Z}^\times} a_k b_{-k} |k|_j^{\frac{\nu-1}{2}}. \tag{5.3.20}$$

The sum of this series is zero unless $\varepsilon_1 + \varepsilon_2 + j \equiv 0 \bmod 2$, as seen by changing k to $-k$, which confirms an observation made immediately after (5.3.13) since the present calculation will lead to the continuous (Eisenstein) part of the spectral decomposition of $\mathcal{M}_j(\mathfrak{N}_1, \mathfrak{N}_2)$ only. Under the condition $\varepsilon_1 + \varepsilon_2 + j \equiv 0 \bmod 2$, one has

$$\sum_{k\in\mathbb{Z}^\times} \alpha_k \beta_{-k} |k|_j^{\frac{-1-\nu_1-\nu_2+\nu}{2}} = 2(-1)^{\varepsilon_2} (\zeta(1-\nu))^{-1} L\left(\frac{1-\nu}{2}, \mathcal{N}_1 \times \mathcal{N}_2\right). \tag{5.3.21}$$

The same holds in the second case, only changing ν to $-\nu$ on both sides.

Assuming from now on that $\varepsilon_1 + \varepsilon_2 \equiv j \bmod 2$, we obtain after the changes of contour have been made the pair of equations

$$f_0(\xi) = \frac{(-1)^{\varepsilon_1}}{8i\pi} \int_{\mathrm{Re}\,\nu = -c} (\zeta(1-\nu))^{-1} L(\frac{1-\nu}{2}, \mathcal{N}_1 \times \mathcal{N}_2)$$
$$\cdot 2^{\frac{-1+\nu_1+\nu_2-\nu}{2}} (-\nu) \Delta_j(\frac{1-\nu_1-\nu_2-\nu}{2}, \nu) |\xi|^{-1-\nu} d\nu \quad (5.3.22)$$

and

$$g_0(x)\,\delta(\xi) = \frac{(-1)^{\varepsilon_1}}{8i\pi} \int_{\mathrm{Re}\,\nu = c} (\zeta(1+\nu))^{-1} L(\frac{1+\nu}{2}, \mathcal{N}_1 \times \mathcal{N}_2)$$
$$\cdot 2^{\frac{-1+\nu_1+\nu_2-\nu}{2}} (-\nu) \frac{\zeta(1-\nu)}{\zeta(\nu)} \Delta_j(\frac{1+\nu_1-\nu_2+\nu}{2}, -\nu) |x|^{-\nu} \delta(\xi)\, d\nu. \quad (5.3.23)$$

If $\mu \in \mathbb{C}$ is such that $\mathrm{Re}\,\mu > 0$ in the first case, $\mathrm{Re}\,\mu > c$ in the second case, one has if one applies the first line of (5.1.33), next a change of contour taking back the line of integration to the line $\mathrm{Re}\,\nu = 0$, the equations

$$M_0(\mu) = \frac{(-1)^{\varepsilon_1}}{2i\pi} \int_{\mathrm{Re}\,\nu = 0} (\zeta(1-\nu))^{-1} L(\frac{1-\nu}{2}, \mathcal{N}_1 \times \mathcal{N}_2)$$
$$\cdot 2^{\frac{-1+\nu_1+\nu_2-\nu}{2}} (-\nu) \Delta_j(\frac{1-\nu_1-\nu_2-\nu}{2}, \nu) \frac{d\nu}{\mu - \nu} \quad (5.3.24)$$

and

$$N_0(\mu) = \frac{(-1)^{\varepsilon_1}}{2i\pi} \int_{\mathrm{Re}\,\nu = 0} (\zeta(1+\nu))^{-1} L(\frac{1+\nu}{2}, \mathcal{N}_1 \times \mathcal{N}_2)$$
$$\cdot 2^{\frac{-1+\nu_1+\nu_2-\nu}{2}} (-\nu) \frac{\zeta(1-\nu)}{\zeta(\nu)} \Delta_j(\frac{1+\nu_1-\nu_2+\nu}{2}, -\nu) \frac{d\nu}{\mu - \nu}. \quad (5.3.25)$$

The continuations of these two functions to the half-plane $\mathrm{Re}\,\mu > 0$ (nothing needs being done in the first case) are holomorphic there. The formulas giving, for $-\mathrm{Re}\,\mu$ sufficiently negative ($\mathrm{Re}\,\mu < -c$ will do in the first case, $\mathrm{Re}\,\mu < 0$ in the second) the functions $M_\infty(\mu)$ and $N_\infty(\mu)$ in place of $M_0(\mu)$ and $N_0(\mu)$ are the same, except for a global change of sign in the right-hand side, in both cases. Again, the continuations of the last two functions to the half-plane $\mathrm{Re}\,\mu < 0$ are holomorphic there.

Recall that we are not really busying ourselves with these integrals, but with a modified version the effect of which is not having to worry about integrability at infinity on vertical ν-lines. We must now compute explicitly the functions $\frac{M_0(\mu)+M_\infty(\mu)}{\zeta(-\mu)}$ and $\frac{N_0(\mu)+N_\infty(\mu)}{\zeta(1-\mu)}$, as defined in Theorem 5.1.4, and show that they agree on the real line. If $\psi = \psi(\lambda)$ is a continuous function on the pure real line, say a $O(\lambda^{-2})$ at infinity (never mind this condition: (b, M)-trick again), one has for every $\lambda' \in \mathbb{R}$

$$\lim_{\varepsilon \to 0} \left[\int_{-\infty}^{\infty} \frac{\psi(\lambda)}{i\lambda' + \varepsilon - i\lambda} \, d\lambda - \int_{-\infty}^{\infty} \frac{\psi(\lambda)}{i\lambda' - \varepsilon - i\lambda} \, d\lambda \right]$$

$$= \lim_{\varepsilon \to 0} \int_{-\infty}^{\infty} \psi(\lambda) \, \frac{2\varepsilon}{\varepsilon^2 + (\lambda' - \lambda)^2} \, d\lambda = 2\pi \, \psi(\lambda'). \quad (5.3.26)$$

We obtain that, at $\mu = i\lambda$ (after the $d\lambda$-integration has been performed, we change the name λ' of the argument to λ), the function $M_0(\mu) + M_\infty(\mu)$ takes the value

$$(-1)^{\varepsilon_1} \, (\zeta(1 - i\lambda))^{-1} \, L(\frac{1 - i\lambda}{2}, \mathcal{N}_1 \times \mathcal{N}_2)$$

$$\cdot 2^{\frac{-1 + \nu_1 + \nu_2 - i\lambda}{2}} \, (-i\lambda) \, \Delta_j(\frac{1 - \nu_1 - \nu_2 - i\lambda}{2}, i\lambda). \quad (5.3.27)$$

Similarly, $N_0(\mu) + N_\infty(\mu)$ takes at $\mu = i\lambda$ the value

$$(-1)^{\varepsilon_1} \, (\zeta(1 + i\lambda))^{-1} \, L(\frac{1 + i\lambda}{2}, \mathcal{N}_1 \times \mathcal{N}_2)$$

$$\cdot 2^{\frac{-1 + \nu_1 + \nu_2 - i\lambda}{2}} \, (-i\lambda) \, \frac{\zeta(1 - i\lambda)}{\zeta(i\lambda)} \, \Delta_j(\frac{1 + \nu_1 - \nu_2 + i\lambda}{2}, -i\lambda). \quad (5.3.28)$$

Setting

$$M_0(i\lambda) + M_\infty(i\lambda) = (-1)^{\varepsilon_1} 2^{\frac{-1 + \nu_1 + \nu_2 - i\lambda}{2}} \, (-i\lambda) \, \phi_1(i\lambda),$$

$$N_0(i\lambda) + N_\infty(i\lambda) = (-1)^{\varepsilon_1} 2^{\frac{-1 + \nu_1 + \nu_2 - i\lambda}{2}} \, (-i\lambda) \, \phi_2(i\lambda), \quad (5.3.29)$$

we have, using the general identity

$$\Delta_j(x, y) = \frac{B_j(x - y)}{B_j(x)} \frac{\zeta(1 - y)}{\zeta(y)} = (-1)^j B_j(x + y) B_j(1 - x) \frac{\zeta(1 - y)}{\zeta(y)}, \quad (5.3.30)$$

the pair of equations

$$\frac{\phi_1(i\lambda)}{\zeta(-i\lambda)} = \frac{(-1)^j}{\zeta(i\lambda)\zeta(-i\lambda)} L\left(\frac{1 - i\lambda}{2}, \mathcal{N}_1 \times \mathcal{N}_2\right)$$

$$\cdot B_j(\frac{1 - \nu_1 - \nu_2 + i\lambda}{2}) B_j(\frac{1 + \nu_1 + \nu_2 + i\lambda}{2}),$$

$$\frac{\phi_2(i\lambda)}{\zeta(1 - i\lambda)} = \frac{(-1)^j}{\zeta(i\lambda)\zeta(-i\lambda)} L\left(\frac{1 + i\lambda}{2}, \mathcal{N}_1 \times \mathcal{N}_2\right)$$

$$\cdot B_j(\frac{1 + \nu_1 - \nu_2 - i\lambda}{2}) B_j(\frac{1 - \nu_1 + \nu_2 - i\lambda}{2}). \quad (5.3.31)$$

These two functions coincide, as shown in (5.2.23).

All conditions making it possible to apply Theorem 5.1.4 to $\mathfrak{T}_j$ (actually, to the symbol denoted as $\mathfrak{R}_M$, a purely notational complication we dispensed with) are satisfied, and the automorphic distribution $\mathfrak{T}_j$ admits an expansion of the type (5.1.4), in which, from what precedes, $C_\infty = C_0 = 0$ and there are no Eisenstein distributions $\mathfrak{E}_{\mu_j}$. Applying Theorem 5.1.4, we obtain (5.3.15).   □

In order to obtain the discrete part of the spectral decomposition of $\mathfrak{T}_j$, some preparation is needed. We start with the following lemma, proved in [39, p.90-91] or [34, p.66-67], which might be compared to Lemma 5.2.1.

Lemma 5.3.3. *Let f be an automorphic function such that f and Δf are square-integrable in the fundamental domain. Assume that its Fourier expansion is*

$$f(x + iy) = \sum_{k \in \mathbb{Z}} A_k(y) e^{2i\pi k x}$$

and that its Roelcke-Selberg expansion reduces to

$$f(z) = \sum_{r \geq 1} \sum_{\ell} \Phi^{r,\ell} \mathcal{M}_{r,\ell}(z). \tag{5.3.32}$$

Given $k \neq 0$, define

$$c_k(\mu) = \frac{1}{8\pi} \int_0^\infty A_k(y) \, y^{-\frac{3}{2}} \frac{(\pi y)^{-\frac{\mu}{2}}}{\Gamma(-\frac{\mu}{2})} \, dy \tag{5.3.33}$$

when $-1 < \operatorname{Re} \mu < 0$. Then, $c_k(\mu)$ extends as a meromorphic function of μ in the half-plane $\operatorname{Re} \mu > -4$: it has no pole with $\operatorname{Re} \mu < 4$ except the pure imaginary points $\pm i\lambda_r$, $r \geq 1$; all its poles are simple. For every r, the projection of f on the eigenspace of Δ in $L^2(\Gamma \backslash \Pi)$ corresponding to the eigenvalue $\frac{1+\lambda_r^2}{4}$ is given as

$$\sum_{\ell} \Phi^{r,\ell} \mathcal{M}_{r,\ell}(z) = y^{\frac{1}{2}} \sum_{k \neq 0} d_k \, K_{\frac{i\lambda_r}{2}} (2\pi |k| y) \, e^{2i\pi k x} \tag{5.3.34}$$

with

$$d_k = -8\pi |k|^{-\frac{i\lambda_r}{2}} \times \operatorname{Res}_{\mu = i\lambda_r} c_k(\mu). \tag{5.3.35}$$

We shall apply Lemma 5.3.3 in two cases, then benefit from the comparison: with $f = \Theta_0 \mathfrak{T}_j$, or with $f^\natural = \mathcal{N}_1 \times \mathcal{N}_2$. Neither function reduces to the discrete part of its spectral decomposition, but the functions c_k or $c_k^\natural$ associated to the continuous part of their spectral decompositions are regular at the points $\pm i\lambda_r$: their only poles in the domain $\operatorname{Re} \mu < 1$ (this will have to be replaced by the domain $\operatorname{Re} \mu < 1 - |\operatorname{Re}(\nu_1 - \nu_2)|$ when, later, we shall substitute for one, or both, Hecke distributions an Eisenstein distribution) are non-trivial zeros of zeta. This argument, easy because the continuous part of the spectral decomposition of $\mathfrak{T}_j$, hence f, or that of $f^\natural$ is already explicit (from Proposition 5.2.2 in the latter case), has been developed in full in [34, theor. 11.1] and recalled in [39, p.116].

Let us start with the function $f^\natural$. Denoting as (a_k) and (b_k) the families of Fourier coefficients of $\mathcal{N}_1$ and $\mathcal{N}_2$, one has

$$
\mathcal{N}_1 \underset{j}{\times} \mathcal{N}_2
$$
$$
= \sum_{k_1, k_2 \in \mathbb{Z}^\times} a_{k_1} b_{k_2} \left[\left(y^{\frac{1}{2}} K_{\frac{\nu_1}{2}} \left(2\pi \left| k_1 \right| y \right) e^{2i\pi k_1 x} \right) \underset{j}{\times} \left(y^{\frac{1}{2}} K_{\frac{\nu_2}{2}} \left(2\pi \left| k_2 \right| y \right) e^{2i\pi k_2 x} \right) \right].
$$
$$
\tag{5.3.36}
$$

If $j = 0$, the coefficient $A_k^\natural(y)$ of the Fourier expansion of this automorphic function is

$$
A_k^\natural(y) = \sum_{k_1 + k_2 = k} a_{k_1} b_{k_2} \, y \, K_{\frac{\nu_1}{2}} \left(2\pi \left| k_1 \right| y \right) K_{\frac{\nu_2}{2}} \left(2\pi \left| k_2 \right| y \right), \tag{5.3.37}
$$

and the function $c_k^\natural(\mu)$ associated to this coefficient by means of (5.3.33) is, using (5.2.11),

$$
c_k^\natural(\mu) = 2^{-6} \pi^{-\frac{3}{2}} \left[\Gamma(-\frac{\mu}{2}) \Gamma(\frac{1 - \mu}{2}) \right]^{-1} \sum_{k_1 + k_2 = k} a_{k_1} b_{k_2} \, |k_1|^{\frac{\mu - \nu_2 - 1}{2}} |k_2|^{\frac{\nu_2}{2}}
$$
$$
\cdot \, \Gamma\left(\frac{1 + \nu_1 + \nu_2 - \mu}{4} \right) \Gamma\left(\frac{1 + \nu_1 - \nu_2 - \mu}{4} \right)
$$
$$
\cdot \, \Gamma\left(\frac{1 - \nu_1 + \nu_2 - \mu}{4} \right) \Gamma\left(\frac{1 - \nu_1 - \nu_2 - \mu}{4} \right)
$$
$$
\cdot \, {}_2F_1 \left(\frac{1 + \nu_1 + \nu_2 - \mu}{4}, \frac{1 - \nu_1 + \nu_2 - \mu}{4}; \frac{1 - \mu}{2}; 1 - \left(\frac{k_2}{k_1} \right)^2 \right) : \tag{5.3.38}
$$

recall that we start from negative values of Re μ. We are only interested in the poles on the pure imaginary line of the continuation of this function of μ. For fixed k, one has

$$
|k_1|^{\frac{\mu - \nu_2 - 1}{2}} |k_2|^{\frac{\nu_2}{2}} = |k_1|^{\frac{\mu - 1}{2}} \left(1 + O\left(\frac{1}{k_1} \right) \right), \quad 1 - \left(\frac{k_2}{k_1} \right)^2 = O\left(\frac{1}{k_1} \right). \tag{5.3.39}
$$

If we had an extra factor $\frac{1}{k_1}$, the series would be absolutely convergent in a domain crossing the pure imaginary line: hence, when applying Lemma 5.3.3 to the present situation, one may replace $c_k^\natural(\mu)$ by its modification on the right-hand side of the equation

$$c_k^\natural(\mu) \sim 2^{-6}\pi^{-\frac{3}{2}} \left[\Gamma(-\frac{\mu}{2})\Gamma(\frac{1-\mu}{2})\right]^{-1} \sum_{k_1 \in \mathbb{Z}^\times} a_{k_1} b_{k-k_1} |k_1|^{\frac{\mu-1}{2}}$$

$$\cdot \Gamma\left(\frac{1+\nu_1+\nu_2-\mu}{4}\right)\Gamma\left(\frac{1+\nu_1-\nu_2-\mu}{4}\right)$$

$$\cdot \Gamma\left(\frac{1-\nu_1+\nu_2-\mu}{4}\right)\Gamma\left(\frac{1-\nu_1-\nu_2-\mu}{4}\right). \tag{5.3.40}$$

Let us introduce, for $j = 0$ or 1, the series

$$S_k^j(\nu_1, \nu_2; \mu) = \sum_{k_1 \in \mathbb{Z}^\times} a_{k_1} b_{k-k_1} |k_1|_j^{\frac{\mu-1}{2}}. \tag{5.3.41}$$

Applying Lemma 5.3.3, one obtains that the series $S_k^0(\nu_1, \nu_2; \mu)$ extends as a meromorphic function in the half-plane Re $\mu > -4$, and that it has no pole with Re $\mu < 4$, except the pure imaginary points $\pm i\lambda_r$, $r \geq 1$: these poles are simple. Finally, since the discrete part of the spectral decomposition of $\mathcal{N}_1\mathcal{N}_2$ is

$$(\mathcal{N}_1\mathcal{N}_2)^{\mathrm{disc}} = \sum_{r \geq 1}\sum_\ell \|\mathcal{N}_{r,\ell}\|^{-2} \left(\mathcal{N}_{r,\ell} \,|\, \mathcal{N}_1\mathcal{N}_2\right)_{L^2(\Gamma\backslash\Pi)} \mathcal{N}_{r,\ell}, \tag{5.3.42}$$

one obtains, this time with $r \in \mathbb{Z}^\times$,

$$\mathrm{Res}_{\mu=i\lambda_r} S_k^0(\nu_1, \nu_2; \mu)$$

$$= -8\pi^{\frac{1}{2}}|k|^{\frac{i\lambda_r}{2}} \frac{\Gamma(-\frac{i\lambda_r}{2})\Gamma(\frac{1-i\lambda_r}{2})}{\Gamma\left(\frac{1+\nu_1+\nu_2-i\lambda_r}{4}\right)\Gamma\left(\frac{1+\nu_1-\nu_2-i\lambda_r}{4}\right)\Gamma\left(\frac{1-\nu_1+\nu_2-i\lambda_r}{4}\right)\Gamma\left(\frac{1-\nu_1-\nu_2-i\lambda_r}{4}\right)}$$

$$\cdot \sum_\ell \|\mathcal{N}_{|r|,\ell}\|^{-2} \left[\int_{\Gamma\backslash\Pi} \overline{\mathcal{N}}_{|r|,\ell}(z)\,\mathcal{N}_1(z)\,\mathcal{N}_2(z)\,dm(z)\right] C_{k,|r|,\ell}, \tag{5.3.43}$$

where $C_{k,|r|,\ell}$ is the kth Fourier coefficient of the Hecke eigenform $\mathcal{N}_{|r|,\ell}$.

We wish now, with $\kappa = 0$ or 1, to compute the coefficients $\Psi_j^{r,\ell}$ in Proposition 5.3.2. Do not confuse the indexes j and κ: the first one refers to the fact that we interest ourselves in the commutative, or anticommutative, part of a sharp product, while the second refers to the fact that we use Θ_κ to move from distributions in the plane to functions in the half-plane. In view of Corollary 5.1.3, these coefficients are given in terms of the coefficients $B_\kappa^{r,\ell}$ ($\kappa = 0$ or 1) in the identities

$$\Theta_\kappa \mathfrak{T}_j = \sum_{r \geq 1}\sum_\ell B_\kappa^{r,\ell} \mathcal{N}_{r,\ell}, \quad \kappa = 0, 1, \tag{5.3.44}$$

by the equations

$$\Psi^{r,\ell} = 2^{-2-\frac{i\lambda_r}{2}}\left(B_0^{|r|,\ell} - \frac{1}{i\lambda_r}B_1^{|r|,\ell}\right), \quad r \in \mathbb{Z}^\times. \tag{5.3.45}$$

We assume that the Hecke distributions $\mathfrak{N}_1$ and $\mathfrak{N}_2$ are linked to the Hecke eigenforms $\mathcal{N}_1$ and $\mathcal{N}_2$ by the equations (2.1.25), in other words

$$\mathcal{N}_1 = 2^{-1-\frac{\nu_1}{2}} \Theta_0 \mathfrak{N}_1, \quad \mathcal{N}_2 = 2^{-1-\frac{\nu_2}{2}} \Theta_0 \mathfrak{N}_2, \tag{5.3.46}$$

so that their Fourier coefficients (α_k) and (β_k), defined so that $\mathfrak{N}_1(x,\xi)$ $= \frac{1}{2} \sum_{k_1 \in \mathbb{Z}^\times} \alpha_{k_1} |\xi|^{-\nu_1-1} \exp\left(2i\pi \frac{k_1 x}{\xi}\right)$ (cf. (5.3.12)) and that a similar equation should hold with $\mathfrak{N}_2$, are given (2.1.26) as $\alpha_k = |k|^{\frac{\nu_1}{2}} a_k$, $\beta_k = |k|^{\frac{\nu_1}{2}} b_k$. Then, one has

$$\mathfrak{T}_j = \sum_{k_1, k_2 \in \mathbb{Z}^\times} \alpha_{k_1} \beta_{k_2} (2i\pi\mathcal{E}) \left(h_{\nu_1, k_1} \underset{j}{\#} h_{\nu_2, k_2} \right) \tag{5.3.47}$$

and, in view of our computation, we may restrict the summation by imposing the additional condition $k_1 + k_2 \neq 0$. It is only the presence of infinitely many terms in this sum that is responsible for the appearance of discrete (Hecke) terms in the spectral decomposition of $\mathfrak{T}_j$, so we may as well, when computing the individual Fourier coefficients of $\Theta_0\mathfrak{T}_j$, throw away all terms with $k_1 k_2 > 0$. Finally, we are left with applying to each remaining term of the sum (5.3.47) the equation (4.5.1) or, preferably, (./.).

Recall (4.6.3) that

$$\Theta_0 \left(\mathcal{F}^{\mathrm{symp}} h_{-\nu, k} \right) = 2^{\frac{3+\nu}{2}} |k|^{\frac{\nu}{2}} \mathcal{W}_{\nu, k}, \quad \Theta_0 h_{\nu, k} = 2^{\frac{3+\nu}{2}} |k|^{-\frac{\nu}{2}} \mathcal{W}_{\nu, k} \tag{5.3.48}$$

if one sets

$$\mathcal{W}_{\pm \nu, k}(x + iy) = y^{\frac{1}{2}} K_{\frac{\nu}{2}}(2\pi |k| y) e^{2i\pi kx}. \tag{5.3.49}$$

It follows that, for $k \neq 0$, the coefficient $A_k(y)$ of $e^{2i\pi kx}$ in the Fourier expansion of $(\Theta_0 \mathfrak{T}_j)(x+iy)$ is, up to terms which will not contribute to the residues of interest,

$$A_k(y) \sim \frac{1}{i} \int_{\mathrm{Re}\,\nu=0} G_k^j(\nu_1, \nu_2; \nu; y) \, d\nu, \tag{5.3.50}$$

with

$$G_k^j(\nu_1, \nu_2; \nu; y) = \frac{(-1)^j}{4\pi} 2^{\frac{2+\nu_1+\nu_2}{2}} (-\nu) y^{\frac{1}{2}} K_{\frac{\nu}{2}}(2\pi |k| y)$$

$$\cdot \sum_{\substack{k_1 + k_2 = k \\ k_1 k_2 < 0}} \alpha_{k_1} \beta_{k_2} |k_1|_j^{\frac{-1-\nu_1-\nu_2}{2}} [H_{k, k_1}(\nu_1, \nu_2; \nu) + H_{k, k_1}(-\nu_1, \nu_2; -\nu)], \tag{5.3.51}$$

and

$$H_{k, k_1}(\nu_1, \nu_2; \nu) = \left| \frac{k_1}{k} \right|^{\frac{\nu}{2}} \Delta_j \left(\frac{1 - \nu_1 - \nu_2 - \nu}{2}, \nu \right)$$

$$\cdot {}_2F_1 \left(\frac{1 - \nu_1 + \nu_2 - \nu}{2}, \frac{1 + \nu_1 + \nu_2 - \nu}{2}; 1 - \nu; \frac{k}{k_1} \right). \tag{5.3.52}$$

Just as in the proof of Proposition 5.3.2, we increase the k_1-summability, replacing the line Re $\nu = 0$ by the line Re $\nu = -c$ (resp. Re $\nu = c$) with c large enough to ensure the convergence of the series

$$\widetilde{S}_k^j(\nu_1, \nu_2; \nu) = \sum_{\substack{k_1 + k_2 = k \\ k_1 k_2 < 0}} \alpha_{k_1} \beta_{k_2} |k_1|_j^{\frac{-1-\nu_1-\nu_2+\nu}{2}} \tag{5.3.53}$$

on the line Re $\nu = -c$: it suffices to take $c > 1$, at the same time taking $c < 2$ so as to avoid unnecessary poles of the functions Δ_j. Observe that, except for the change $\mu \mapsto \nu$, this series is, because of (2.1.26), an inessential modification of the series $S_k^j(\nu_1, \nu_2; \nu)$ introduced in (5.3.41): indeed, deleting finitely many terms will not change the poles of residues of its analytic continuation. We are currently dealing with the case of two Hecke distributions, but the same condition $1 < c < 2$ can be arranged, as will be needed later, when we deal instead with two Eisenstein distributions $\frac{1}{2}\mathfrak{E}_{\nu_1}$ and $\frac{1}{2}\mathfrak{E}_{\nu_2}$ and assume that $|\text{Re}\,(\nu_1 \pm \nu_2)| < 1$, since in that case $\alpha_{k_1} = \sigma_{\nu_1}(|k_1|)$.

A Taylor expansion of the hypergeometric functions near the value 0 of the argument shows that, in view of our present investigation, we can replace these functions by 1: for, if saving another factor k_1^{-1} were possible, no change of contour whatsoever would be necessary to ensure the convergence of the series (5.3.53). In order to apply Lemma 5.3.3, we use (again) the integral [22, p.91], in which we assume that $-\text{Re}\,\mu$ is large,

$$\frac{1}{8\pi} \int_0^\infty K_{\frac{\nu}{2}}(2\pi |k| y) \frac{(\pi y)^{-\frac{\mu}{2}}}{\Gamma(-\frac{\mu}{2})} \frac{dy}{y} = \frac{1}{32\pi} \frac{|k|^{\frac{\mu}{2}}}{\Gamma(-\frac{\mu}{2})} \Gamma(\frac{-\mu+\nu}{4}) \Gamma(\frac{-\mu-\nu}{4}). \tag{5.3.54}$$

Set, for $r \in \mathbb{Z}^\times$,

$$\text{pr}_{i\lambda_r} \mathfrak{T}_j = \sum_\ell \Psi_j^{r,\ell} \mathfrak{N}_{r,\ell}. \tag{5.3.55}$$

Using (5.3.35) together with (5.3.51) and (5.3.54), we obtain that the kth Fourier coefficient (i.e., the coefficient of $y^{\frac{1}{2}} K_{\frac{i\lambda_r}{2}}(2\pi |k| y) e^{2i\pi k x}$) of the image under Θ_0 of $\text{pr}_{i\lambda_r} \mathfrak{T}_j$ is the residue at $\mu = i\lambda_r$ of the continuation of the function

$$\frac{(-1)^{j+1}}{i\pi} \frac{2^{-3+\frac{\nu_1+\nu_2}{2}}}{\Gamma(-\frac{\mu}{2})} |k|^{\frac{\mu-i\lambda_r}{2}} [F(\nu_1, \nu_2; \mu) - F(-\nu_1, \nu_2; \mu)], \tag{5.3.56}$$

with

$$F(\nu_1, \nu_2; \mu) = \int_{\text{Re}\,\nu = -c} K(\nu_1, \nu_2; \nu; \mu)\, d\nu := \int_{\text{Re}\,\nu = -c} \Gamma(\frac{-\mu+\nu}{4}) \Gamma(\frac{-\mu-\nu}{4})$$

$$\cdot (-\nu) |k|^{-\frac{\nu}{2}} \Delta_j \left(\frac{1-\nu_1-\nu_2-\nu}{2}, \nu \right) S_k^j(\nu_1, \nu_2; \nu)\, d\nu \tag{5.3.57}$$

(we have changed ν to $-\nu$ and c to $-c$ in the second integral). Note that the integrand is rapidly decreasing at infinity on the vertical lines involved. It is of course the factor $S_k^j(\nu_1, \nu_2; \nu)$ that involves "all the arithmetic" there.

From this point on, we assume that $j = 0$, and it is only in this case that we have completed our application of Lemma 5.3.3 to $\mathcal{N}_1 \times \mathcal{N}_2$. Calculations in the case when $j = 1$ are similar, though more complicated, and have been prepared by the proof of Proposition 5.2.2 since, as already used, the argument of each hypergeometric function involved after we have used the Weber-Schafheitlin integral can be taken to the value 0. However, we shall not impose these new calculations to the reader and we shall instead give later a short incorrect argument, which will at the same time provide some verification of the formula obtained when $j = 0$.

As we have obtained the analytic continuation (at least in the domain $-4 < \text{Re } \nu < 4$) of the function $S_k^0(\nu_1, \nu_2; \mu)$ and its residues in (5.3.43), it is possible to perform changes of contour: we analyse the integral $F(\nu_1, \nu_2; \mu)$ first. It is an analytic function of μ for $\text{Re } \mu < -c$, a domain we may start from: recall that $1 < c < 2$. With $c < c' < 2$, we use the modification γ of the line $\text{Re } \nu = -c$, obtained for some $\varepsilon > 0$ when replacing the segment from $-c + i\lambda_r - i\varepsilon$ to $-c + i\lambda_r + i\varepsilon$ by the "long" piecewise straight line from $-c + i\lambda_r - i\varepsilon$ to $c' + i\lambda_r - i\varepsilon$, next to $c' + i\lambda_r - i\varepsilon$ to $c' + i\lambda_r + i\varepsilon$, finally to $-c + i\lambda_r + i\varepsilon$. Denote as $\mathcal{D}$ the domain, including the initial domain for μ, on the left of γ. For ε small enough, neither γ nor $\mathcal{D}$ will contain any point $i\lambda_s$ with $s \in \mathbb{Z}^\times$, $s \neq r$. We may choose ε so that the poles $1 \pm (\nu_1 + \nu_2)$ of the Delta function in the integrand $K(\nu_1, \nu_2; \nu; \mu)$ of the $d\nu$-integral (5.3.57) defining $F(\nu_1, \nu_2; \mu)$ will not lie on γ: one at most of the two may, however, lie in $\mathcal{D}$, and one may assume that the points ± 1 do not lie in D. When μ moves on the segment from $-c + i\lambda_r$ to $i\lambda_r$, $\pm\mu$ never lies on γ and the two Gamma factors reach no singularity, for ν on γ. As a consequence, the formula obtained after the change of contour has taken place will provide the continuation of the function $F(\nu_1, \nu_2; \mu)$ to some domain containing the initial domain $\{\mu\colon \text{Re } \mu < -c\}$ and reaching the point $i\lambda_r$. So as to obtain the residue of the continuation of $F(\nu_1, \nu_2; \mu)$ at $\mu = i\lambda_r$, we may thus replace the $d\nu$-integral (5.3.57) by $-2i\pi$ times the sum of residues of $K(\nu_1, \nu_2; \nu; \mu)$ at poles inside $\mathcal{D}$.

One has

$$\text{Res}_{\nu=i\lambda_r} K(\ldots) = (-i\lambda_r)\,\Gamma(\frac{-\mu + i\lambda_r}{4})\Gamma(\frac{-\mu - i\lambda_r}{4})\,\Gamma(-\frac{\mu}{2})|k|^{-\frac{i\lambda_r}{2}}$$
$$\cdot \Delta_0\left(\frac{1 - \nu_1 - \nu_2 - i\lambda_r}{2}, i\lambda_r\right)\,\text{Res}_{\nu=i\lambda_r} S_k^0(\nu_1, \nu_2; \nu). \quad (5.3.58)$$

On the other hand, if one of the two points $1 \pm (\nu_1 + \nu_2)$ lies in $\mathcal{D}$, the residue of $K(\nu_1, \nu_2; \nu; \mu)$ at this point is, as a function of μ, just a multiple of $\Gamma(\frac{-\mu+1\pm(\nu_1+\nu_2)}{4})\Gamma(\frac{-\mu-1\mp(\nu_1+\nu_2)}{4})$: this function of μ will thus be regular at $\mu = i\lambda_r$. Finally, the residue of the continuation of $F(\nu_1, \nu_2; \mu)$ at the point $i\lambda_r$ is $-2i\pi$

times the residue at $\mu = i\lambda_r$ of the expression (5.3.58), to wit

$$8i\pi\,(-i\lambda_r)\,\Gamma(-\frac{i\lambda_r}{2})\,|k|^{-\frac{i\lambda_r}{2}}\Delta_0\left(\frac{1-\nu_1-\nu_2-i\lambda_r}{2},\,i\lambda_r\right)\mathrm{Res}_{\nu=i\lambda_r}S_k^0(\nu_1,\nu_2;\nu). \tag{5.3.59}$$

Using (5.3.56), we obtain that the kth Fourier coefficient of $\Theta_0\left(\mathrm{pr}_{i\lambda_r}\mathfrak{T}_0\right)$ is

$$-2^{\frac{\nu_1+\nu_2}{2}}(-i\lambda_r)\,|k|^{-\frac{i\lambda_r}{2}}\,\mathrm{Res}_{\nu=i\lambda_r}S_k^0(\nu_1,\nu_2;\nu)$$
$$\cdot\left[\Delta_0\left(\frac{1-\nu_1-\nu_2-i\lambda_r}{2},\,i\lambda_r\right)-\Delta_0\left(\frac{1+\nu_1-\nu_2-i\lambda_r}{2},\,i\lambda_r\right)\right]: \tag{5.3.60}$$

we have used the fact, obvious from (5.3.43) that changing ν_1 to its negative does not change the residue of interest here. Using (5.3.43), we find that the coefficient of $\mathcal{N}_{|r|,\ell}$ in the spectral decomposition of $\Theta_0\left(\mathrm{pr}_{i\lambda_r}\mathfrak{T}_0\right)$ is

$$2^{\frac{3+\nu_1+\nu_2}{2}}\,\pi^{\frac{1}{2}}(-i\lambda_r)\,\frac{\Gamma(-\frac{i\lambda_r}{2})\Gamma(\frac{1-i\lambda_r}{2})}{\prod_{\eta_1,\eta_2=\pm1}\Gamma(\frac{1+\eta_1\nu_1+\eta_2\nu_2-i\lambda_r}{4})}\,\frac{\left(\mathcal{N}_{|r|,\ell}\,|\,\mathcal{N}_1\mathcal{N}_2\right)_{L^2(\Gamma\backslash\Pi)}}{\|\mathcal{N}_{|r|,\ell}\|^2}$$
$$\cdot\left[\Delta_0\left(\frac{1-\nu_1-\nu_2-i\lambda_r}{2},\,i\lambda_r\right)-\Delta_0\left(\frac{1+\nu_1-\nu_2-i\lambda_r}{2},\,i\lambda_r\right)\right]. \tag{5.3.61}$$

If considering $\Theta_1\left(\mathrm{pr}_{i\lambda_r}\mathfrak{T}_0\right)$ instead, the following modifications are needed. When applying $(./.)$, we must insert there the extra factor $-\nu$, which will appear on the right-hand side of (5.3.51). But let us not forget that, going from the integrals there to the one in (5.3.57), we changed ν to $-\nu$ in the second integral only (so as to have two integrals both taken on the line $\mathrm{Re}\,\nu=-c$): hence, in (5.3.56), we must insert respectively $-\nu$ and ν as factors of $F(\nu_1,\nu_2;\mu)$ and $F(-\nu_1,\nu_2;\mu)$. To obtain the kth Fourier coefficient of $\Theta_1\left(\mathrm{pr}_{i\lambda_r}\mathfrak{T}_0^1\right)$, we must thus, simply, accompany the two Δ_0-terms in (5.3.60) by the factors $-i\lambda_r$ and $i\lambda_r$ respectively. Ultimately, the coefficient of $\mathcal{N}_{|r|,\ell}$ in the spectral decomposition of $\Theta_1\left(\mathrm{pr}_{i\lambda_r}\mathfrak{T}_0\right)$ is given by the modification of (5.3.60) obtained in the way just described.

Using (5.3.45), we obtain, for $r\in\mathbb{Z}^\times$,

$$\Psi^{r,\ell}=2^{\frac{\nu_1+\nu_2-i\lambda_r}{2}}\,\pi^{\frac{1}{2}}(-i\lambda_r)\,\frac{\Gamma(-\frac{i\lambda_r}{2})\Gamma(\frac{1-i\lambda_r}{2})}{\prod_{\eta_1,\eta_2=\pm1}\Gamma(\frac{1+\eta_1\nu_1+\eta_2\nu_2-i\lambda_r}{4})}$$
$$\cdot\frac{\left(\mathcal{N}_{|r|,\ell}\,|\,\mathcal{N}_1\mathcal{N}_2\right)_{L^2(\Gamma\backslash\Pi)}}{\|\mathcal{N}_{|r|,\ell}\|^2}\,\Delta_0\left(\frac{1-\nu_1-\nu_2-i\lambda_r}{2},\,i\lambda_r\right) \tag{5.3.62}$$

or, making the Δ_0 function explicit ((4.4.3) and (1.1.1)),

$$\Psi^{r,\ell}=2^{\frac{\nu_1+\nu_2-i\lambda_r}{2}}\,\pi\,(-i\lambda_r)\,\frac{\Gamma(-\frac{i\lambda_r}{2})\Gamma(\frac{i\lambda_r}{2})}{\prod_{\eta_1,\eta_2=\pm1}\Gamma(\frac{1+\eta_1\nu_1+\eta_2\nu_2+\eta_1\eta_2 i\lambda_r}{4})}$$
$$\cdot\frac{\left(\mathcal{N}_{|r|,\ell}\,|\,\mathcal{N}_1\mathcal{N}_2\right)_{L^2(\Gamma\backslash\Pi)}}{\|\mathcal{N}_{|r|,\ell}\|^2}. \tag{5.3.63}$$

As already indicated, rather than redoing in the case when $j = 1$ the lengthy computations that precede (the question is purely computational: we have shown why similar computations *could* be carried in the new case), we shall give a grossly incorrect very short argument. This will provide at the same time some measure of verification of the result obtained for $j = 0$. One starting point is the identity (3.2.7), and the observation that, in the case when a function f in Π is the sum of a (discrete) series of generalized eigenfunctions of Δ for real eigenvalues $\geq \frac{1}{4}$, say

$$f(z) = \sum_{r \geq 1} g_r(z) \quad \text{with} \quad \Delta g_r = \frac{1 + \lambda_r^2}{4} \text{ and } \lambda_r > 0, \tag{5.3.64}$$

the identity is preserved provided we set

$$f_{\frac{1+\lambda^2}{4}}(z)\, d\lambda = \sum_{r \geq 1} \frac{1}{\pi} \frac{\Gamma(\frac{i\lambda_r}{2})\Gamma(-\frac{i\lambda_r}{2})}{\Gamma(\frac{1+i\lambda_r}{2})\Gamma(\frac{1-i\lambda_r}{2})}\, g_r(z)\, \delta(\lambda - \lambda_r). \tag{5.3.65}$$

Another starting point is Theorem 4.6.6, which asserts that if $h^1 = h_{\nu_1,k_1}$ and $h^2 = h_{\nu_2,k_2}$ with $|\mathrm{Re}\,(\nu_1 \pm \nu_2)| < 1$ and $k_1 k_2 (k_1 + k_2) \neq 0$, one has

$$\Theta_0 \left(h^1 \underset{j}{\#} h^2 \right) = \int_{-\infty}^{\infty} R_j(\nu_1, \nu_2; i\lambda) \left[\Theta_0 h^1 \underset{j}{\times} \Theta_0 h^2 \right]_{\frac{1+\lambda^2}{4}} d\lambda, \tag{5.3.66}$$

if one sets

$$R_j(\nu_1, \nu_2; \nu) = (-i)^j \pi^2\, \frac{\Gamma(\frac{1+\nu}{2})\Gamma(\frac{1-\nu}{2})}{\prod_{\eta_1,\eta_2 = \pm 1} \Gamma(\frac{1 + \eta_1\nu_1 + \eta_2\nu_2 + \eta_1\eta_2\nu + 2j}{4})}. \tag{5.3.67}$$

The identity, a consequence of (5.3.66),

$$\Theta_0 \mathcal{M}_j \left(h^1, h^2 \right) = \int_{-\infty}^{\infty} R_j(\nu_1, \nu_2; i\lambda) \left[\Theta_0 h^1 \underset{j}{\times} \Theta_0 h^2 \right]_{\frac{1+\lambda^2}{4}} (-i\lambda)\, d\lambda \tag{5.3.68}$$

is then valid without having to assume that $k_1 + k_2 \neq 0$.

In view of Theorem 5.3.1, given two Hecke distributions $\mathfrak{N}_1$ and $\mathfrak{N}_2$, the automorphic distribution $\mathcal{M}_j(\mathfrak{N}_1, \mathfrak{N}_2)$ can be defined as a series of functions $\mathcal{M}_j(h_{\nu_1,k_1}, h_{\nu_2,k_2})$ with the appropriate coefficients, which is weakly convergent in $\mathcal{S}'(\mathbb{R}^2)$: moreover, as Θ_0 is a continuous map from $\mathcal{S}'(\mathbb{R}^2)$ to $C^\infty(\Pi)$, we obtain $\Theta_0 \mathcal{M}_j(\mathfrak{N}_1, \mathfrak{N}_2)$ in the form of a convergent series in $C^\infty(\Pi)$. However, this does not imply that the identity (5.3.68) remains valid after one has substituted $\mathfrak{N}_1$ and $\mathfrak{N}_2$ for h^1 and h^2: for the subscript $\frac{1+\lambda^2}{4}$ would now allude to the spectral theory in $L^2(\Gamma \backslash \Pi)$, not that in $L^2(\Pi)$ (which has of course only a purely continuous spectrum). Still, let us finish our calculation along these lines, keeping in mind that this is just to avoid the purely computational part of a calculation the feasibility of which has been properly established.

Still setting $\mathfrak{T}_j = \mathcal{M}_j(\mathfrak{N}_1, \mathfrak{N}_2)$, we may rewrite the extension of (5.3.68) taken for granted, together with the analogous Θ_1-version, as the single equation (where $\kappa = 0$ or 1)

$$\Theta_\kappa \mathfrak{T}_j = 2^{1 + \frac{\nu_1 + \nu_2}{2}} \int_0^\infty (-i\lambda)^{1+\kappa}$$

$$\cdot \left[R_j(\nu_1, \nu_2; i\lambda) + (-1)^\kappa R_j(\nu_1, \nu_2; -i\lambda) \right] \left[\mathcal{N}_1 \underset{j}{\times} \mathcal{N}_2 \right]_{\frac{1+\lambda^2}{4}} d\lambda. \quad (5.3.69)$$

We can then quickly complete Proposition 5.3.2, making in the case of two Hecke distributions the coefficients of the identity (5.3.8) explicit.

Theorem 5.3.4. *Let $\mathfrak{N}_1$ and $\mathfrak{N}_2$ be two Hecke distributions, with the degrees of homogeneity $-1 - \nu_1$ and $1 - \nu_2$ and the parities ε_1 and ε_2; let $j = 0$ or 1. Set $\mathcal{N}_1 = \Theta_0 \mathfrak{N}_1^{\mathrm{resc}}, \mathcal{N}_2 = \Theta_0 \mathfrak{N}_2^{\mathrm{resc}}$ (2.1.25). The symbol $\mathfrak{T}_j = \mathcal{M}_j(\mathfrak{N}_1, \mathfrak{N}_2)$ is the image under the operator $2i\pi\mathcal{E}$ of an automorphic distribution, to be denoted as $\mathrm{Sharp}_j(\mathfrak{N}_1, \mathfrak{N}_2)$, which admits a decomposition of the kind (5.3.8), to wit*

$$\mathrm{Sharp}_j(\mathfrak{N}_1, \mathfrak{N}_2) = \frac{1}{8\pi} \int_{-\infty}^\infty \Omega_{i\lambda}(\mathfrak{N}_1, \mathfrak{N}_2) \, \mathfrak{E}_{i\lambda}^{\mathrm{resc}} \, \frac{d\lambda}{\zeta(i\lambda)\zeta(-i\lambda)}$$

$$+ \frac{1}{2} \sum_{r \in \mathbb{Z}^\times} \sum_\ell \Omega_{r,\ell}^j(\mathfrak{N}_1, \mathfrak{N}_2) \frac{\Gamma(\frac{i\lambda_r}{2})\Gamma(-\frac{i\lambda_r}{2})}{\|\mathcal{N}_{|r|,\ell}\|^2} \mathfrak{N}_{r,\ell}^{\mathrm{resc}}, \quad (5.3.70)$$

in which the coefficients are given as follows: $\Omega_{i\lambda}(\mathfrak{N}_1, \mathfrak{N}_2) = 0$ unless $j \equiv \varepsilon_1 + \varepsilon_2$ mod 2, in which case

$$\Omega_{i\lambda}(\mathfrak{N}_1, \mathfrak{N}_2)$$
$$= (-1)^{\varepsilon_2} 2^{\frac{\nu_1+\nu_2}{2}} B_j\left(\frac{1 + \nu_1 + \nu_2 + i\lambda}{2}\right) B_j\left(\frac{1 - \nu_1 - \nu_2 + i\lambda}{2}\right) L\left(\frac{1 - i\lambda}{2}, \mathcal{N}_1 \times \mathcal{N}_2\right), \quad (5.3.71)$$

and

$$\Omega_{r,\ell}^j(\mathfrak{N}_1, \mathfrak{N}_2) = 2^{2 + \frac{\nu_1+\nu_2}{2}} \pi \prod_{\eta_1, \eta_2 = \pm 1} \left[\Gamma\left(\frac{1 + \eta_1\nu_1 + \eta_2\nu_2 + \eta_1\eta_2 i\lambda_r + 2j}{4}\right) \right]^{-1}$$

$$\cdot \int_{\Gamma \backslash \Pi} \overline{\mathcal{N}}_{|r|,\ell} \left(\mathcal{N}_1 \underset{j}{\times} \mathcal{N}_2 \right) dm. \quad (5.3.72)$$

Proof. The first part follows immediately from (5.3.15), using $\mathfrak{E}_{i\lambda}^{\mathrm{resc}} = 2^{\frac{-1-i\lambda}{2}} \mathfrak{E}_{i\lambda}$. Note that the factor $-i\lambda$, present in (5.3.15), has disappeared to be replaced by an application of the operator $2i\pi\mathcal{E}$.

So far as the discrete part $(\mathfrak{T}_j)^{\mathrm{disc}}$ of the spectral decomposition (5.3.14) of $\mathfrak{T}_j$ is concerned, let us start with the observation that it is no longer true (as was the case when dealing with the continuous part of the decomposition) that it reduces to zero unless $j \equiv \varepsilon_1 + \varepsilon_2$ mod 2 : instead, only Hecke distributions of even type (under the transformation $(x, \xi) \mapsto (-x, \xi)$) will show up on the right-hand side of (5.3.14) if such is the case, and only Hecke distributions of odd type if $j \equiv 1 + \varepsilon_1 + \varepsilon_2$ mod 2. But is is the same conditions that ensure the presence of Hecke eigenforms of a given parity in the spectral decomposition of the automorphic function $\mathcal{N}_1 \underset{j}{\times} \mathcal{N}_2$. Recall, again, that $\Theta_0 \mathfrak{N}_{r,\ell}^{\mathrm{resc}} = \mathcal{N}_{|r|,\ell}$. The

coefficients on the right-hand side of (5.3.14) are characterized by the pair of equations

$$\Theta_0 \left[(\mathfrak{T}_j)^{\mathrm{disc}} \right] = \sum_{r \in \mathbb{Z}^\times} \sum_\ell \Psi_j^{r,\ell} \, 2^{1 + \frac{i\lambda_r}{2}} \mathcal{N}_{|r|,\ell},$$

$$\Theta_1 \left[(\mathfrak{T}_j)^{\mathrm{disc}} \right] = \sum_{r \in \mathbb{Z}^\times} \sum_\ell (-i\lambda_r) \Psi_j^{r,\ell} \, 2^{1 + \frac{i\lambda_r}{2}} \mathcal{N}_{|r|,\ell} \tag{5.3.73}$$

or, reducing the summation to the set $\{r = 1, 2, \dots\}$ and setting $\kappa = 0$ or 1,

$$\Theta_\kappa \, (\mathfrak{T}_j)^{\mathrm{disc}} = \sum_{r \geq 1} \sum_\ell (-i\lambda_r)^\kappa \left[2^{\frac{i\lambda_r}{2}} \Psi_j^{r,\ell} + (-1)^\kappa 2^{-\frac{i\lambda_r}{2}} \Psi_j^{1,-r,\ell} \right] \mathcal{N}_{r,\ell}. \tag{5.3.74}$$

On the other hand, the discrete part of the spectral decomposition of the automorphic function $\mathcal{N}_1 \underset{j}{\times} \mathcal{N}_2$ (present in (2.1.25)) is

$$\left(\mathcal{N}_1 \underset{j}{\times} \mathcal{N}_2 \right)^{\mathrm{disc}} = \sum_{r \geq 1} \sum_\ell \frac{\mathcal{N}_{r,\ell}}{\| \mathcal{N}_{r,\ell} \|^2} \int_{\Gamma \backslash \Pi} \overline{\mathcal{N}}_{r,\ell} \left(\mathcal{N}_1 \underset{j}{\times} \mathcal{N}_2 \right) dm. \tag{5.3.75}$$

Taking advantage of (5.3.65) and setting

$$T_j(\nu_1, \nu_2; \nu) = G_j(\nu_1, \nu_2; \nu) \times \frac{1}{\pi} \frac{\Gamma(\frac{\nu}{2})\Gamma(-\frac{\nu}{2})}{\Gamma(\frac{1+\nu}{2})\Gamma(\frac{1-\nu}{2})}$$

$$= (-i)^j \pi \frac{\Gamma(\frac{\nu}{2})\Gamma(-\frac{\nu}{2})}{\prod_{\eta_1, \eta_2 = \pm 1} \Gamma(\frac{1 + \eta_1 \nu_1 + \eta_2 \nu_2 + \eta_1 \eta_2 \nu + 2j}{4})}, \tag{5.3.76}$$

one can rewrite the discrete part of the identity (5.3.69) as

$$\Theta_\kappa \, (\mathfrak{T}_j)^{\mathrm{disc}} = 2^{1 + \frac{\nu_1 + \nu_2}{2}} \sum_{r \geq 1} \sum_\ell \frac{\mathcal{N}_{r,\ell}}{\| \mathcal{N}_{r,\ell} \|^2} (-i\lambda_r)^{1+\kappa}$$

$$\cdot [T_j(\nu_1, \nu_2; i\lambda_r) + (-1)^\kappa T_j(\nu_1, \nu_2; -i\lambda_r)] \int_{\Gamma \backslash \Pi} \overline{\mathcal{N}}_{|r|,\ell} \left(\mathcal{N}_1 \underset{j}{\times} \mathcal{N}_2 \right) dm. \tag{5.3.77}$$

Comparing this to (5.3.74), one obtains, for $r \in \mathbb{Z}^\times$,

$$\Psi_j^{r,\ell} = 2^{\frac{\nu_1 + \nu_2 - i\lambda_r}{2}} \pi \, (i\lambda_r)$$

$$\cdot T_j(\nu_1, \nu_2; i\lambda_r) \| \mathcal{N}_{|r|,\ell} \|^{-2} \int_{\Gamma \backslash \Pi} \overline{\mathcal{N}}_{|r|,\ell} \left(\mathcal{N}_1 \underset{j}{\times} \mathcal{N}_2 \right) dm. \tag{5.3.78}$$

Equation (5.3.72) follows. Also, in the case when $j = 0$, equation (5.3.63) is confirmed.

Note that one gets rid of the coefficient $2^{2+\frac{\nu_1+\nu_2}{2}}$ if one replaces $\mathfrak{N}_1$ and $\mathfrak{N}_2$ by their rescaled versions, a choice we shall make, in the next section, when dealing with the sharp product of two Eisenstein distributions. $\qquad\square$

5.4 The case of two Eisenstein distributions

What remains to be done is extending Theorem 5.3.4 to the cases when one of the two Hecke distributions, or both, is replaced by an Eisenstein distribution: rather, we shall take half Eisenstein distributions, for a reason explained in Remark 2.1.1(iii). In the first case, there are very small differences only, and we shall not rewrite the theorem, only indicate the modifications to be made to its proof. When aiming at the calculation of $\mathcal{M}_j\left(\mathfrak{N}_1, \frac{1}{2}\mathfrak{E}_{\nu_2}\right)$, where $\mathfrak{N}_1$ is a Hecke distribution (same notation as in (5.3.12)), setting $\beta_k = \sigma_\nu(|k|)$, we observe that the series (5.3.7) will converge for some $a < 2$ (using again the bound $a_k = \mathrm{O}\left(|k|^{\frac{5}{28}+\varepsilon}\right)$ for the Fourier coefficients of the Hecke eigenform $\mathcal{N}_1$) provided that $|\mathrm{Re}\ \nu| < \frac{9}{14}$, in which case Theorem 5.3.1 will be applicable.

The analysis starts exactly as in the case of two Hecke distributions, with the analogue of (5.3.16). There are, however, new terms that must be added to the right-hand side, to wit those which originate from coupling one of the two terms $\frac{1}{2}\zeta(-\nu_2)|\xi|^{-\nu_2-1}$ and $\frac{1}{2}\zeta(1-\nu_2)|x|^{-\nu_2}\delta(\xi)$ of the Fourier expansion (1.1.38) of $\frac{1}{2}\mathfrak{E}_{\nu_2}(x,\xi)$ with any term of the expansion of $\mathfrak{N}_1$. However, we do not get in this way any contribution to the functions denoted as $f_0(\xi)$ and $g_0(x)$ in (5.3.18) and (5.3.19): as a consequence, nothing is changed in the integral term in Theorem 5.3.4, apart from the fact that the function $L\left(\frac{1-i\lambda}{2}, \mathcal{N}_1 \times E^*_{\frac{1-\nu_2}{2}}\right)$ can be expressed if so desired (5.2.28) as the product of two values of the L-function relative to $\mathcal{N}_1$. The new terms will not contribute, either, to the residues the computation of which is the key to that of the discrete part of the spectral decomposition of $\mathcal{M}_j\left(\mathfrak{N}_1, \frac{1}{2}\mathfrak{E}_{\nu_2}\right)$, for the simple reason that, as k_2 is now fixed at the value 0, k_1 is fixed too when $k = k_1 + k_2$ is.

We may thus leave the analysis of $\mathcal{M}_j\left(\mathfrak{N}_1, \frac{1}{2}\mathfrak{E}_{\nu_2}\right)$ at this point, only stating that Theorem 5.3.4 extends without modification in this case. We consider now two Eisenstein distributions $\mathfrak{E}_{\nu_1}$ and $\mathfrak{E}_{\nu_2}$: we have already mentioned that Theorem 5.3.1 applies to this pair provided that $|\mathrm{Re}\ (\nu_1 \pm \nu_2)| < 1$. Keeping the notation (α_k) (resp. (β_k)) for the sequence of coefficients of the Fourier series expansions (1.2.32) of the modular distributions $\frac{1}{2}\mathfrak{E}_{\nu_1}$ and $\frac{1}{2}\mathfrak{E}_{\nu_2}$ (again, the factor $\frac{1}{2}$ gives these coefficients the same role as in the expansion (5.3.12) of a Hecke distribution), we have for $k \in \mathbb{Z}^\times$ the equations $\alpha_k = \sigma_{\nu_1}(|k|)$ and $\beta_k = \sigma_{\nu_2}(|k|)$.

Again, we start from the analogue of (5.3.16), and shall be satisfied with mentioning that the condition $|\mathrm{Re}\,(\nu_1 \pm \nu_2)| < 1$ makes it possible to perform all the changes of contour used in the case of two Hecke distributions: we made such a checking, for instance, immediately before (5.3.36), or shortly after (5.3.53). Before continuing, let us make the observation that the anticommutative part of Sharp $\left(\frac{1}{2}\,\mathfrak{E}_{\nu_1},\, \frac{1}{2}\,\mathfrak{E}_{\nu_2}\right)$ will reduce to a series of Hecke distributions of odd type under the symmetry $(x,\xi) \mapsto (-x,\xi)$: all the other terms of the spectral decomposition we are aiming at will arise from the commutative part (the one associated to the index $j = 0$). There are now some modifications to be made. Let us first eliminate anything that might concern the calculation of the discrete part of the spectral decomposition of $\mathcal{M}_j\left(\frac{1}{2}\,\mathfrak{E}_{\nu_1},\, \frac{1}{2}\,\mathfrak{E}_{\nu_2}\right)$: there is no change at this point, for the trivial reason that fixing $k = k_1 + k_2$ and, at the same time, k_1 or k_2 to the value 0 does not leave us with any series, so that no new poles will occur when applying Lemma 5.3.3. There are 4 new terms, originating from combining any of the two special terms of the Fourier expansion of $\frac{1}{2}\,\mathfrak{E}_{\nu_1}$ with any of the two special terms of the Fourier expansion of $\frac{1}{2}\,\mathfrak{E}_{\nu_2}$. These new terms, as given by Theorem 4.5.5, are the products of $-\nu$, where $-1-\nu$ is the total degree of homogeneity of the distribution under consideration, by the following expressions:

$$\frac{1}{4}\,\zeta(-\nu_1)\zeta(-\nu_2)\,|\xi|^{-2-\nu_1-\nu_2},$$

$$\frac{1}{4}\,\zeta(-\nu_1)\,2^{1+\nu_1}\,\frac{\zeta(\nu_2)\zeta(2+\nu_1-\nu_2)}{\zeta(-1-\nu_1+\nu_2)}\,|x|^{1+\nu_1-\nu_2}\delta(\xi),$$

$$\frac{1}{4}\,\zeta(-\nu_2)\,2^{1+\nu_2}\,\frac{\zeta(\nu_1)\zeta(2-\nu_1+\nu_2)}{\zeta(-1+\nu_1-\nu_2)}\,|x|^{1-\nu_1+\nu_2}\delta(\xi),$$

$$\frac{1}{4}\,2^{\nu_1+\nu_2}\zeta(\nu_1)\zeta(\nu_2)\,|\xi|^{-2+\nu_1+\nu_2}. \qquad (5.4.1)$$

But there is another major difference with the situation dealt with before. In contrast to the function $L\left(\frac{1-\nu}{2}, \mathcal{N}_1 \times \mathcal{N}_2\right)$ which, according to Proposition 5.2.2, is a holomorphic function in the complex plane, the function

$$L\left(\frac{1-\nu}{2},\, \frac{1}{2}E^*_{\frac{1-\nu_1}{2}} \times \frac{1}{2}E^*_{\frac{1-\nu_2}{2}}\right) = \zeta\left(\frac{1+\nu_1+\nu_2-\nu}{2}\right)\zeta\left(\frac{1+\nu_1-\nu_2-\nu}{2}\right)$$
$$\cdot\,\zeta\left(\frac{1-\nu_1+\nu_2-\nu}{2}\right)\zeta\left(\frac{1-\nu_1-\nu_2-\nu}{2}\right) \qquad (5.4.2)$$

(cf. (5.2.29)), which occurs in the equations to take the place of (5.3.22) or (5.3.24), has poles $\nu = -1\pm\nu_1\pm\nu_2$ not to be disregarded when making the change of contour involved between the two equations: similarly, the poles $\nu = 1\pm\nu_1\pm\nu_2$ must be considered when making the change of contour between the equations taking the place of (5.3.23) and (5.3.25). Let us list the poles really present. First, in the part of the proof (taking the place of that of Proposition 5.3.2) concerned with the analysis of the distributions $f_0(\xi)$ and $g_0(x)\,\delta(\xi)$, we may assume that in the change of contour, from $\mathrm{Re}\,\nu = 0$ to $\mathrm{Re}\,\nu = -c$, that ensured the convergence of the series (5.3.20), c, while still < 2, has been chosen $> 1 + |\text{Re } \nu_1| + |\text{Re } \nu_2|$: then, the 4 poles $\nu = -1 \pm \nu_1 \pm \nu_2$ are to be considered in the move from (5.3.22) to (5.3.24), and the 4 poles $\nu = 1 \pm \nu_1 \pm \nu_2$ in the move from (5.3.23) to (5.3.25). Not all, though, for the function $\Delta_0 \left(\frac{1-\nu_1-\nu_2-\nu}{2}, \nu \right)$ vanishes at $\nu = -1 \pm (\nu_1+\nu_2)$: let us not forget that, when aiming at the Eisenstein part (continuous and exceptional) of the spectral decomposition of a sharp product of Eisenstein distributions, we may assume that $j = 0$, as observed immediately after (5.2.37). Hence, only the poles $-1 \pm (\nu_1 - \nu_2)$ are to be considered in the move from (5.3.22) to (5.3.24), and only the poles $1 \pm (\nu_1 + \nu_2)$ in the move from (5.3.23) to (5.3.25). So as to avoid double poles, we shall assume that $\nu_1 \pm \nu_2 \neq 0$, but the remaining cases, assuming only that ν_1 and ν_2 are not both zero, will be treated in Remark 5.4.1 at the very end of the chapter.

We obtain a collection of 4 residues: we shall compute only one of them, say the one at $\nu = -1 - \nu_1 + \nu_2$ since, as will be seen presently, this is really only a verification. Let us write in full the term taking now the place of (5.3.22). It is

$$f_0(\xi) = \frac{1}{8i\pi} \int_{\text{Re } \nu = -c} (\zeta(1-\nu))^{-1} L \left(\frac{1-\nu}{2}, \frac{1}{2} E^*_{\frac{1-\nu_1}{2}} \times \frac{1}{2} E^*_{\frac{1-\nu_2}{2}} \right)$$
$$\cdot 2^{\frac{-1+\nu_1+\nu_2-\nu}{2}} (-\nu) \Delta_0 (\frac{1-\nu_1-\nu_2-\nu}{2}, \nu) |\xi|^{-1-\nu} d\nu. \quad (5.4.3)$$

Provided that $\nu_1 \neq \nu_2$, the residue of the function (5.4.2) there is $-2 \zeta(1+\nu_1)\zeta(1+\nu_1-\nu_2)\zeta(1-\nu_2)$, so that the residue of the integrand on the right-hand side of (5.4.3) is the product of 2ν, taken at $\nu = -1 - \nu_1 + \nu_2$, by the product of 3 values of zeta just computed, and by

$$(\zeta(2+\nu_1-\nu_2))^{-1} 2^{\nu_1} \Delta_0(-\nu_2, -1-\nu_1+\nu_2) |\xi|^{\nu_1-\nu_2}$$
$$= (\zeta(2+\nu_1-\nu_2))^{-1} 2^{\nu_1} \frac{B_0(-\nu_1)}{B_0(1-\nu_2)B_0(-1-\nu_1+\nu_2)} |\xi|^{\nu_1-\nu_2}$$
$$= 2^{\nu_1} \frac{\zeta(-\nu_1)}{\zeta(1+\nu_1)} \frac{\zeta(\nu_2)}{\zeta(1-\nu_2)\zeta(-1-\nu_1+\nu_2)} |\xi|^{\nu_1-\nu_2}. \quad (5.4.4)$$

Overall, moving the line of integration, in the equation taking the place of (5.3.22), from Re $\nu = -c$ to Re $\nu = 0$, we get the extra term

$$(1 + \nu_1 - \nu_2) \times 2^{\nu_1-1} \frac{\zeta(-\nu_1)\zeta(\nu_2)\zeta(1+\nu_1-\nu_2)}{\zeta(-1-\nu_1+\nu_2)} |\xi|^{\nu_1-\nu_2}, \quad (5.4.5)$$

which will contribute to $M_0(\mu)$ the extra term

$$(1 + \nu_1 - \nu_2) \times 2^{\nu_1+1} \frac{\zeta(-\nu_1)\zeta(\nu_2)\zeta(1+\nu_1-\nu_2)}{(\mu+1+\nu_1-\nu_2)\zeta(-1-\nu_1+\nu_2)}, \quad (5.4.6)$$

while this addition to $M_0(i\lambda) + M_\infty(i\lambda)$ will reduce to 0 (trivially). On the other hand, the second term in the list (5.4.1) contributes to $N_0(\mu)$ the extra term

$$(1 + \nu_1 - \nu_2) \times 2^{\nu_1+1} \frac{\zeta(-\nu_1)\zeta(\nu_2)\zeta(2+\nu_1-\nu_2)}{(\mu+1+\nu_1-\nu_2)\zeta(-1-\nu_1+\nu_2)} : \quad (5.4.7)$$

again, the contribution to $N_0(i\lambda) + N_\infty(i\lambda)$ will reduce to zero. It follows from Theorem 5.1.4 that neither of the two additional terms under study can contribute to the continuous part of the spectral decomposition of $\mathcal{M}\left(\frac{1}{2}\,\mathfrak{E}_{\nu_1}, \frac{1}{2}\,\mathfrak{E}_{\nu_2}\right)$. However, both expressions (5.4.6) and (5.4.7) have poles at $\mu = -1 - \nu_1 + \nu_2$, to be denoted as μ_{-1} if one wishes to follow strictly the instructions of Theorem 5.1.4: moreover, as required in (5.1.35), one observes that $\frac{1}{4\zeta(-\mu_{-1})}$ times the residue of (5.4.6) and $\frac{1}{4\zeta(1-\mu_{-1})}$ times the residue of (5.4.7) agree to the common value

$$C_{-1} = (1 + \nu_1 - \nu_2) \times 2^{\nu_1 - 1} \frac{\zeta(-\nu_1)\zeta(\nu_2)}{\zeta(-1 - \nu_1 + \nu_2)}. \tag{5.4.8}$$

It follows that the two terms under consideration, taken together, ultimately contribute to $\mathcal{M}\left(\mathfrak{E}_{\nu_1}, \frac{1}{2}\,\mathfrak{E}_{\nu_2}\right)$ the image under $2i\pi\mathcal{E}$ (this stands for the factor $1 + \nu_1 - \nu_2$) of the distribution $C_{-1}\mathfrak{E}_{-1-\nu_1+\nu_2}$.

The coincidence expressed by the condition (5.1.35) is no accident: it arose from the fact that the term $2^{\nu_1-1} \frac{\zeta(-\nu_1)\zeta(\nu_2)\zeta(1+\nu_1-\nu_2)}{\zeta(-1-\nu_1+\nu_2)} |\xi|^{\nu_1-\nu_2}$ in (5.4.5) is exactly the one needed to complete the second term from the list (5.4.1) so that the two will combine to give the two "special" terms of a multiple of some Eisenstein distribution, to wit $C_{-1}\mathfrak{E}_{-1-\nu_1+\nu_2}$. This had of course to be expected, since $\mathcal{M}\left(\frac{1}{2}\,\mathfrak{E}_{\nu_1}, \frac{1}{2}\,\mathfrak{E}_{\nu_2}\right)$ is an automorphic distribution. The 3 more extra terms originating from residues at the other poles of the function in (5.4.2) produce, ultimately, terms which just complete in the same way the first, third and last term from the list (5.4.1). We obtain the following theorem, in which the simplest coefficients are found in connection with the pair $\mathfrak{E}_{\nu_1}^{\mathrm{resc}}, \mathfrak{E}_{\nu_2}^{\mathrm{resc}}$: one might compare it to [35, p.170] (the function denoted there as $\mathfrak{F}_\nu^\sharp$ is here denoted as $\mathfrak{E}_\nu^{\mathrm{resc}}$), not forgetting, however, that the meaning of the same equation in this reference was much weaker than the one needed here, while its proof was totally different.

Theorem 5.4.1. *Assume that* $|\mathrm{Re}\,(\nu_1 \pm \nu_2)| < 1$ *and* $\nu_1 \pm \nu_2 \neq 0$. *The symbol* $\mathcal{M}\left(\mathfrak{E}_{\nu_1}^{\mathrm{resc}}, \mathfrak{E}_{\nu_2}^{\mathrm{resc}}\right)$ *is the image under the operator* $2i\pi\mathcal{E}$ *of an automorphic distribution, to be denoted as* $\mathrm{Sharp}\left(\mathfrak{E}_{\nu_1}^{\mathrm{resc}}, \mathfrak{E}_{\nu_2}^{\mathrm{resc}}\right)$, *which admits a decomposition*

$$\mathrm{Sharp}\left(\mathfrak{E}_{\nu_1}^{\mathrm{resc}}, \mathfrak{E}_{\nu_2}^{\mathrm{resc}}\right) = \text{exceptional terms}$$
$$+ \frac{1}{8\pi} \int_{-\infty}^{\infty} \Omega_{i\lambda}\left(\mathfrak{E}_{\nu_1}^{\mathrm{resc}}, \mathfrak{E}_{\nu_2}^{\mathrm{resc}}\right) \mathfrak{E}_{i\lambda}^{\mathrm{resc}} \frac{d\lambda}{\zeta(i\lambda)\zeta(-i\lambda)}$$
$$+ \frac{1}{2} \sum_{r\in\mathbb{Z}^\times} \sum_\ell \Omega_{r,\ell}\left(\mathfrak{E}_{\nu_1}^{\mathrm{resc}}, \mathfrak{E}_{\nu_2}^{\mathrm{resc}}\right) \frac{\Gamma(\frac{i\lambda_r}{2})\Gamma(-\frac{i\lambda_r}{2})}{\|\mathcal{N}_{|r|,\ell}\|^2} \mathfrak{N}_{r,\ell}^{\mathrm{resc}},$$
$$\tag{5.4.9}$$

in which the exceptional terms add up to

$$\text{Exc:} = \frac{\zeta(-\nu_1)\zeta(-\nu_2)}{\zeta(-1-\nu_1-\nu_2)}\, \mathfrak{E}^{\text{resc}}_{1+\nu_1+\nu_2} + \frac{\zeta(-\nu_1)\zeta(\nu_2)}{\zeta(-1-\nu_1+\nu_2)}\, \mathfrak{E}^{\text{resc}}_{-1-\nu_1+\nu_2}$$

$$+ \frac{\zeta(\nu_1)\zeta(-\nu_2)}{\zeta(-1+\nu_1-\nu_2)}\, \mathfrak{E}^{\text{resc}}_{-1+\nu_1-\nu_2} + \frac{\zeta(\nu_1)\zeta(\nu_2)}{\zeta(-1+\nu_1+\nu_2)}\, \mathfrak{E}^{\text{resc}}_{1-\nu_1-\nu_2} \quad (5.4.10)$$

and the coefficients are given as follows:

$$\Omega_{i\lambda}\left(\mathfrak{E}^{\text{resc}}_{\nu_1}, \mathfrak{E}^{\text{resc}}_{\nu_2}\right) = 2 \prod_{\eta_1,\eta_2=\pm 1} \zeta\left(\frac{1+\eta_1\nu_1+\eta_2\nu_2+\eta_1\eta_2 i\lambda}{2}\right) \quad (5.4.11)$$

and, denoting as $j = 0$ or 1 the indicator of the parity of $\mathcal{N}_{|r|,\ell}$,

$$\Omega_{r,\ell}\left(\mathfrak{E}^{\text{resc}}_{\nu_1}, \mathfrak{E}^{\text{resc}}_{\nu_2}\right) = \frac{4\pi}{ij} L^{\natural}\left(\frac{1+\nu_1-\nu_2}{2}, \mathfrak{N}_{r,\ell}\right) L^{\natural}\left(\frac{1-\nu_1-\nu_2}{2}, \mathfrak{N}_{-r,\ell}\right),$$

$$(5.4.12)$$

where the function $L^{\natural}$ associated to a Hecke distribution has been defined in (1.2.8).

Proof. Apart from the exceptional terms, the calculation of which has just been detailed, it is just a matter of recopying Theorem 5.3.4, paying attention to the fact that replacing $\frac{1}{2}\mathfrak{E}_\nu$ by $\mathfrak{E}^{\text{resc}}_\nu$ amounts to multiplying it by $2^{\frac{1-\nu}{2}}$: also, one should remember that the image under Θ_0 of $\mathfrak{E}^{\text{resc}}_\nu$ is $E^*_{\frac{1-\nu}{2}}$. Let us give some details in the case of the discrete coefficients, first remembering that only Hecke distributions of the parity associated to j occur in $\mathcal{M}_j\left(\frac{1}{2}\mathfrak{E}_{\nu_1}, \frac{1}{2}\mathfrak{E}_{\nu_2}\right)$. Recopying (5.3.72), one has now

$$\Omega_{r,\ell}\left(\frac{1}{2}\mathfrak{E}_{\nu_1}, \frac{1}{2}\mathfrak{E}_{\nu_2}\right)$$

$$= 2^{2+\frac{\nu_1+\nu_2}{2}} \pi \prod_{\eta_1,\eta_2=\pm 1}\left[\Gamma\left(\frac{1+\eta_1\nu_1+\eta_2\nu_2+\eta_1\eta_2 i\lambda_r + 2j}{4}\right)\right]^{-1}$$

$$\cdot \int_{\Gamma\backslash\Pi} \mathcal{N}_{|r|,\ell}\left(\frac{1}{2}E^*_{\frac{1-\nu_1}{2}}, \times_j \frac{1}{2}E^*_{\frac{1-\nu_2}{2}}\right) dm. \quad (5.4.13)$$

Then, one uses (5.2.39). Simplifying the quotient of two products of four Gamma factors, one is left with the product of $L\left(\frac{1+\nu_1-\nu_2}{2}, \mathcal{N}_{|r|,\ell}\right) L\left(\frac{1-\nu_1-\nu_2}{2}, \mathcal{N}_{|r|,\ell}\right)$ by the quotient of two products of two Gamma factors, which one identifies with a product of two B_j factors with the help of (1.1.1), obtaining the equation

$$\Omega_{r,\ell}\left(\mathfrak{E}^{\text{resc}}_{\nu_1}, \mathfrak{E}^{\text{resc}}_{\nu_2}\right) = \frac{\pi}{ij} B_j\left(\frac{1-\nu_1+\nu_2-i\lambda_r}{2}\right) B_j\left(\frac{1+\nu_1+\nu_2+i\lambda_r}{2}\right)$$

$$\cdot L\left(\frac{1+\nu_1-\nu_2}{2}, \mathcal{N}_{|r|,\ell}\right) L\left(\frac{1-\nu_1-\nu_2}{2}, \mathcal{N}_{|r|,\ell}\right). \quad (5.4.14)$$

With the help of (1.2.8), one transforms (5.4.14) to (5.4.12), one advantage of which is its simpler functional equation (1.2.9). $\qquad\square$

Proposition 5.4.2. *In the case when* $|\mathrm{Re}\,(\nu_1 \pm \nu_2)| < 1$ *and* $\nu_1 \pm \nu_2 = 0$*, Theorem 5.1.4 and the decomposition (5.4.9) remain valid, with the following modifications. If* $\nu_1 - \nu_2 = 0$ *and* $\nu_1 \neq 0$*, the sum of the 2nd and 3rd term of the list* (5.4.10) *must be replaced by*

$$\mathrm{Res}_{\nu=-1}\left[\frac{(\zeta(\frac{1-\nu}{2}))^2\zeta(\frac{1+2\nu_1+\nu}{2})\zeta(\frac{1-2\nu_1+\nu}{2})}{\zeta(\nu)}\,\frac{\mathfrak{E}_\nu^{\mathrm{resc}}}{\zeta(-\nu)}\right]. \tag{5.4.15}$$

If $\nu_1 + \nu_2 = 0$ *and* $\nu_1 \neq 0$*, the sum of the first and 4th term of the list* (5.4.10) *must be replaced by*

$$-\mathrm{Res}_{\nu=1}\left[\frac{(\zeta(\frac{1+\nu}{2}))^2\zeta(\frac{1+2\nu_1-\nu}{2})\zeta(\frac{1-2\nu_1-\nu}{2})}{\zeta(-\nu)}\,\frac{\mathfrak{E}_\nu^{\mathrm{resc}}}{\zeta(\nu)}\right]. \tag{5.4.16}$$

If $\nu_1 = \nu_2 = 0$*, both replacements have to be made.*

Proof. We had to discard the cases when $\nu_1 + \nu_2 = 0$ or $\nu_1 - \nu_2 = 0$, in which the presence of double poles prevents an application of Theorem 5.1.4. But the condition $\nu_1 = \pm\nu_2$ does not prevent the application of Theorem 5.3.1, a consequence of which is also that, as a tempered distribution, $\mathcal{M}\left(\mathfrak{E}_{\nu_1}^{\mathrm{resc}}, \mathfrak{E}_{\nu_2}^{\mathrm{resc}}\right)$ is an analytic function of ν_1, ν_2 in the domain where $|\mathrm{Re}\,(\nu_1 \pm \nu_2)| < 1$. Consider the case when $\nu_1 - \nu_2 = 0$ (the two cases are totally similar). Then, the second (or third) exceptional term in (5.4.1) is $-2^{\nu_1}\pi^2\zeta(-\nu_1)\zeta(\nu_1)\,|x|\,\delta(\xi)$. A look at (1.2.32) shows that $\zeta(2)\,|x|\,\delta(\xi)$ would be the second term in the expansion of $\mathfrak{E}_{-1}(x, \xi)$ if such an Eisenstein distribution did exist: but it does not, since the first term of its expansion would be an infinite constant. Let us start from the decomposition (5.4.9), assuming again that $|\mathrm{Re}\,(\nu_1 \pm \nu_2)| < 1$ and $\nu_1 - \nu_2 \neq 0$ and making the integral term explicit as

$$\frac{1}{4i\pi}\int_{\mathrm{Re}\,\nu=0}\prod_{\eta_1,\eta_2=\pm1}\zeta\left(\frac{1+\eta_1\nu_1+\eta_2\nu_2+\eta_1\eta_2\,\nu}{2}\right)\mathfrak{E}_\nu^{\mathrm{resc}}\,\frac{d\nu}{\zeta(\nu)\zeta(-\nu)}. \tag{5.4.17}$$

Setting $\beta = 2\,|\mathrm{Im}\,(-\nu_1+\nu_2)|$ and choosing c with $1 + |\mathrm{Re}\,(\nu_1-\nu_2)| < c < 2$, let us change the line of integration $i\mathbb{R}$ to the piecewise straight line γ obtained when replacing the part of $i\mathbb{R}$ from $-i\beta$ to $i\beta$ by the line from $-i\beta$ to $-c - i\beta$ to $-c + i\beta$ to $i\beta$. If $|\mathrm{Im}\,(-\nu_1+\nu_2)|$ is small enough, neither the function $\zeta(\nu)$ (of course) nor the function $\zeta(-\nu)$ can vanish at any point of the rectangular region delimited by γ. The poles of the integrand of (5.4.17) there are $1 - \nu_1 + \nu_2$ and $-1 + \nu_1 - \nu_2$: they are simple, as long as $\nu_1 \neq \nu_2$. At the pole $-1 - \nu_1 + \nu_2$, the residue of the factor $\zeta(\frac{1-\nu_1+\nu_2-\nu}{2})$ is -2 and the other zeta factors are regular. The residue there (times $2i\pi$) which has to be considered during the move kills the second exceptional term (5.4.10), while the residue (times $2i\pi$) at $-1+\nu_1-\nu_2$ kills the third exceptional term. We may thus replace the line $\mathrm{Re}\,\nu = 0$ in (5.4.9) by the line γ, provided that we drop the second and third exceptional terms from the list (5.4.10).

The new decomposition of Sharp $(\mathfrak{E}_{\nu_1}^{\mathrm{resc}}, \mathfrak{E}_{\nu_2}^{\mathrm{resc}})$ into homogeneous components is then valid also in the case when $\nu_1 = \nu_2 \neq 0$, in which its integral part becomes

$$\frac{1}{4i\pi} \int_{\gamma} \left[\zeta(\frac{1-\nu}{2})\right]^2 \zeta(\frac{1+2\nu_1+\nu}{2})\zeta(\frac{1-2\nu_1+\nu}{2}) \, \mathfrak{E}_{\nu}^{\mathrm{resc}} \, \frac{d\nu}{\zeta(\nu)\zeta(-\nu)}. \qquad (5.4.18)$$

The integrand has now a double pole at $\nu = -1$. The distribution $(\zeta(-\nu))^{-1}\mathfrak{E}_{\nu}^{\mathrm{resc}}$ is a regular function of ν at $\nu = -1$, where it coincides with the constant 1. One moves back the line of integration to the line $\operatorname{Re} \nu = 0$, obtaining the result expressed by (5.4.15).

Note that the residue under examination, while still automorphic, ceases to be a modular (i.e., homogeneous) distribution, being instead a linear combination of a constant and of the distribution $\frac{d}{d\nu}\Big|_{\nu=-1} \left[(\zeta(-\nu))^{-1}\mathfrak{E}_{\nu}^{\mathrm{resc}}\right]$: it does not lie in the nullspace of the operator $2i\pi\mathcal{E} - 1$ (which would make it a constant), but in the nullspace of the square of this operator. $\qquad \square$

Chapter 6

The operator with symbol $\mathfrak{E}_\nu$

6.1 Extending the validity of the spectral decomposition of a sharp product

Theorem 5.4.1 has been established under the assumptions that $|\mathrm{Re}\,(\nu_1 \pm \nu_2)| < 1$ and $\nu_1 \pm \nu_2 \neq 0$. The first condition occurred already when we justified that $\mathcal{M}\left(\mathfrak{E}^{\mathrm{resc}}_{\nu_1}, \mathfrak{E}^{\mathrm{resc}}_{\nu_2}\right)$ is a meaningful distribution. We wish, however, to consider now the case when $0 < \mathrm{Re}\,\nu_1 < 1$, $0 < \mathrm{Re}\,\nu_2 < 1$.

Theorem 6.1.1. *Set*

$$R = (2i\pi\mathcal{E})^2, \quad \mathfrak{R} = -4\pi^2\,[\mathrm{mad}(P \wedge Q)]^2 : \tag{6.1.1}$$

the first operator, which acts on symbols, and the second, which acts on operators, correspond to each other under the Weyl calculus. The symbol $(R-4)$ $\cdot\, \mathcal{M}\left(\mathfrak{E}^{\mathrm{resc}}_{\nu_1}, \mathfrak{E}^{\mathrm{resc}}_{\nu_2}\right)$, initially a well-defined tempered distribution in the domain $|\mathrm{Re}\,(\nu_1 \pm \nu_2)| < 1$, extends as an analytic function of ν_1, ν_2 to the domain defined by the conditions $0 < \mathrm{Re}\,\nu_1 < 1$, $0 < \mathrm{Re}\,\nu_2 < 1$. In the case when $0 < \mathrm{Re}\,\nu_1 < 1$, $0 < \mathrm{Re}\,\nu_2 < 1$, $\nu_1 \neq \nu_2$ and $\mathrm{Re}\,(\nu_1 + \nu_2) > 1$, the symbol $(R-4)\,\mathcal{M}\left(\mathfrak{E}^{\mathrm{resc}}_{\nu_1}, \mathfrak{E}^{\mathrm{resc}}_{\nu_2}\right)$ is the image under the operator $(R-4)(2i\pi\mathcal{E})$ of the distribution $\mathrm{Sharp}\left(\mathfrak{E}^{\mathrm{resc}}_{\nu_1}, \mathfrak{E}^{\mathrm{resc}}_{\nu_2}\right)$ given by the same equations as in Theorem 5.4.1, save for the fact that the 4th term of (5.4.10) must be deleted.

Proof. Recall from (1.1.38) that, for $k \neq 0$, the kth Fourier coefficient α_k of $\mathfrak{E}^{\mathrm{resc}}_{\nu_1}$ is $2^{\frac{-1-\nu_1}{2}}\sigma_{\nu_1}(|k|)$: it is thus a $\mathrm{O}\left(|k|^{\frac{\delta_1}{2}}\right)$ provided that $\delta_1 > 2\,\mathrm{Re}\,\nu_1$. The first sentence is then proved just as Theorem 5.3.1, taking benefit of Proposition 4.5.4, together with the fact that $\iota > \delta_1 + \delta_2 - \mathrm{Re}\,(\nu_1 + \nu_2)$ will be satisfied if $\iota \geq 2$.

It is now a matter of continuing analytically the image under $(R-4)(2i\pi\mathcal{E})$ of the right-hand side of the equation (5.4.9) giving when $|\mathrm{Re}\,(\nu_1 \pm \nu_2)| < 1$ the distribution $\mathrm{Sharp}\left(\mathfrak{E}^{\mathrm{resc}}_{\nu_1}, \mathfrak{E}^{\mathrm{resc}}_{\nu_2}\right)$. We do it first in the part U of the domain

$0 < \operatorname{Re} \nu_1 < 1, 0 < \operatorname{Re} \nu_2 < 1, \nu_1 \neq \nu_2$ defined by the conditions $\nu_1 + \nu_2 \neq 1$, together with the fact that $\nu_1 + \nu_2 - 1$ should not be a zero of zeta.

Let us continue analytically the various terms of the decomposition (5.4.9) one at a time. Despite the singularity (when $j = 0$) of the second B_j factor in (5.4.14) if $1 - \nu_1 - \nu_2 - i\lambda_r = 0$, this is not a singularity of the product of this factor by the corresponding L-factor (a consequence of (1.2.10), unless one prefers to remember that the two B_j factors originated from (5.2.39)). Looking at the exceptional terms (5.4.10), we must worry about the zeros of the zeta functions in the denominators, not forgetting either that $\mathfrak{E}_\mu$ is meaningless for $\mu = \pm 1$. This leads to the condition $\zeta(-1 + \nu_1 + \nu_2) \neq 0$ on one hand, to the condition $\nu_1 \neq \nu_2$ on the other hand. Let us, finally, rewrite in full the integral term of the decomposition: it is

$$\frac{1}{4i\pi} \int_{\operatorname{Re} \nu = 0} \prod_{\varepsilon_1, \varepsilon_2 = \pm 1} \zeta\left(\frac{1 + \varepsilon_1 \nu_1 + \varepsilon_2 \nu_2 + \varepsilon_1 \varepsilon_2 \nu}{2}\right) \mathfrak{E}_\nu^{\mathrm{resc}} \frac{d\nu}{\zeta(\nu)\zeta(-\nu)}. \qquad (6.1.2)$$

Given any real number t, we now show that (6.1.2) extends analytically to the part of the domain U defined by the extra condition $|\operatorname{Im}(\nu_1 + \nu_2) + t| < \varepsilon$ and we make its continuation to the subdomain defined by the condition $\operatorname{Re}(\nu_1 + \nu_2) > 1$ explicit. One may choose ε so that there are no zeros of zeta with the imaginary part $t \pm \varepsilon$ or $-t \pm \varepsilon$.

Starting with the assumption that $0 < \operatorname{Re} \nu_1, 0 < \operatorname{Re} \nu_2$ and $\operatorname{Re}(\nu_1 + \nu_2) < 1$, we make a change of contour in (6.1.2), replacing the segment from $i(t - \varepsilon)$ to $i(t + \varepsilon)$ by the piecewise straight line from $i(t - \varepsilon)$ to $1 + i(t - \varepsilon)$, next to $1 + i(t + \varepsilon)$, finally to $i(t + \varepsilon)$. Within the rectangle $\mathcal{R}$ three sides of which have just been defined, the poles of the integrand are the zeros ρ_m of zeta there (if any, finitely many in any case) and the point $\nu = 1 - \nu_1 - \nu_2$; let us not worry about the pole at $\nu = 1$ of $\mathfrak{E}_\nu^{\mathrm{resc}}$, which is taken care of by the factor $\zeta(\nu)$ in the denominator. One can thus replace the initial contour $\operatorname{Re} \nu = 0$ by the contour γ, provided one adds to the new integral $-2i\pi$ times the sum of residues at the points $\nu = \rho_m$ and at $\nu = 1 - \nu_1 - \nu_2$. As will be seen presently, there is no need to make the ones of the first species explicit. The new term to add is exactly $-\frac{\zeta(\nu_1)\zeta(\nu_2)}{\zeta(-1+\nu_1+\nu_2)} \mathfrak{E}_{1-\nu_1-\nu_2}^{\mathrm{resc}}$, so that it will cancel the 4th term in the list (5.4.10). Consider now, under the assumption that $|\operatorname{Im}(\nu_1 + \nu_2) + t| < \varepsilon$, the integral on the contour γ: the argument of any of the 4 functions in the numerator cannot be 1, either because $|\operatorname{Im}(\nu_1 + \nu_2) + t| \geq \varepsilon$ or, if this is not the case, $\operatorname{Re} \nu = 1$ (one has $\operatorname{Re}(\nu_1 + \nu_2) > 0$ and $|\operatorname{Re}(\nu_1 - \nu_2)| < 1$). If follows that this integral is an analytic function in the domain $\operatorname{Re} \nu_1 > 0$, $\operatorname{Re} \nu_2 > 0$, $|\operatorname{Im}(\nu_1 + \nu_2) + t| < \varepsilon$. Finally, in the case when, on top of that, $\operatorname{Re}(\nu_1 + \nu_2) > 1$, the points ρ_m still lie inside the rectangle $\mathcal{R}$, but this is no longer the case for the point $\nu = 1 - \nu_1 - \nu_2$. Changing back the contour γ to the line $\operatorname{Re} \nu = 0$, one ends up with the integral (6.1.2) one started with, but one must now delete the 4th term from the list (5.4.10).

When Re $(\nu_1 + \nu_2) > 1$, we may, using analyticity, drop the condition that $\zeta(\nu_1 + \nu_2 - 1) \neq 0$. Indeed, the sole reason for this condition was to ensure that the 4th term from the list (5.4.10) was meaningful: but this term is no longer present in the decomposition.   □

6.2 The odd-odd part of $\mathrm{Op}(\mathfrak{E}_\nu)$ when $|\mathrm{Re}\ \nu| < \frac{1}{2}$

When $|\mathrm{Re}\ (\nu_1 \pm \nu_2)| < 1$ and $\nu_1 \pm \nu_2 \neq 0$ (the latter condition can be dispensed with, as proved in Proposition 5.4.2), Theorem 5.4.1 shows that, given a pair v, u of functions in $\mathcal{S}(\mathbb{R})$ such that the Wigner function $W(v, u)$ lies in the image of $\mathcal{S}(\mathbb{R}^2)$ under $2i\pi\mathcal{E}$, the operator $\mathrm{Op}\left(\mathfrak{E}_{\nu_1}^{\mathrm{sharp}}\right) \mathrm{Op}\left(\mathfrak{E}_{\nu_2}^{\mathrm{sharp}}\right)$ can be tested on the pair v, u, and that

$$\left(v \,\middle|\, \mathrm{Op}\left(\mathfrak{E}_{\nu_1}^{\mathrm{sharp}}\right) \mathrm{Op}\left(\mathfrak{E}_{\nu_2}^{\mathrm{sharp}}\right) u\right) = \left\langle \mathrm{Sharp}\left(\mathfrak{E}_{\nu_1}^{\mathrm{resc}}, \mathfrak{E}_{\nu_2}^{\mathrm{resc}}\right),\, W(v, u)\right\rangle. \tag{6.2.1}$$

The following lemma, reproduced from [39, p.260] in view of our present interest in the question, shows that the condition demanded from $W(v, u)$ will be satisfied if u and v are odd.

Lemma 6.2.1. *If* $u, v \in \mathcal{S}_{\mathrm{odd}}(\mathbb{R})$, *the function* $W(v, u)$ *lies in the space image of* $\mathcal{S}(\mathbb{R}^2)$ *under* $2i\pi\mathcal{E}$. *If* $u \neq 0$ *lies in* $\mathcal{S}_{\mathrm{even}}(\mathbb{R})$, *the function* $W(u, u)$ *never lies in that space.*

Proof. A symbol $h \in \mathcal{S}(\mathbb{R}^2)$ lies in the image of $\mathcal{S}(\mathbb{R}^2)$ under $2i\pi\mathcal{E}$ if and only if $\int_0^\infty h(tx, t\xi)\, dt = 0$ for every $(x, \xi) \neq 0$ or, what amounts to the same, if $\int_0^\infty h(ts, t)\, dt = 0$ for every $s \neq 0$. Indeed, the equation $h = \left(x\frac{\partial}{\partial x} + \xi\frac{\partial}{\partial \xi} + 1\right) f$ is equivalent to $h(tx, t\xi) = \frac{d}{dt}(t\, f(tx, t\xi))$. Starting then from

$$W(v, u)(x, \xi) = 2 \int_{-\infty}^\infty \overline{v}(x + r)\, u(x - r)\, e^{4i\pi r\xi}\, dr \tag{6.2.2}$$

and using the fact that, if u and v have the same parity, $W(v, u)$ is an even function in $\mathbb{R}^2$, one obtains, for $s \neq 0$,

$$\int_0^\infty W(v, u)(ts, t)\, dt = \int_{-\infty}^\infty dt \int_{-\infty}^\infty \overline{v}(ts + r)\, u(ts - r)\, e^{4i\pi tr}\, dr$$

$$= \frac{1}{2|s|} \int_{\mathbb{R}^2} \overline{v}(y)\, u(x)\, \exp\left(\frac{i\pi(y^2 - x^2)}{s}\right)\, dx\, dy. \tag{6.2.3}$$

This is zero if u and v are odd, while if $v = u$ is even, it is $\frac{2}{|s|}\left|\int_{-\infty}^\infty u(x)e^{\frac{-i\pi x^2}{s}}\, dx\right|^2$ and can therefore not be identically zero unless $u = 0$.   □

Theorem 6.2.2. *When* $0 < |\mathrm{Re}\ \nu| < \frac{1}{2}$ *and* $\nu \notin \mathbb{R}$, *the operator* $\mathrm{Op}\left(\mathfrak{E}_\nu^{\mathrm{resc}}\right)$ *sends the space* $\mathcal{S}_{\mathrm{odd}}(\mathbb{R})$ *to the space* $L^2_{\mathrm{odd}}(\mathbb{R})$. *If* $u \in \mathcal{S}_{\mathrm{odd}}(\mathbb{R})$ *and* $\langle \mathfrak{E}_2^{\mathrm{resc}}, W(u, u)\rangle \neq 0$, *the norm* $\|\mathrm{Op}\left(\mathfrak{E}_\nu^{\mathrm{resc}}\right) u\|$ *goes to infinity as* $\mathrm{Re}\ \nu \to \frac{1}{2}$: *the same holds as* $\mathrm{Re}\ \nu \to -\frac{1}{2}$ *if* $\langle \mathfrak{E}_0^{\mathrm{resc}}, W(u, u)\rangle \neq 0$.

Proof. It $u \in \mathcal{S}_{\text{odd}}(\mathbb{R})$, one has $\hat{u}(\xi) = O(|\xi|)$ as $\xi \to 0$, so that the operator with symbol $|\xi|^{-\nu-1}$, the first term of the decomposition into homogeneous components of $\mathfrak{E}_\nu(x,\xi)$, sends u to some element of $L^2_{\text{odd}}(\mathbb{R})$ if Re $\nu < \frac{1}{2}$. The same is true, if Re $\nu > -\frac{1}{2}$, in connection with the second term $|x|^{-\nu}\delta(\xi)$ of the decomposition, since the $\mathcal{G}$-transform of this symbol is the symbol $2^\nu B_0(\nu)|\xi|^{\nu-1}$, so that

$$\text{Op}\left(|x|^{-\nu}\delta(\xi)\right)u = 2^\nu B_0(\nu)\,\text{Op}\left(|\xi|^{\nu-1}\right)\overset{\vee}{u}. \tag{6.2.4}$$

Finally, it follows from (4.1.12) that each operator $\text{Op}(h_{\nu,k})$ with $k \in \mathbb{Z}^\times$ sends $\mathcal{S}_{\text{odd}}(\mathbb{R})$ to $L^2_{\text{odd}}(\mathbb{R})$. Given $N = 1, 2, \ldots$, denote as $\mathfrak{E}_\nu^{[N]}$ the truncation of $\mathfrak{E}_\nu^{\text{resc}}$ obtained when retaining only the terms $h_{\nu,k}$ of the decomposition of $\mathfrak{E}_\nu^{\text{resc}}$ such that $|k| \leq N$. As $N \to \infty$, the image under $2i\pi\mathcal{E}$ of the distribution Sharp $(\mathfrak{E}_{\bar{\nu}}^{\text{resc}}, \mathfrak{E}_\nu^{\text{resc}})$ is the weak limit, in $\mathcal{S}'(\mathbb{R}^2)$, of the distribution Sharp $\left(\mathfrak{E}_{\bar{\nu}}^{[N]}, \mathfrak{E}_\nu^{[N]}\right)$. Then, for $u \in \mathcal{S}_{\text{odd}}(\mathbb{R})$,

$$\|\text{Op}\left(\mathfrak{E}_\nu^{[N]}\right)u\|^2 = \langle\text{Sharp}\left(\mathfrak{E}_{\bar{\nu}}^{[N]}, \mathfrak{E}_\nu^{[N]}\right), W(u,u)\rangle \tag{6.2.5}$$

goes as $N \to \infty$ to $\langle\text{Sharp}(\mathfrak{E}_{\bar{\nu}}^{\text{resc}}, \mathfrak{E}_\nu^{\text{resc}}), W(u,u)\rangle$, a finite number. From a weak compactness argument, it follows that $\text{Op}(\mathfrak{E}_\nu^{\text{resc}})u \in L^2(\mathbb{R})$.

Setting $\sigma = \text{Re }\nu$ with $0 < |\sigma| < \frac{1}{2}$, it follows from Theorem 5.4.1 that

$$\text{Sharp}(\mathfrak{E}_{\bar{\nu}}^{\text{resc}}, \mathfrak{E}_\nu^{\text{resc}})$$

$$= \frac{|\zeta(-\nu)|^2}{\zeta(-1-2\sigma)}\mathfrak{E}_{1+2\sigma}^{\text{resc}} + \frac{|\zeta(\nu)|^2}{\zeta(-1+2\sigma)}\mathfrak{E}_{1-2\sigma}^{\text{resc}}$$

$$+ \frac{\zeta(-\bar{\nu})\zeta(\nu)}{\zeta(-1-\bar{\nu}+\nu)}\mathfrak{E}_{-1-\bar{\nu}+\nu}^{\text{resc}} + \frac{\zeta(\bar{\nu})\zeta(-\nu)}{\zeta(-1+\bar{\nu}-\nu)}\mathfrak{E}_{-1+\bar{\nu}-\nu}^{\text{resc}}$$

$$+ \frac{1}{4\pi}\int_{-\infty}^\infty \prod_{\eta_1,\eta_2=1}\zeta\left(\frac{1+\eta_1\bar{\nu}+\eta_2\nu+\eta_1\eta_2 i\lambda}{2}\right)\mathfrak{E}_{i\lambda}^{\text{resc}}\frac{d\lambda}{\zeta(i\lambda)\zeta(-i\lambda)}$$

$$+ \frac{1}{2}\sum_{r\in\mathbb{Z}^\times}\sum_\ell \Omega_{r,\ell}(\mathfrak{E}_{\bar{\nu}}^{\text{resc}}, \mathfrak{E}_\nu^{\text{resc}})\frac{\Gamma(\frac{i\lambda_r}{2})\Gamma(-\frac{i\lambda_r}{2})}{\|\mathcal{N}_{|r|,\ell}\|^2}\mathfrak{N}_{r,\ell}^{\text{resc}}, \tag{6.2.6}$$

where the coefficients of the last (discrete) sum have been made explicit in (5.4.14). Testing this identity on $W(u,u)$, one obtains the second part of Theorem 6.2.2 since $\zeta(-2) = 0$. $\square$

To go further, we shall specialize u as a Hermite function.

6.3 The harmonic oscillator

Recall that $P = \frac{1}{2i\pi}\frac{d}{dx}$ and that Q is the operator of multiplication by x. If A is a linear operator from $\mathcal{S}(\mathbb{R})$ to $\mathcal{S}'(\mathbb{R})$, one has

$$(P+iQ)A(Q-iP) - (Q-iP)A(P-iQ) = 2(PAQ - QAP): \tag{6.3.1}$$

in other words, if one introduces, as done in Physics, the annihilation operator $C = \pi^{\frac{1}{2}}(P - iQ)$ and the creation operator $C^* = \pi^{\frac{1}{2}}(P + iQ)$,

$$\text{mad}(P \wedge Q) = \frac{1}{2i\pi}\,\text{mad}(C \wedge C^*). \tag{6.3.2}$$

Elementary developments regarding these operators, and the associated harmonic oscillator

$$L = \pi\left(Q^2 + P^2\right) = \text{Op}\left(\pi(x^2 + \xi^2)\right) \tag{6.3.3}$$

can be found in considerably many references, in particular as the first example in Physics textbooks concerned with the rudiments of quantum mechanics.

The function $\phi^{(0)}(x) = 2^{\frac{1}{4}}e^{-\pi x^2}$ is the so-called ground state of L, to wit its normalized eigenstate with smallest eigenvalue $\frac{1}{2}$. Using the identities

$$C^*C = L - \frac{1}{2}, \quad CC^* = L + \frac{1}{2}, \tag{6.3.4}$$

a consequence of the commutator relation $[P, Q] = \frac{1}{2i\pi}$, it is immediate by induction that if one sets

$$\phi^{(m+1)} = (m + 1)^{-\frac{1}{2}}C^*\,\phi^{(m)}, \quad m = 0, 1, \ldots, \tag{6.3.5}$$

$\phi^{(m)}$ is normalized in $L^2(\mathbb{R})$ for every m and that $L\phi^{(m)} = \left(m + \frac{1}{2}\right)\phi^{(m)}$: the function $\phi^{(0)}$ (resp. $\phi^{(1)}$) is the same as the one denoted as ϕ_i^0 (resp. ϕ_i^1) in (3.1.17). Finally, we note that $\left(L - \frac{1}{2}\right)\phi^{(0)} = 0$ and that, for $m \geq 1$,

$$C\phi^{(m)} = m^{-\frac{1}{2}}CC^*\phi^{(m-1)} = m^{-\frac{1}{2}}\left(L + \frac{1}{2}\right)\phi^{(m-1)} = m^{\frac{1}{2}}\phi^{(m-1)}. \tag{6.3.6}$$

The set $\left(\phi^{(m)}\right)_{m=0,1,\ldots}$ is an orthonormal basis of $L^2(\mathbb{R})$, and one can define, for $\alpha > 0$, the operator $\exp\left(-\alpha L\right)$ by the equation

$$\exp\left(-\alpha L\right)u = \sum_{m=0}^{\infty} e^{-\alpha(m+\frac{1}{2})}\left(\phi^{(m)}\,|\,u\right)\phi^{(m)} \tag{6.3.7}$$

(all scalar products in this section are in $L^2(\mathbb{R})$). One has [30, p.204]

$$\exp\left(-\alpha L\right) = \text{Op}(F_\alpha) \quad \text{with} \quad F_\alpha(x, \xi) = \frac{1}{\cosh\frac{\alpha}{2}}\exp\left(-2\pi\tanh\frac{\alpha}{2}\,(x^2 + \xi^2)\right). \tag{6.3.8}$$

Then, it is immediate from (2.1.13) that

$$\left(\mathcal{G}\,F_\alpha\right)(x, \xi) = \frac{1}{\sinh\frac{\alpha}{2}}\exp\left(-\frac{2\pi}{\tanh\frac{\alpha}{2}}\,(x^2 + \xi^2)\right). \tag{6.3.9}$$

In view of the meaning, given immediately before (3.1.15), of the operator $\mathcal{G}$ when acting on symbols, one has for $u \in \mathcal{S}(\mathbb{R})$ the identity

$$\operatorname{Op}(\mathcal{G}\,F_\alpha)\,u = \sum_{m=0}^{\infty} (-1)^m e^{-\alpha(m+\frac{1}{2})} \left(\phi^{(m)}\,|\,u\right) \phi^{(m)}, \qquad (6.3.10)$$

since $\phi^{(m)}$ is even or odd according to the parity of m. The fact that the symbol $e^{-2\pi\varepsilon\,(x^2+\xi^2)}$ is that of a positive-definite operator if $0 < \varepsilon < 1$, but is the product of a positive-definite operator by the check operator if $\varepsilon > 1$, served in ([30]) as the basis of a new approach to the question of continuity of pseudodifferential operators: but no arithmetic occurred there. We also set $G_\alpha = \frac{1}{2}(F_\alpha - \mathcal{G}\,F_\alpha)$, so that, for $u \in \mathcal{S}(\mathbb{R})$,

$$\operatorname{Op}(G_\alpha)\,u = \sum_{n=0}^{\infty} e^{-\alpha(2n+\frac{3}{2})} \left(\phi^{(2n+1)}\,|\,u\right) \phi^{(2n+1)}. \qquad (6.3.11)$$

Lemma 6.3.1. *Recall that* $R = (2i\pi\mathcal{E})^2$. *Let* h *be a radial symbol in* $\mathcal{S}(\mathbb{R}^2)$. *It lies in the image of* $\mathcal{S}(\mathbb{R}^2)$ *under the operator* $R - 4$ *if and only if*

$$\int_{\mathbb{R}^2} (x^2+\xi^2)^{\frac{1}{2}}\, h(x,\xi)\,dx\,d\xi = 0 \quad \text{and} \quad \int_{\mathbb{R}^2} (x^2+\xi^2)^{-\frac{3}{2}}\, [h(x,\xi) - h(0,0)]\,dx\,d\xi = 0.$$
$$(6.3.12)$$

Proof. Let us solve the equations $(R - 4)\,f = 4h$. Averaging under the action of the rotation group, it is no loss of generality to assume that f is also radial. Setting, with $r = (x^2+\xi^2)^{\frac{1}{2}}$, $f(x,\xi) = f_1(r)$, $h(x,\xi) = h_1(r)$, we solve the equation $\left[\left(r\frac{d}{dr} + 1\right)^2 - 4\right] f_1 = 4h_1$ by the method of "variation of constants", obtaining that the solution of this equation which is rapidly decreasing at infinity is

$$f_1(r) = r^{-3} \int_r^\infty t^2 h_1(t)\,dt - r \int_r^\infty t^{-2} h_1(t)\,dt. \qquad (6.3.13)$$

The question is whether this function of $r > 0$ is, near $r = 0$, an analytic function of r^2. Simply writing $\int_r^\infty = \int_0^\infty - \int_0^r$, one sees that the pair of conditions (6.3.12) suffices to that effect. The two conditions are also necessary, as seen if one observes that the two terms of (6.3.13) are the solutions rapidly decreasing at infinity of the equations $\left(r\frac{d}{dr} + 3\right) f_1 = -h_1$ and $\left(r\frac{d}{dr} - 1\right) f_1 = h_1$. $\qquad \square$

A better understanding of what is really meant by the fact that, when $\operatorname{Re}(\nu_1 + \nu_2) > 1$ (and $0 < \operatorname{Re}\nu_1 < 1, 0 < \operatorname{Re}\nu_2 < 1$), one can only test $\mathcal{M}\left(\mathfrak{E}_{\nu_1}^{\mathrm{resc}}, \mathfrak{E}_{\nu_2}^{\mathrm{resc}}\right)$ on symbols in the image of $\mathcal{S}(\mathbb{R}^2)$ under the operator $R-4$ will be obtained from the following calculations. Consider, for $m = 0, 1, \ldots$, the rank-one projection operator p_m such that $p_m u = (\phi^{(m)}\,|\,u)\,\phi^{(m)}$, and recall that its symbol is the Wigner function $W(\phi^{(m)}, \phi^{(m)})$: for convenience, define also $p_{-1} = 0$. One has for $m \geq 0$,

$$
\begin{aligned}
2i\pi\, [\mathrm{mad}&(P \wedge Q)\, p_m]\, u \\
&= C p_m C^* u - C^* p_m C\, u \\
&= (\phi^{(m)} \,|\, C^* u)\, C\phi^{(m)} - (\phi^{(m)} \,|\, Cu)\, C^*\phi^{(m)} \\
&= (C\phi^{(m)} \,|\, u)\, C\phi^{(m)} - (C^*\phi^{(m)} \,|\, u)\, C^*\phi^{(m)} \\
&= m(\phi^{(m-1)} \,|\, u)\, \phi^{(m-1)} - (m+1)(\phi^{(m+1)} \,|\, u)\, \phi^{(m+1)},
\end{aligned}
\tag{6.3.14}
$$

hence

$$
2i\pi\, \mathrm{mad}(P \wedge Q)\, p_m = m\, p_{m-1} - (m+1)\, p_{m+1}. \tag{6.3.15}
$$

A first consequence is that the symbol of p_1, next the symbol of p_{2n+1} for $n = 0, 1, \ldots$, lies in the image of $\mathcal{S}(\mathbb{R}^2)$ under $2i\pi\mathcal{E}$: this is of course a special case of Lemma 6.2.1. So far as projections on even states of the harmonic oscillator are concerned, it is only the symbols of differences $(2n+2)\, p_{2n+2} - (2n+1)\, p_{2n}$ that lie in that space.

Next, one has, with $\mathfrak{R}$ as defined in (6.1.1),

$$
(\mathfrak{R} - 4)\, p_m = m(m-1)\, p_{m-2} - (2m^2 + 2m + 5)\, p_m + (m+1)(m+2)\, p_{m+2}. \tag{6.3.16}
$$

In particular, taking $m = 1$, $2p_3 - 3p_1$ lies in the image of the operator $\mathfrak{R} - 4$. The same is generally true, by induction, for the operator $(2n+2)\, p_{2n+3} - (2n+3)\, p_{2n+1}$, since it follows from (6.3.16) that

$$
\begin{aligned}
(\mathfrak{R} - 4)&\, p_{2n+1} \\
&= (2n+1)(2n)\, p_{2n-1} - (8n^2 + 12n + 9)\, p_{2n+1} + (2n+2)(2n+3)\, p_{2n+3} \\
&= 2n\, [(2n+1)\, p_{2n-1} - 2n\, p_{2n+1}] + (2n+3)\, [(2n+2)\, p_{2n+3} - (2n+3)\, p_{2n+1}].
\end{aligned}
\tag{6.3.17}
$$

Something similar can be said about projections on even states of the harmonic oscillator, starting with the fact that the symbol of $2p_2 - 5p_0$ lies in $(R-4)\mathcal{S}(\mathbb{R}^2)$, but this is not as useful since this symbol does not lie in the image of $\mathcal{S}(\mathbb{R}^2)$ under $2i\pi\mathcal{E}$: it is the linear combination $2p_2 - p_0$ that does.

Let us compute now the coefficient α_{2n+1} of x in the polynomial $e^{\pi x^2} \cdot \phi^{(2n+1)}(x)$. More generally, the polynomial $e^{\pi x^2} \phi^{(m)}(x)$ is related to the Hermite polynomials H_m, as defined in [22, p.250], by the equation

$$
e^{\pi x^2} \phi^{(m)}(x) = (-1)^{[\frac{m}{2}]} \left(\frac{2^{\frac{1}{2}-m}}{m!} \right)^{\frac{1}{2}} H_m(x\sqrt{2\pi}), \tag{6.3.18}
$$

as can be checked from the normalization condition

$$
\int_{-\infty}^{\infty} e^{-x^2} H_m(x) H_n(x)\, dx = \pi^{\frac{1}{2}} 2^m m!\, \delta_{n,m} \tag{6.3.19}
$$

together with an examination of the sign of the term of highest degree in the polynomials concerned. Since [22, p.250], one has explicitly

$$H_{2n+1}(x\sqrt{2\pi}) = (2n+1)!\sum_{k=0}^{n}\frac{(-1)^k}{k!}\frac{(x\sqrt{8\pi})^{2n+1-2k}}{(2n+1-2k)!},\qquad(6.3.20)$$

it follows that

$$\alpha_{2n+1} = 2^{-n+\frac{5}{4}}\pi^{\frac{1}{2}}\frac{\sqrt{(2n+1)!}}{n!},\qquad(6.3.21)$$

so that $(2n+2)\,\alpha_{2n+3}^2 = (2n+3)\,\alpha_{2n+1}^2$.

The difference $(2n+2)\,p_{2n+3} - (2n+3)\,p_{2n+1}$, just as each of its two terms, can be expanded as a linear combination of rank-one operators of the kind $u\mapsto \left(x^{2k+1}e^{-\pi x^2}\,|\,u\right)x^{2j+1}e^{-\pi x^2}$. The benefit of taking precisely the coefficients $2n+2$ and $2n+3$ (recall that the linear combination $(2n+2)\,p_{2n+3} - (2n+3)\,p_{2n+1}$ is the one that lies in $\mathfrak{R}-4$) is that there is no remaining term with $j=k=0$: at least one of the two factors $x^{2k+1}e^{-\pi x^2}$ and $x^{2j+1}e^{-\pi x^2}$ has to be divisible by x^3. This confirms the benefit of dealing with functions on the line with a certain degree of flatness at 0, as was indicated in Section 3.3. But it is not true that the Wigner function of the pair of functions both equal to $x^3 e^{-\pi x^2}$ lies in the image of $R-4$: to manage this, one would have to use a higher-level Weyl calculus Op^p with $p\geq 2$, as introduced in [35, sections 7,9] and briefly mentioned at the end of Chapter 3.

That we have not done so is due to the complications one comes across when developing the automorphic p-Weyl calculus, even though the equation (9.54) in [35] provides a way to relate the p-Weyl calculus to the $(p-1)$-Weyl calculus when $p\geq 1$. Its first case deals with the Op^1-calculus and its easy part is the fact that if $h\in\mathcal{S}'_{\mathrm{even}}(\mathbb{R}^2)$, $\mathrm{Op}^1(h)$ agrees with $\mathrm{Op}(h)$ on functions in $\mathcal{S}_{\mathrm{odd}}(\mathbb{R})$. But, if $h=(1-i\pi\mathcal{E})f$ with $f\in\mathcal{S}'_{\mathrm{even}}(\mathbb{R}^2)$ and if one denotes as $Q^2\mathcal{S}_{\mathrm{even}}(\mathbb{R})$ the space which is the image of $\mathcal{S}_{\mathrm{even}}(\mathbb{R})$ under the multiplication by x^2, the operator $\mathrm{Op}^1(h)$ agrees on that space with the operator $\Lambda\,(\mathrm{Op}(f))$, where $\Lambda(A) = \frac{1}{2}\left[QAQ^{-1}+Q^{-1}AQ-i\pi(PAQ-QAP)\right]$. This equation does not make it possible to reduce the automorphic Op^1-calculus to the automorphic Op-calculus: but such a possibility could not be expected, since, in terms of what was recalled in Section 3.3, the representations Met_p, with $p=0,1,\ldots$, are pairwise inequivalent.

What could be realized from these reminders is that, with the use of the Weyl calculus, one can have a good understanding of the odd-odd parts of operators with Eisenstein distributions $\mathfrak{E}_\nu$ for symbols only when $|\mathrm{Re}\,\nu|<\frac{1}{2}$. To understand their even-even parts in the same domain for ν, one should replace the Weyl calculus by the Op^1-calculus, while using an Op^p-calculus with $p\geq 2$ would be necessary to understand both parts of these operators when $|\mathrm{Re}\,\nu|<1$, even more so if wanting to move to higher values of Re ν. Dealing with symbols as singular as those which present themselves in the automorphic situation forces one to change the Weyl calculus, up to some point, to a calculus of operators acting in a natural way on functions with some degree of flatness at 0.

6.4 The square of zeta on the critical line; non-critical zeros

We start with two lemmas.

Lemma 6.4.1. *For $m = 0, 1, \ldots$, $z \neq 1, 2, \ldots$ and $z \neq -m - 1, -m - 2, \ldots$, one has*

$$\langle (x^2 + \xi^2)^{-z}, W(\phi^{(m)}, \phi^{(m)}) \rangle = \frac{m!}{\Gamma(m+1+z)} (2\pi)^z \, {}_2F_1 (z, m+1; m+1+z; -1). \tag{6.4.1}$$

Proof. The condition $z \neq 1, 2, \ldots$ ensures that $(x^2 + \xi^2)^{-z}$ is a well-defined distribution (it is characterized by the way it is tested on radial functions). Starting from the equation

$$(x^2 + \xi^2)^{-z} = \frac{(2\pi)^z}{\Gamma(z)} \int_0^\infty e^{-2\pi\delta(x^2 + \xi^2)} \delta^{z-1} d\delta, \quad \text{Re } z > 0, \tag{6.4.2}$$

and setting $\delta = \tanh \frac{\alpha}{2}$, so that $d\delta = \frac{1}{2} \frac{d\alpha}{\left(\cosh \frac{\alpha}{2} \right)^2}$, one obtains from (6.3.8) the equation

$$\mathrm{Op}\left((x^2 + \xi^2)^{-z} \right) = \frac{1}{2} \frac{(2\pi)^z}{\Gamma(z)} \int_0^\infty \frac{\left(\tanh \frac{\alpha}{2} \right)^{z-1}}{\cosh \frac{\alpha}{2}} \exp(-\alpha L) \, d\alpha. \tag{6.4.3}$$

Testing against the pair $(\phi^{(m)}, \phi^{(m)})$ and using the fact that $L\phi^{(m)} = (m+\frac{1}{2}) \phi^{(m)}$, one obtains

$$\left(\phi^{(m)} \,|\, \mathrm{Op}\left((x^2 + \xi^2)^{-z} \right) \phi^{(m)} \right) = \frac{1}{2} \frac{(2\pi)^z}{\Gamma(z)} \int_0^\infty \frac{\left(\tanh \frac{\alpha}{2} \right)^{z-1}}{\cosh \frac{\alpha}{2}} e^{-\alpha(m+\frac{1}{2})} \, d\alpha. \tag{6.4.4}$$

Setting $e^{-\alpha} = t$, so that $\tanh \frac{\alpha}{2} = \frac{1-t}{1+t}$, $\cosh \frac{\alpha}{2} = \frac{1}{2} t^{-\frac{1}{2}} (1 + t)$ and $d\alpha = \frac{dt}{t}$, one finds if $z \neq -m - 1, -m - 2, \ldots$, using [22, p.64],

$$\left(\phi^{(m)} \,|\, \mathrm{Op}\left((x^2 + \xi^2)^{-z} \right) \phi^{(m)} \right) = \frac{(2\pi)^z}{\Gamma(z)} \int_0^1 t^m (1 - t)^{z-1} (1 + t)^{-z} dt$$

$$= \frac{m!}{\Gamma(m+1+z)} (2\pi)^z \, {}_2F_1 (z, m+1; m+1+z; -1). \tag{6.4.5}$$

Note that the dt-integral converges for Re $z > 0$ and the formula is first established under the assumption that Re $z > 0$ together with $z \neq 1, 2, \ldots$ and $z \neq -m - 1, -m-2, \ldots$: it remains valid, by analytic continuation, if one drops the condition Re $z > 0$. $\square$

Lemma 6.4.2. *If* $\mathfrak{S} \in \mathcal{S}'(\mathbb{R}^2)$ *is homogeneous of degree* $-1 - \mu$, *with* $\mu \neq 1, 3, \ldots$ *and* $\mu \neq -3, -5, \ldots$, *one has*

$$\langle \mathfrak{S}, W(\phi^{(m)}, \phi^{(m)}) \rangle = \frac{m!}{\left(\frac{3+\mu}{2}\right)_m} \frac{{}_2F_1\left(\frac{1+\mu}{2}, m+1; m+\frac{3+\mu}{2}; -1\right)}{{}_2F_1\left(\frac{1+\mu}{2}, 1; \frac{3+\mu}{2}; -1\right)} (\Theta_0\mathfrak{S})(i).$$
(6.4.6)

Proof. Setting $t = e^{-\alpha}$, one has, with F_α as introduced in (6.3.8),

$$F_\alpha(x, \xi) = \frac{2t^{\frac{1}{2}}}{1+t} \exp\left(-2\pi(x^2 + \xi^2)\frac{1-t}{1+t}\right)$$

$$= \frac{2t^{\frac{1}{2}}}{1+t}\left(\frac{1-t}{1+t}\right)^{i\pi\mathcal{E}-\frac{1}{2}}\left(e^{-2\pi(x^2+\xi^2)}\right).$$
(6.4.7)

In particular, if $\mathfrak{S}$ is homogeneous of degree $-1 - \mu$,

$$\langle \mathfrak{S}, F_\alpha \rangle = \frac{t^{\frac{1}{2}}}{1+t}\left\langle\left(\frac{1-t}{1+t}\right)^{-i\pi\mathcal{E}-\frac{1}{2}}\mathfrak{S}, 2\,e^{-2\pi(x^2+\xi^2)}\right\rangle$$

$$= t^{\frac{1}{2}}(1-t)^{\frac{\mu-1}{2}}(1+t)^{\frac{-\mu-1}{2}}(\Theta_0\mathfrak{S})(i).$$
(6.4.8)

Since

$$F_\alpha = \sum_{m\geq 0} t^{m+\frac{1}{2}}W(\phi^{(m)}, \phi^{(m)}),$$
(6.4.9)

one has

$$\langle \mathfrak{S}, W(\phi^{(m)}, \phi^{(m)}) \rangle = A_m(\mu)(\Theta_0\mathfrak{S})(i),$$
(6.4.10)

with

$$A_m(\mu) = \frac{1}{m!}\frac{d^m}{dt^m}\bigg|_{t=0}\left[(1-t)^{\frac{\mu-1}{2}}(1+t)^{\frac{-\mu-1}{2}}\right].$$
(6.4.11)

What remains to be done is computing this coefficient, which can be done by using the special distribution $\mathfrak{S} = (x^2 + \xi^2)^{\frac{-1-\mu}{2}}$. In this case, according to (6.4.5), one has if $\mu \neq -2m - 3, -2m - 5, \ldots$

$$\langle \mathfrak{S}, W(\phi^{(m)}, \phi^{(m)}) \rangle = \frac{m!}{\Gamma(m+\frac{3+\mu}{2})}(2\pi)^{\frac{1+\mu}{2}}{}_2F_1\left(\frac{1+\mu}{2}, m+1; m+\frac{3+\mu}{2}; -1\right).$$
(6.4.12)

Applying also the case $m = 0$ of this identity, together with the equation $(\Theta_0\mathfrak{S})(i) = \langle \mathfrak{S}, W(\phi^{(0)}, \phi^{(0)}) \rangle$, one obtains the lemma. $\square$

In each of the two parts of the domain $0 < \mathrm{Re}\, \nu_1 < 1$, $0 < \mathrm{Re}\, \nu_2 < 1$ defined by the conditions $\mathrm{Re}\,(\nu_1 + \nu_2) < 1$ and $\mathrm{Re}\,(\nu_1 + \nu_2) > 1$, we have introduced a certain quasi-distribution (in the sense of Remark 4.1.1(ii)) $\mathfrak{T} = \mathrm{Sharp}\,(\mathfrak{E}_{\nu_1}, \mathfrak{E}_{\nu_2})$. It was to be expected that $(2i\pi\mathcal{E})\,\mathfrak{T}$ would be a genuine distribution when $\mathrm{Re}\,(\nu_1 + \nu_2) < 1$, while $\left[(2i\pi\mathcal{E})^2 - 4\right](2i\pi\mathcal{E})\,\mathfrak{T} = (R-4)(2i\pi\mathcal{E})\,\mathfrak{T}$ would be a genuine distribution when $\mathrm{Re}\,(\nu_1 + \nu_2) < 1$ or $\mathrm{Re}\,(\nu_1 + \nu_2) > 1$. From Theorem 5.4.1, Proposition 5.4.2 and Theorem 6.1.1, it turned out that it did in both cases, after all, coincide with a genuine distribution. But the identity that justified its introduction only allowed to test in on functions in the image of $\mathcal{S}(\mathbb{R}^2)$ under $2i\pi\mathcal{E}$ in the first case, under $(R-4)(2i\pi\mathcal{E})$ in the second one.

When ν is on the critical line $\mathrm{Re}\,\nu = \frac{1}{2}$, then it follows from (6.2.6) that $\mathrm{Sharp}\,(\mathfrak{E}_{\bar\nu}^{\mathrm{resc}}, \mathfrak{E}_{\nu}^{\mathrm{resc}})$ is not a genuine distribution, as already observed in the proof of the second part of Theorem 6.2.2. One can give the square of zeta, on the critical line, an interpretation as a discontinuity.

Proposition 6.4.3. *The function*

$$H(\nu) = \langle (R-4)\,\mathrm{Sharp}\,(\mathfrak{E}_{\bar\nu}^{\mathrm{resc}}, \mathfrak{E}_{\nu}^{\mathrm{resc}}),\, W(\phi^1, \phi^1)\rangle, \qquad (6.4.13)$$

well-defined as a real-analytic function of ν in each of the two components of the domain $0 < \mathrm{Re}\,\nu < 1$, $\mathrm{Re}\,\nu \neq \frac{1}{2}$, has well-defined limits at any point ω on the critical line, from the left or from the right. One has

$$H(\omega + 0) - H(\omega - 0) = -\frac{16}{3}\,\frac{{}_2F_1\left(\frac{1}{2}, 2; \frac{5}{2}; -1\right)}{{}_2F_1\left(\frac{1}{2}, 1; \frac{3}{2}; -1\right)}\, E_{\frac{1}{2}}^*(i)\,|\zeta(\omega)|^2. \qquad (6.4.14)$$

Proof. We use (6.2.6). The operator $R-4$ acts on $\mathfrak{E}_{1+2\sigma}^{\mathrm{resc}}$ as the multiplication by $(-1 - 2\sigma)^2 - 4 = (2\sigma - 1)(2\sigma + 3)$ while the function $\frac{1}{\zeta(-1-2\sigma)}$ has a simple pole at $\sigma = \frac{1}{2}$, just killed by the preceding factor. None of the other terms of the decomposition (6.2.6) has any singularity in the strip $0 < \mathrm{Re}\,\nu < 1$, but the second term, present when $\sigma = \mathrm{Re}\,\nu < \frac{1}{2}$, is absent when $\mathrm{Re}\,\nu > \frac{1}{2}$. The jump of the function H at a point ω on the critical line is thus (since $R-4$ acts on $\mathfrak{E}_0^{\mathrm{resc}}$ as the multiplication by -4)

$$\frac{4\,|\zeta(\omega)|^2}{\zeta(0)}\,\langle \mathfrak{E}_0^{\mathrm{resc}},\, W(\phi^{(1)}, \phi^{(1)})\rangle. \qquad (6.4.15)$$

According to (6.4.2),

$$\langle \mathfrak{E}_0^{\mathrm{resc}},\, W(\phi^{(1)}, \phi^{(1)})\rangle = \frac{2}{3}\,\frac{{}_2F_1\left(\frac{1}{2}, 2; \frac{5}{2}; -1\right)}{{}_2F_1\left(\frac{1}{2}, 1; \frac{3}{2}; -1\right)}\,(\Theta_0\mathfrak{E}_0^{\mathrm{resc}})(i), \qquad (6.4.16)$$

where

$$(\Theta_0\mathfrak{E}_0^{\mathrm{resc}})(i) = E_{\frac{1}{2}}^*(i) \qquad (6.4.17)$$

(a well-defined nonzero number). The proposition follows.

Lemma 6.4.2 provides of course a similar formula in which $W(\phi^{(1)}, \phi^{(1)})$ is replaced by $W(\phi^{(2n+1)}, \phi^{(2n+1)})$: only the coefficient of $|\zeta(\omega)|^2$ is changed. One could also consider $W(\phi^{(2n)}, \phi^{(2n)})$ instead, but this would demand replacing the operator $R - 4$ by the operator $(R - 4)(2i\pi\mathcal{E})$: recall that we have taken benefit of the fact that $W(u, u)$ lies in the image of $\mathcal{S}(\mathbb{R}^2)$ under $2i\pi\mathcal{E}$ in the case when $u \in \mathcal{S}_{\text{odd}}(\mathbb{R})$, a fact never true when u is even (Lemma 6.2.1).

One should emphasize that, while (from (6.3.17)), one has

$$H(\nu) = 3\left[-2\,\|\operatorname{Op}\left(\mathfrak{E}_\nu^{\text{resc}}\right)\phi^1\|^2 + 3\,\|\operatorname{Op}\left(\mathfrak{E}_\nu^{\text{resc}}\right)\phi^3\|^2\right] \tag{6.4.18}$$

when $\operatorname{Re}\nu < \frac{1}{2}$, this equation ceases to be valid when $\operatorname{Re}\nu > \frac{1}{2}$: neither term on the right-hand side is meaningful. What remains valid in this case, with the notation in the proof of Theorem 6.2.2, is that $H(\nu)$ is the limit as $N \to \infty$ of the expression $3\left[-2\,\|\operatorname{Op}\left(\mathfrak{E}_\nu^{[N]}\right)\phi^1\|^2 + 3\,\|\operatorname{Op}\left(\mathfrak{E}_\nu^{[N]}\right)\phi^3\|^2\right]$. $\qquad\square$

In connection with (6.2.6), this gives when $|\operatorname{Re}\nu| < \frac{1}{2}$, $\nu \neq 0$ an expression of $\|\operatorname{Op}\left(\mathfrak{E}_\nu^{\text{resc}}\right)\phi^{(2n+1)}\|^2$. One can also test the right-hand side of the identity (6.2.6) on the function $G_\alpha \in \mathcal{S}(\mathbb{R}^2)$.

Lemma 6.4.4. *With* $\delta = \tanh\frac{\alpha}{2}$, *one has if* $\mu \neq \pm 1$

$$\langle\mathfrak{E}_\mu^{\text{resc}}, G_\alpha\rangle = \frac{1}{2}\delta^{-\frac{1}{2}}(1-\delta^2)^{\frac{1}{2}}\left[\delta^{\frac{\mu}{2}} - \delta^{-\frac{\mu}{2}}\right]E_{\frac{1-\mu}{2}}^*(i) \tag{6.4.19}$$

and, for every pair r, ℓ *with* $r \in \mathbb{Z}^\times$,

$$\langle\mathfrak{N}_{r,\ell}^{\text{resc}}, G_\alpha\rangle = \frac{1}{2}\delta^{-\frac{1}{2}}(1-\delta^2)^{\frac{1}{2}}\left[\delta^{\frac{i\lambda_r}{2}} - \delta^{-\frac{i\lambda_r}{2}}\right]\mathcal{N}_{|r|,\ell}(i). \tag{6.4.20}$$

Proof. Associating (6.4.8) with (6.3.8), (6.3.9), one obtains if $\mathfrak{S}$ is homogeneous of degree $-1 - \mu$ and $t = e^{-\alpha}$,

$$\langle\mathfrak{S}, \mathcal{G}F_\alpha\rangle = t^{\frac{1}{2}}(1-t)^{\frac{-\mu-1}{2}}(1+t)^{\frac{\mu-1}{2}}(\Theta_0\mathfrak{S})(i) \tag{6.4.21}$$

and

$$\langle\mathfrak{S}, G_\alpha\rangle = \frac{1}{2}t^{\frac{1}{2}}\left[(1-t)^{\frac{\mu-1}{2}}(1+t)^{\frac{-\mu-1}{2}} - (1-t)^{\frac{-\mu-1}{2}}(1+t)^{\frac{\mu-1}{2}}\right](\Theta_0\mathfrak{S})(i). \tag{6.4.22}$$

As $t^{\frac{1}{2}}(1-t^2)^{-\frac{1}{2}} = \frac{1}{2}\delta^{-\frac{1}{2}}(1-\delta^2)^{\frac{1}{2}}$, the two equations are consequences of the first and last lines of Proposition 2.1.1. $\qquad\square$

Theorem 6.4.5. *Given* $\nu = \sigma + it$ *with* $0 < \sigma < \frac{1}{2}$, *set, for* $0 < \delta < 1$,

$$\sum_{n=0}^{\infty} e^{-(2n+\frac{3}{2})\alpha}\|\operatorname{Op}\left(\mathfrak{E}_\nu^{\text{resc}}\right)\phi^{(2n+1)}\|^2$$

$$= (2\pi)^{\frac{1}{2}+\sigma}\frac{\Gamma(\frac{1}{2} - \sigma)}{4(1 + 2\sigma)}|\zeta(-\nu)|^2 E_{1+\sigma}(i) \times \alpha^{-1-\sigma} + \operatorname{Err}(\nu; \alpha). \tag{6.4.23}$$

If $\zeta(\nu) \neq 0$, the error term $\mathrm{Err}(\nu; \alpha)$ is a $O\left(\alpha^{-1}\right)$ as $\delta \to 0$, and is not a $O\left(\alpha^{-\theta}\right)$ for any $\theta < 1$. If $\zeta(\nu) = 0$, the error term is a $O\left(\alpha^{-\frac{1}{2}}\right)$.

Proof. Let us start by remarking that $E_{1+\sigma}(i) = \frac{1}{2}\sum_{(m,n)=1}(m^2 + n^2)^{-1-\sigma}$ is positive, while the factor $\frac{\Gamma(\frac{1}{2}-\sigma)}{4(1+2\sigma)}$ is positive if $0 < \sigma < \frac{1}{2}$, negative if $\frac{1}{2} < \sigma < 1$. The main term is positive since we are in the first case, as it should be. But it would not remain so in the second case, so that it is not for technical reasons only that we had to assume that $|\mathrm{Re}\,(\nu_1 + \nu_2)| < 1$ in Theorem 5.4.1.

Setting $\delta = \tanh \frac{\alpha}{2}$, we examine one at a time the results of testing on G_α the various terms on the right-hand side of (6.2.6). According to (6.4.19), one has

$$\langle \mathfrak{E}^{\mathrm{resc}}_{1+2\sigma}, G_\alpha \rangle = \frac{1}{2}\,\delta^{-\frac{1}{2}}(1 - \delta^2)^{\frac{1}{2}}\left[\delta^{\frac{1}{2}+\sigma} - \delta^{-\frac{1}{2}-\sigma}\right]E^*_{-\sigma}(i), \tag{6.4.24}$$

and

$$E^*_{-\sigma}(i) = E^*_{1+\sigma}(i) = \zeta^*(2 + 2\sigma)\,E_{1+\sigma}(i) = \zeta^*(-1 - 2\sigma)\,E_{1+\sigma}(i). \tag{6.4.25}$$

It follows after some elementary calculations that

$$\frac{|\zeta(-\nu)|^2}{\zeta(-1 - 2\sigma)}\,\langle \mathfrak{E}^{\mathrm{resc}}_{1+2\sigma}, G_\alpha \rangle = \frac{1}{8}\,\pi^{\frac{1}{2}+\sigma}\Gamma(-\frac{1}{2} - \sigma)\,|\zeta(-\nu)|^2$$
$$\cdot\,\delta^{-\frac{1}{2}}(1 - \delta^2)^{\frac{1}{2}}\left[\delta^{\frac{1}{2}+\sigma} - \delta^{-\frac{1}{2}-\sigma}\right]E_{1+\sigma}(i) \tag{6.4.26}$$

is the sum of the main term on the right-hand side of (6.4.23) and of a $O\left(\delta^{1-\sigma}\right)$.

The same formula, only replacing ν by $-\nu$, shows that the result of testing on G_α the second (exceptional) term on the right-hand side of (6.2.6) is a $O\left(\delta^{-1+\sigma}\right)$ and no better than that, unless of course $\zeta(\nu) = 0$, in which case this term is zero.

The 3rd and 4th (exceptional) terms on the right-hand side of (6.2.6), when tested on G_α, combine, recalling that $\nu = \frac{1}{2} + it$, to

$$2\,\mathrm{Re}\left[\frac{\zeta(-\bar{\nu})\zeta(\nu)}{\zeta(-1 - \bar{\nu} + \nu)}\,\langle \mathfrak{E}^{\mathrm{resc}}_{-1-\bar{\nu}+\nu}, G_\alpha \rangle\right]$$
$$= \mathrm{Re}\left(\frac{\zeta(-\bar{\nu})\zeta(\nu)}{\zeta(-1 - \bar{\nu} + \nu)}\,E_{\frac{2+\bar{\nu}-\nu}{2}}(i)\,\delta^{-1+it}\right) + O(1), \tag{6.4.27}$$

which is as $\delta \to 0$ a $O\left(\delta^{-1}\right)$ but not a $O\left(\delta^{-\theta}\right)$ for any $\theta < 1$, unless $\zeta(\nu) = 0$. We must recall, at this point, that $E_{1-it}(i) \neq 0$ for $t \in \mathbb{R}$: starting from the identity $E_s(i) = \frac{1}{2}\sum_{(m,n)=1}(m^2 + n^2)^{-s} = \frac{2\zeta(s)}{\zeta(2s)}L(\chi_4, s)$ for $\mathrm{Re}\,s \geq 1$, where χ_4 is the non-trivial Dirichlet character modulo 4, this is just a well-known extension [27, p.373] of Hadamard's theorem (the non-vanishing of zeta on the boundary of the critical strip).

So as to bound by $C\,\delta^{-\frac{1}{2}}$ the result of testing on G_α the integral term on the right-hand side of (6.2.6), it suffices, in view of (6.4.19), to obtain the bound $|E^*_{\frac{1-i\lambda}{2}}(i)| \le C\,(1+|\lambda|)^{-N}$ for an arbitrary N. This follows from the Fourier series expansion (2.1.18)

$$E^*_{\frac{1-i\lambda}{2}}(i) = \zeta^*(1-i\lambda) + \zeta^*(1+i\lambda) + 2\sum_{k\neq 0} |k|^{-\frac{i\lambda}{2}} \sigma_{i\lambda}(|k|)\, K_{\frac{i\lambda}{2}}(2\pi\,|k|), \quad (6.4.28)$$

and the estimate obtained by going halfway between the estimate (4.6.9) and the esimate [22, p.85]

$$K_{\frac{i\lambda}{2}}(2\pi|k|) = \int_0^\infty e^{-2\pi|k|\,\cosh t}\cos\frac{\lambda t}{2}\,dt \le e^{-\pi|k|}\int_0^\infty e^{-2\pi(\cosh t-\frac{1}{2})}dt, \quad (6.4.29)$$

so as to get an estimate of $K_{\frac{i\lambda}{2}}(2\pi\,|k|)$ making it possible to save simultaneously powers of $|\lambda|$ and of $|k|$. Exactly the same works with the series of Hecke distributions on the last line of (6.2.6), only replacing (6.4.28) by (2.1.19).

Finally, in view of the size of the remainder of interest, one may replace $\delta^{-1-\sigma} = \left(\frac{\alpha}{2}\right)^{-1-\sigma}(1+\mathrm{O}(\alpha))$ by $2^{1+\sigma}\alpha^{-1-\sigma}$, and the same goes with the other powers of δ involved in the statement of Theorem 6.4.5. $\square$

Remarks 6.4.1. (i) In the beginning of the proof of Theorem 6.4.5, we have observed that the main term in the expansion of the series $\sum_{m\ \mathrm{odd}} e^{-\alpha(m+\frac{1}{2})}\|\operatorname{Op}(\mathfrak{E}^{\mathrm{resc}}_\nu)\,\phi^{(m)}\|^2$ (as a function of $\alpha \to 0$) was of course positive. While the right-hand side of (6.2.6) can certainly be tested on the symbol $\frac{1}{2}(F_\alpha + \mathcal{G}\,F_\alpha)$ in place of G_α (which is the difference, rather than the sum, of the two terms), it will then yield when $0 < \operatorname{Re}\nu < \frac{1}{2}$ and α is small enough a negative result, hence it cannot provide an expression of a sum such as $\sum_{m\ \mathrm{even}} e^{-\alpha(m+\frac{1}{2})}\|\operatorname{Op}(\mathfrak{E}^{\mathrm{resc}}_\nu)\,\phi^{(m)}\|^2$. But this fits with the fact that, as soon as in Lemma 4.1.2, we have seen that if $A_1 = \operatorname{Op}(h_{\nu_1,k_1})$, $A_2 = \operatorname{Op}(h_{\nu_2,k_2})$ and $k_1+k_2 = 0$, the composition $A = A_1 A_2$ does not act from $\mathcal{S}(\mathbb{R})$ to $\mathcal{S}'(\mathbb{R})$, while each of the two terms of the difference $PAQ - QAP$ does. In particular, even when $0 < \operatorname{Re}\nu < \frac{1}{2}$, $\operatorname{Op}(\mathfrak{E}^{\mathrm{resc}}_\nu)\,\phi^{(m)}$ does not lie in $L^2(\mathbb{R})$ when m is even, and the terms of the series just alluded to are meaningless. Note from (6.3.15) that, despite (6.3.2), it is not true, with the notation just used, that each of the two terms of the difference $CAC^* - C^*AC$ acts from $\mathcal{S}(\mathbb{R})$ to $\mathcal{S}'(\mathbb{R})$.

(ii) If $\mathfrak{N}$ is a Hecke distribution, it follows from Theorem 5.3.4 and arguments similar to the ones in the proof of Theorem 6.4.5 that

$$\sum_{n=0}^\infty e^{-\alpha(2n+\frac{3}{2})}\|\operatorname{Op}(\mathfrak{N})\,\phi^{(2n+1)}\|^2$$

is a $\mathrm{O}\left(\alpha^{-\frac{1}{2}}\right)$ as $\alpha \to 0$: there is no "main term" in this estimate.

Chapter 7

From non-holomorphic to holomorphic modular forms

This expository chapter has the following two aims. First, we wish to indicate what a proper generalization of pseudodifferential analysis, combining it with representation theory and, ultimately, with arithmetic, could be (and should be, in our opinion). This will be done under the heading "Quantization", a traditional, and quite appropriate, wording originating with the forefathers of quantum mechanics, and stressing the fact that the measurement process requires making operators (quantum observables) out of functions on some phase space (classical observables). Quantization theory can be combined with representation theory in two reciprocal directions: one can, as done in Kirillov's theory, build a process (starting with polarizations) enabling one to construct irreducible representations from the geometric action of a Lie group G on one of its coadjoint orbits $\mathcal{X}$; or, which is one way to approach some of the generalizations we have in mind, one may wish, starting from an irreducible representation π of G in a Hilbert space H, to build a symbolic calculus of (partially defined in general) operators on H by means of symbols living on $\mathcal{X}$.

In the first section to follow, we shall suggest a slightly more general program, then specialize it immediately to the case when $\mathcal{X} = \mathbb{R}^2$ and the quantization rule is the Weyl calculus, a situation involving already many more possibilities than what one might expect, depending on which group G and which space of symbols are concerned. In each reported case, we shall be led to a sharp composition formula (the composition of symbols corresponding to the composition of operators): as will be seen, there are at least three essentially distinct such formulas.

Some readers, especially the ones with a broader knowledge of arithmetic, cannot fail to ask whether anything comparable to what has been done in this book can also be done if one replaces non-holomorphic modular forms by modular forms of the holomorphic type. We shall indicate in Section 7.2 that this is indeed the case: our presentation will be a (compressed) summary of previous work on anaplectic representation and alternative pseudodifferential analysis, the substitutes for the metaplectic representation and Weyl pseudodifferential analysis called for by this aim.

7.1   Quantization theory and composition formulas

The basic frame we consider as appropriate for piecing pseudodifferential analysis (in a broad sense) and representation theory together requires the following data: a manifold $\mathcal{X}$ (the so-called phase space), a linear space $\mathcal{C}$, often consisting of functions on some space the dimension of which is half that of $\mathcal{X}$, a representation π of some group G in $\mathcal{C}$, finally a "quantization rule" Op associating to any element $\mathfrak{S}$ in some specified space of functions or distributions in $\mathcal{X}$ (a "symbol") an operator $\mathrm{Op}(\mathfrak{S})$ in $\mathcal{C}$, possibly partially defined only. These data imply the definition of a representation $\widetilde{\pi}$ of G in the space of symbols, to wit the one defined by the identity

$$\pi(g)\,\mathrm{Op}(\mathfrak{S})\,\pi(g)^{-1} = \mathrm{Op}\left(\widetilde{\pi}(g)^{-1}\mathfrak{S}\right), \qquad (7.1.1)$$

a truly meaningful formula only in the case when all operators $\mathrm{Op}(\mathfrak{S})$ under consideration have a common $\pi(G)$-invariant domain.

   This is a potentially very rich structure in view of the following possibilities, none of which can, however, be implemented without a great amount of work in the more interesting cases. First, the usual composition of operators combines with the quantization rule to produce a (partially defined) sharp composition rule of symbols. Next, the representation $\widetilde{\pi}$ can sometimes be decomposed into irreducibles: combining the sharp composition of irreducible symbols with it, one reaches what we consider as being the right concept of composition formula in the given pseudodifferential analysis. This will be made explicit presently in a few quite distinct cases. Finally, given a subgroup Γ of G (typically an "arithmetic" subgroup of a Lie group), one can specialize in the consideration of symbols invariant under all elements $\widetilde{\pi}(g)$ with $g \in \Gamma$, still asking for the decomposition of the sharp product $\mathfrak{S}_1 \# \mathfrak{S}_2$ of two such symbols into "irreducible" terms, a word to be taken in a sense no longer referring, in general, to a group action. Still within the general scheme, we may reserve the name of "geometric quantization" to situations in which the representation $\widetilde{\pi}$ is given by the equation $\widetilde{\pi}(g)\,\mathfrak{S} = \mathfrak{S} \circ g^{-1}$ in terms of an action of G by diffeomorphisms of the phase space. In Section 7.2, we shall come across a first example of quantization lying outside this geometric scheme.

   The cases considered up to this point include those in which $\mathcal{X} = \mathbb{R}^2$, $\mathcal{C} = L^2(\mathbb{R})$ and the quantization rule Op is that defining the Weyl calculus, while $G = SL(2, \mathbb{R})$ (resp. $\mathbb{R}^2$), the projective representation π being the metaplectic representation (resp. the projective representation τ in (3.1.7)): recall that one obtains a genuine representation if replacing $SL(2, \mathbb{R})$ (resp. $\mathbb{R}^2$) by its twofold cover (resp. by the central extension of $\mathbb{R}^2$ known as the Heisenberg group). The distribution $\mathfrak{S}$ can be any element of $\mathcal{S}'(\mathbb{R}^2)$ since we have allowed only partially defined operators: but, then, the sharp composition of symbols is only a partially defined operation too. One could decide to allow only symbols in $L^2(\mathbb{R}^2)$ or even $\mathcal{S}(\mathbb{R}^2)$: all operators are then of Hilbert-Schmidt type, and the sharp composition of symbols is everywhere defined. We have also considered, which was what this book mostly consists of, the case of Γ-invariant symbols with $\Gamma = SL(2,\mathbb{Z})$: it led, as was to be expected, to great difficulties.

We wish to show, first, that the above concepts defining what we regard as a proper sharp composition formula lead, in the case of the Weyl calculus, to all known special cases of such a formula. In the case in which the basic space of symbols is $L^2(\mathbb{R}^2)$ and the representation π is the Heisenberg representation of the Heisenberg group in $L^2(\mathbb{R})$, the representation $\widetilde{\pi}$ is the same as the action by translations of $\mathbb{R}^2$ in $L^2(\mathbb{R}^2)$ (3.1.10). The space $L^2(\mathbb{R}^2)$ decomposes as a continuous superposition of irreducible subspaces, each being of multiplicity 1 and characterized by a "generalized" (i.e., not in $L^2(\mathbb{R}^2)$) function $(x,\xi) \mapsto e^{2i\pi(x\eta - y\xi)}$. Now, this function is just the symbol of the operator $\tau_{y,\eta}$ in (3.1.7). Our program thus calls, in this case, for the decomposition of a product $\tau_{y,\eta}\tau_{y',\eta'}$ into a superposition of operators $\tau_{y'',\eta''}$: the answer, given in (3.1.8), shows that just one term, rather than a superposition, is needed, and this formula is sometimes called the Weyl exponential version of the Heisenberg commutation relation. Combining this formula with the symplectic Fourier transformation in $\mathcal{S}'(\mathbb{R}^2)$, one obtains the fully equivalent integral formula (3.3.1), all practitioners of pseudodifferential analysis are familiar with.

Let us consider now the case when, still using the Weyl calculus, we put the emphasis not on the Heisenberg representation, but on the metaplectic representation: then, (3.1.11) substitutes for (3.1.10), but there are still several possible choices for the given space of symbols, the main demand being that it should be invariant under linear transformations of the variable (x,ξ) associated to matrices in $G = SL(2,\mathbb{R})$. Let us take for it, first, the space of all polynomials in (x,ξ): the irreducible spaces for the action of G under consideration consist of the spaces of polynomials globally homogeneous of a given degree. The composition formula, according to our general program, thus consists in expressing the sharp product $h_1 \# h_2$ of two homogeneous polynomials as a linear combination of homogeneous polynomials. It reads

$$(h_1 \# h_2)(x,\xi) = \sum \frac{(-1)^j}{j!\,k!} \left(\frac{1}{4i\pi}\right)^{j+k} \frac{\partial^{j+k}}{\partial x^j \partial \xi^k} h_1(x,\xi) \frac{\partial^{j+k}}{\partial x^k \partial \xi^j} h_2(x,\xi) \quad (7.1.2)$$

and is, again, universally known to practitioners of pseudodifferential analysis. Note that such a formula would be worthless when dealing with automorphic symbols, since not a single term of this series would in general be meaningful: automorphic distributions are just too singular.

Exactly the same formula would do, had we taken for space of symbols the space of C^∞ symbols, polynomial with respect to the variable ξ only: but this would require changing the group G to the subgroup of matrices $\{\left(\begin{smallmatrix} a & b \\ 0 & a^{-1} \end{smallmatrix}\right)\}$. This space of symbols is important, since the associated operators are exactly the differential operators with C^∞ coefficients. Pseudodifferential analysis started, half a century ago, with applications to P.D.E of situations in which symbols with a certain type of behaviour at infinity (especially with respect to ξ) were helpful: then, the formula (7.1.2), without being an exact one, is still often valid as giving an asymptotic expansion of the symbol $h_1 \# h_2$. Do not confuse "asymptotic" with "formal": proving that the remainders are the symbols of increasingly "good" operators is essential here. This may have led many people to the misconception that asymptotic (or even formal) expansions are a permanent part of (generalized) pseudodifferential analysis.

Consider again the Weyl calculus, choosing $\mathcal{C} = \mathcal{S}(\mathbb{R})$ and taking the space $\mathcal{S}(\mathbb{R}^2)$ as a space of symbols (then, the sharp composition is always well-defined), finally using the metaplectic representation again. The irreducible spaces for the action $\widetilde{\pi}$ (the action in (3.1.11), extended to $L^2(\mathbb{R}^2)$) make up a continuous family of irreducible spaces with infinite multiplicity, characterized by a pure imaginary number $i\lambda$ together with an index $\delta = 0$ or 1: the space associated to the pair $(i\lambda, \delta)$ consists of all functions h in the plane, globally homogeneous of degree $-1 - i\lambda$ and of parity defined by δ, such that the function $s \mapsto h(s, 1)$ lies in $L^2(\mathbb{R})$. Our program calls for the decomposition of the sharp product of two such functions as an integral superposition of such functions. A complete answer was given in [39, p.31] and, for simplicity, we have reproduced here, in equations (3.3.2)-(3.3.5), the case when δ is fixed to the value 0: we only considered here globally even distributions in the plane and, accordingly, parity-preserving operators from $\mathcal{S}(\mathbb{R})$ to $\mathcal{S}'(\mathbb{R})$.

Contrary to the preceding composition formulas, the last one (first introduced in [34, section 5]) is generally not known. It was crucial, however, in the present book, in which the case of automorphic symbols was discussed. Note that this fits again with our general scheme, the irreducible terms of decompositions of automorphic distributions being Eisenstein and Hecke distributions: however, the space of symbols is no longer acted upon by a non-trivial group, and irreducible terms refer instead to decompositions relative to the spectral theory of a family of commuting operators (the Euler and Hecke operators).

We do not wish, here, to expand these views on quantization theory, but we cannot, so as to prepare for the next section, avoid mentioning how the way the metaplectic representation relates to the discrete series of representations of the twofold cover of $SL(2, \mathbb{R})$ is construed in connection with (generalized) pseudodifferential analysis. Recall [18] that the discrete series $(\mathcal{D}_{\tau+1})_{\tau > -1}$ of the universal cover of $SL(2, \mathbb{R})$ is defined by means of operators $\mathcal{D}_{\tau+1}(g)$, $g = \left(\begin{smallmatrix} a & b \\ c & d \end{smallmatrix}\right)$ acting on spaces of holomorphic functions in the hyperbolic half-plane Π, defined up to scalar factors in the group $\exp{(2i\pi\tau\mathbb{Z})}$ by the equation

$$(\mathcal{D}_{\tau+1}(g)f)(z) = (-cz + a)^{-\tau-1} f\left(\frac{dz - b}{-cz + a}\right), \qquad z \in \Pi. \tag{7.1.3}$$

When $\tau > 0$, one realizes $\mathcal{D}_{\tau+1}$ as a unitary representation in some easily defined Hilbert space of functions f. Taking a Laplace transformation as an intertwining operator, one may realize the representation as a representation $\pi_{\tau+1}$ defined with the help of integral kernels of Bessel type in the space $L^2((0, \infty); t^{-\tau}dt)$. One of the advantages of the realization $\pi_{\tau+1}$ is that it extends in an easier way to the case when $\tau > -1$, retaining its unitarity property.

When $2\tau \in \mathbb{Z}$, the representation $\pi_{\tau+1}$ may be regarded as a representation of the twofold cover of $SL(2, \mathbb{R})$, another name for the metaplectic group. While the metaplectic representation is not irreducible, since the space $L^2(\mathbb{R})$ decomposes as the sum of invariant subspaces $L^2_{\text{even}}(\mathbb{R}) \oplus L^2_{\text{odd}}(\mathbb{R})$, every representation $\pi_{\tau+1}$ is. A fundamental fact is that the even (resp. odd) part of the metaplectic representation is unitarily equivalent to $\pi_{\frac{1}{2}}$ (resp. $\pi_{\frac{3}{2}}$), the intertwining operator being, up to normalization, just the quadratic change of variable $x \mapsto x^2$ from the line to the half-line. Details were given in [35, p.64], in a way of interest to us for several reasons. We considered there, in place of the (metaplectic) representation Met $\sim$ $\pi_{\frac{1}{2}} \oplus \pi_{\frac{3}{2}}$, a more general representation $\text{Met}_p \sim \pi_{p+\frac{1}{2}} \oplus \pi_{p+\frac{3}{2}}$, with $p = 0, 1, \ldots$. The representation Met_p is precisely the p-metaplectic representation alluded to in Section 3.3 and at the end of Section 6.3.

We have made use, in Chapter 6, of the harmonic oscillator $\pi(P^2 + Q^2)$, to be denoted here as Λ. Its position in relation to the metaplectic representation is central, since it is the infinitesimal operator of this representation associated to the element $\left(\begin{smallmatrix} 0 & 1 \\ -1 & 0 \end{smallmatrix}\right)$ of the Lie algebra $\mathfrak{sl}(2, \mathbb{R})$, in other words the infinitesimal generator, in the sense of Stone's theorem on one-parameter groups of unitary operators, of the image under Met of the subgroup $SO(2)$ of $SL(2, \mathbb{R})$. Now, such an operator, to be denoted as L_p, can be defined in relation to the representation $\pi_{p+\frac{1}{2}}$, and is just as central there. But it can no longer be written as the sum of squares of two first-order differential operators (on the half-line). Our interest in piecing together the two representations $\pi_{p+\frac{1}{2}}$ and $\pi_{p+\frac{3}{2}}$ originated from the fact that if one identifies, by means of two properly normalized versions of the quadratic change of variable $x \mapsto x^2$ from the line to the half-line, pairs of functions on the half-line with functions on the line, decomposed into their even and odd parts, one recovers the possibility of decomposing as a sum of squares not the operator L_p, but the direct sum $\Lambda_p = \left(\begin{smallmatrix} L_p & 0 \\ 0 & L_{p+1} \end{smallmatrix}\right)$: this was just a generalization of the observation that, under the unitary equivalence Met $\sim \pi_{\frac{1}{2}} \oplus \pi_{\frac{3}{2}}$, the harmonic oscillator Λ becomes $\left(\begin{smallmatrix} L_0 & 0 \\ 0 & L_1 \end{smallmatrix}\right)$. In the general case, one obtains the decomposition $\Lambda_p = \pi(P_p^2 + Q^2)$, where

$$P_p = \frac{1}{2i\pi}\begin{pmatrix} 0 & \frac{d}{dx} + \frac{p}{x} \\ \frac{d}{dx} - \frac{p}{x} & 0 \end{pmatrix}: \tag{7.1.4}$$

the decomposition of $L^2(\mathbb{R})$ into its even and odd parts is to be made in the order associated to the parity of p: the operator P_p happens to be the simplest member of the class of so-called Dunkl operators [8].

Now that we dispose, for general p, of a pair (Q, P_p), it is a natural thing to define a p-Weyl calculus generalizing the Weyl calculus (its $p = 0$ case), defining the quantization rule by the demand that, given $(y, \eta) \in \mathbb{R}^2$, the operator with symbol $(x, \xi) \mapsto e^{2i\pi(x\eta - y\xi)}$ should be the operator $\exp(2i\pi(\eta Q - \xi P_p))$ (Stone's theorem again: the operator $\eta Q - \xi P_p$ is self-adjoint). This was done, along these lines, in [35, sections 7,9], and we have already mentioned, in Section 3.3 and at the end of Section 6.3, that the main difficulties (regarding the possibility to define the composition of operators with automorphic symbols) we came across in the Weyl calculus would cease to be present, had we used instead the p-Weyl calculus with p large enough.

Unfortunately but not unpredictably, the p-Weyl calculus leads to more complicated formulas throughout its development. The main reason is that the pair (Q, P_p) does not fit, unless $p = 0$, among the infinitesimal operators of a unitary representation of a finite-dimensional Lie group (just iterate the bracket operation, starting from the given pair, to see this). Apart from this fact, there are relatively few differences. One must replace the space $\mathcal{S}(\mathbb{R})$ by its image $\mathcal{S}_p(\mathbb{R})$ under the operator of multiplication by the function $x \mapsto x^p$, obtaining again a space of nice functions preserved under the p-metaplectic representation. Of course, the whole construction has been made so that the p-Weyl calculus should satisfy the analogue of the covariance property (3.1.11), just replacing there the representation Met by Met_p.

Let us come back to a situation alluded to in the introduction of this chapter, in which we take for π an irreducible representation of $G = SL(2, \mathbb{R})$: it is then a natural thing, in view of Kirillov's theory, that, so as to build a (generalized) pseudodifferential calculus of operators acting on the corresponding Hilbert space, one should let symbols live on a one-sheeted hyperboloid (resp. one sheet of a two-sheeted hyperboloid) in the dual of the Lie algebra of $SL(2, \mathbb{R})$ if dealing with a representation π from the principal (resp. the discrete) series. In the second case, a quantizing map could be the Berezin one [1, 2], or the one defined in [31], for which, in contrast to the case of the Berezin calculus, a sharp composition formula of integral type (in the style of (3.3.1)) is available [32]. In the case of the one-sheeted hyperboloid, it was developed in [33], where it was shown that a sharp composition formula, somewhat similar to (7.1.2), existed: but, in place of the so-called "Moyal brackets" (the homogeneous terms of the expansion just mentioned), it was necessary to use Rankin-Cohen brackets, a notion of interest in holomorphic modular form theory introduced in [6]. A role of these brackets was reached, independently, in [7], a paper in which formal series in terms of a parameter, satisfying some associativity property, and looking somewhat like the right-hand side of (7.1.2), are built: note that, despite its title, there are no (generalizations of) pseudodifferential operators in that paper, which fits more properly within the so-called star-product theory.

The main idea that led to the construction of the pair consisting of the p-metaplectic representation and p-Weyl calculus, to wit the fact that one should interest oneself in the direct sum of two irreducible representations, will be helpful too in building the pair consisting of the anaplectic representation and alternative pseudodifferential analysis: we shall satisfy ourselves, however, with building an alternative to the Weyl calculus proper, not to the p-Weyl calculus. The phase space will be, again, the plane $\mathbb{R}^2$ but the representation $\widetilde{\pi}$ corresponding to the "covariance formula" which will take the place of (3.1.10) or of (3.1.11) will cease to be of a geometric type, i.e., will cease to be defined by means of changes of coordinates in the plane. This may be the right place to remark that the Weyl calculus is characterized, up to multiplication by a unique constant, by its two covariance properties (3.1.10) and (3.1.11). For any other calculus with these two properties would be linked to the Weyl calculus by means of an operator, acting on symbols, commuting with the pair of operators $\frac{\partial}{\partial x}, \frac{\partial}{\partial \xi}$ as well as with the Euler operator on $\mathbb{R}^2$: such an operator must be scalar. Just the same will be valid with the alternative pseudodifferential analysis, which may well be, in this sense, the only "one-dimensional" competitor of the Weyl calculus.

7.2 Anaplectic representation and pseudodifferential analysis

It is a very exotic world we are entering now: but this is unavoidable in view of our project, which consists in building an alternative to the automorphic Weyl calculus, in which the symbols (again, functions in the plane) of the required species will be decomposable as series of holomorphic modular forms instead of non-holomorphic ones. The aim of this short section is to provide a few of the main ideas, but it could not be considered as an introduction to the subject: the introduction of the book [38] and its Remark 3.1.2 give already some more information, while the book proper covers in detail what will be summed up in this last section.

We start with a description of the first representation $\widetilde{\pi}_1$ of a pair, the first one dealing with the group $SL(2, \mathbb{R})$ (the second one will deal with $\mathbb{R}^2$). There is nothing surprising about it, and people with some experience with harmonic analysis or arithmetic will immediately realize that it is just a special case of more general well-known constructions, involving more linear algebra and a possibly more knowledgeable terminology. We decompose $L^2(\mathbb{R}^2)$ according to the representation, by linear changes of coordinates, of the rotation group, obtaining $L^2(\mathbb{R}^2) = \oplus_{m \in \mathbb{Z}} L_m^2(\mathbb{R}^2)$, where a function h lies in $L_m^2(\mathbb{R}^2)$ if it satisfies the identity

$$h \circ \left(\begin{smallmatrix} \cos\theta & -\sin\theta \\ \sin\theta & \cos\theta \end{smallmatrix} \right) = e^{-im\theta} h. \tag{7.2.1}$$

The action $\widetilde{\pi}_1$ of $SL(2,\mathbb{R})$ on "symbols" (again, functions in $\mathbb{R}^2$) is meant to preserve this decomposition. One defines it on a family of generators of $SL(2,\mathbb{R})$, setting

$$\left[\widetilde{\pi}_1\left(\left(\begin{smallmatrix} 1 & 0 \\ c & 1 \end{smallmatrix}\right)\right)u\right](x,\xi) = u(x,\xi)\, e^{i\pi c(x^2+\xi^2)}, \quad \widetilde{\pi}_1\left(\left(\begin{smallmatrix} 0 & 1 \\ -1 & 0 \end{smallmatrix}\right)\right)u = -i\,\mathcal{F}u,$$

$$\left[\widetilde{\pi}_1\left(\left(\begin{smallmatrix} a & 0 \\ 0 & a^{-1} \end{smallmatrix}\right)\right)u\right](x,\xi) = a^{-1}u(a^{-1}x, a^{-1}\xi), \quad a > 0 \quad (7.2.2)$$

(this time, $\mathcal{F}$ is the usual Euclidean Fourier transformation in the plane, not the symplectic one). This group of transformations does not normalize the action of the additive group of $\mathbb{R}^2$ by translations. To recover such a property, we shall define an entirely different (non-geometric, again) action $\widetilde{\pi}_2$ of this latter group. Given an entire function h in $\mathbb{C}^2$, set, for $(\alpha,\beta) \in \mathbb{R}^2$ (or even $\mathbb{C}^2$),

$$(\widetilde{\pi}_2(\alpha,\beta)h)(x,\xi) = e^{-2\pi\beta(x-i\xi)}h(x - i\alpha, \xi - \alpha). \qquad (7.2.3)$$

This is the action we are interested in, and the basic space $\mathcal{S}^A(\mathbb{R}^2)$ consists of symbols $h \in \mathcal{S}(\mathbb{R}^2)$ which extend as entire functions, and such that the functions $\widetilde{\pi}_2(\alpha,\beta)h$ obtained by letting the pair (α,β) remain in a bounded subset of $\mathbb{C}^2$ make up a bounded subset of $\mathcal{S}(\mathbb{R}^2)$.

One can identify each "isotypic" space $L^2_m(\mathbb{R}^2)$ with $m \neq 0$ as a space of holomorphic functions in the hyperbolic half-plane, as follows. If $m = 1, 2, \ldots$, set $c_m = (2\pi)^{\frac{m-1}{2}}((m-1)\,!)^{-\frac{1}{2}}$. Given $h \in L^2_{\pm m}(\mathbb{R}^2)$, set

$$(\Theta_{\pm m}h)(z) = z^{-m-1}\int_{\mathbb{R}^2}(x\pm i\xi)^m \exp\left(-i\pi\frac{x^2+\xi^2}{z}\right)h(x,\xi)\,dx, \quad z \in \Pi, \;(7.2.4)$$

so that the map $c_m\Theta_m$ is an isometry from $L^2_{\pm m}(\mathbb{R}^2)$ onto the Hilbert space $\mathcal{H}_{m+1}$ consisting of holomorphic functions in Π, square-summable with respect to the measure $(\mathrm{Im}\,z)^{m-1}d\mathrm{Re}\,z\,d\mathrm{Im}\,z$. The map Θ_m intertwines the restriction to $L^2_{\pm m}(\mathbb{R}^2)$ of the representation (7.2.2) with the representation $\mathcal{D}_{m+1}$ as made explicit in (7.1.3). This map has some analogy with the restriction of the map Θ_0 (2.1.5) to the space of even functions in the plane of a given degree of homogeneity, which has an important role in this book. Despite the fact that the representation π to be considered presently is more linked to the full principal series of G than to the discrete series, this occurrence of Π (as opposed to a one-sheeted hyperboloid) does not contradict the last but one paragraph of Section 7.2, in which coherence with Kirillov's method of orbits was emphasized: for we are no longer dealing here with geometric quantization.

We proceed now toward a construction of the anaplectic representation. Just as the metaplectic representation is the sum of two irreducible representations from the discrete series of $G = SL(2,\mathbb{R})$, it will be made of two representations *almost* taken from the full set of unitary representations of g. With Knapp's notation [18], the first one is a representation $\mathcal{C}_{-\frac{1}{2}}$ taken from the complementary series of G.

We need, however, to introduce also a signed version of it, no longer unitarizable but pseudo-unitarizable (more about it presently). To cover both cases, we set [38, p.25] if ρ is real, $0 < |\rho| < 1$, and $\varepsilon = 0$ or 1,

$$(\mathcal{C}_{\rho,\varepsilon}(g)w)(\sigma) = |-b\sigma + d|_\varepsilon^{-1-\rho} w\left(\frac{a\sigma - c}{-b\sigma + d}\right) \quad \text{if } g = \left(\begin{smallmatrix} a & b \\ c & d \end{smallmatrix}\right). \tag{7.2.5}$$

Here, w is a function on the line, and the representation $\pi_{\rho,0}$ is unitary for the scalar product associated to the norm such that

$$\||w\||_{\rho,0}^2 = \int_{-\infty}^{\infty} \overline{w}(\sigma)\,(|D|^\rho w)\,(\sigma)\,d\sigma, \tag{7.2.6}$$

where $|D|^\rho$ stands for the operator of convolution by the Fourier transform of the function $s \mapsto |s|^\rho$. The anaplectic representation is just the sum $\mathcal{C}_{-\frac{1}{2},0} \oplus \mathcal{C}_{\frac{1}{2},1}$. Like the metaplectic representation, it has higher-dimensional analogues but, in contrast to the metaplectic case, the construction of these is much more difficult [36].

The (one-dimensional) anaplectic representation has a pleasant realization as a space of functions u on the line, which we now briefly describe. In usual analysis, there are functions, such as the Hermite functions, which are both regular (say, analytic) and rapidly decreasing at infinity. In the space $\mathfrak{A}$ basic in the anaplectic theory, these two properties will have to be shared between two functions, linked to each other in a one-to-one way. To do so, it is handy to characterize functions of interest by a set of 4 functions, only two of which are independent. Say that a function f of a real variable is nice if it extends as an entire function of $z \in \mathbb{C}$ bounded by some exponential $C \exp\left(R|z|^2\right)$, and its restriction to the positive half-line is bounded by some exponential $C \exp\left(-\varepsilon x^2\right)$ with $\varepsilon > 0$. The space $\mathfrak{A}$ consists of all entire functions u of one variable with the property that there exists a (necessarily unique, a consequence of the Phragmén-Lindelöf theorem) 4-tuple

$$\boldsymbol{f} = (f_0,\, f_1,\, f_{i,0},\, f_{i,1}) \tag{7.2.7}$$

of nice functions such that

$$f_{i,0}(z) = \frac{1-i}{2}\left(f_0(iz) + i\,f_0(-iz)\right),$$
$$f_{i,1}(z) = \frac{1+i}{2}\left(f_1(iz) - i\,f_1(-iz)\right), \tag{7.2.8}$$

and such that the even part u_{even} of u coincides with the even part of f_0, while the odd part u_{odd} of u coincides with the odd part of f_1. An example of function in $\mathfrak{A}$, both typical and fundamental, consists of the function

$$\phi(x) = (\pi\,|x|)^{\frac{1}{2}}\,I_{-\frac{1}{4}}(\pi\,x^2). \tag{7.2.9}$$

Its $\mathbb{C}^4$-realization is the function $\boldsymbol{f} = (\psi, 0, \psi, 0)$, with

$$\psi(x) = 2^{\frac{1}{2}} \, \pi^{-\frac{1}{2}} \, x^{\frac{1}{2}} \, K_{\frac{1}{4}}(\pi \, x^2) = (\pi \, x)^{\frac{1}{2}} \, [\, I_{-\frac{1}{4}}(\pi \, x^2) - I_{\frac{1}{4}}(\pi \, x^2)\,], \quad x > 0. \quad (7.2.10)$$

The space $\mathfrak{A}$ is stable under the usual operators P and Q, as well as under the exponential versions of these; the Heisenberg representation is defined in the usual way. Since functions in $\mathfrak{A}$ may increase, at infinity, like the inverse of a Gaussian function, the usual concepts of integral (over the real line) or of L^2-norm cannot subsist: but they have perfect substitutes, most easily defined with the help of the "$\mathbb{C}^4$-realization" of functions in $\mathfrak{A}$. One defines the "integral" by the equation

$$\text{Int}\,[u] = 2^{\frac{1}{2}} \int_0^\infty (f_0(x) + f_{i,0}(x)) \, dx, \qquad (7.2.11)$$

obtaining indeed a linear form invariant under translations.

Next, if $\boldsymbol{f} = (f_0, f_1, f_{i,0}, f_{i,1})$ and $\boldsymbol{h} = (h_0, h_1, h_{i,0}, h_{i,1})$ are the $\mathbb{C}^4$-realizations of two functions u and v in $\mathfrak{A}$, one sets

$$(v \,|\, u)$$
$$= 2^{\frac{1}{2}} \int_0^\infty \left(\bar{h}_0(x) f_0(x) + \bar{h}_1(x) f_1(x) + \bar{h}_{i,0}(x) f_{i,0}(x) - \bar{h}_{i,1}(x) f_{i,1}(x) \right) \, dx.$$
$$(7.2.12)$$

This is a pseudo-scalar product, the same as a scalar product except for positivity: but it is still non-degenerate. The operators $\exp(2i\pi(\eta Q - yP))$, defined in the usual way, preserve the space $\mathfrak{A}$ even for $(y, \eta) \in \mathbb{C}^2$; the operators obtained when $(y, \eta) \in \mathbb{R}^2$ preserve also the pseudoscalar product.

One can now define a Fourier transformation $\mathcal{F}_{\text{ana}}$ in $\mathfrak{A}$ in a "usual" way, setting

$$(\mathcal{F}_{\text{ana}} u)\,(x) = \text{Int}\,(u \, e_x), \qquad (7.2.13)$$

with $e_x(y) = e^{-2i\pi x y}$. The function ϕ in (7.2.9) has in anaplectic analysis the role played in usual analysis by the standard Gaussian function: in particular, it is (pseudo-)normalized, and invariant under $\mathcal{F}_{\text{ana}}$. One can finally define the anaplectic representation as follows [38, p.19].

It is the unique representation Ana of $SL(2, \mathbb{R})$ in the space $\mathfrak{A}$ with the following properties:

(i) if $g = \left(\begin{smallmatrix} 1 & 0 \\ c & 1 \end{smallmatrix} \right)$, one has $(\text{Ana}(g)\,u)(x) = u(x)\,e^{i\pi c x^2}$;

(ii) if $g = \left(\begin{smallmatrix} a & 0 \\ 0 & a^{-1} \end{smallmatrix} \right)$ with $a > 0$, one has $(\text{Ana}(g)\,u)(x) = a^{-\frac{1}{2}}\,u(a^{-1}x)$;

(iii) one has $\text{Ana}\left(\left(\begin{smallmatrix} 0 & 1 \\ -1 & 0 \end{smallmatrix} \right) \right) = \mathcal{F}_{\text{ana}}$.

This representation is pseudo-unitary, i.e., it preserves the scalar product introduced in (7.2.12). It combines with the usual Heisenberg representation in the way characterized by the equation

$$\text{Ana}(g)\, e^{2i\pi\,(\eta Q - yP)}\, \text{Ana}(g^{-1}) = e^{2i\pi\,(\eta' Q - y' P)} \tag{7.2.14}$$

if $g = \left(\begin{smallmatrix} a & b \\ c & d \end{smallmatrix}\right) \in SL(2,\mathbb{R})$ and $g\left(\begin{smallmatrix} y \\ \eta \end{smallmatrix}\right) = \left(\begin{smallmatrix} y' \\ \eta' \end{smallmatrix}\right)$.

The operators

$$A^* = \pi^{\frac{1}{2}}\left(x - \frac{1}{2\pi}\frac{d}{dx}\right) \quad \text{and} \quad A = \pi^{\frac{1}{2}}\left(x + \frac{1}{2\pi}\frac{d}{dx}\right), \tag{7.2.15}$$

the analogues of the creation and annihilation operators of usual analysis, will be called the raising and lowering operators in anaplectic analysis: for each of them is actually invertible. Starting from the function ϕ, one can then easily, with their help, build a sequence of functions playing in anaplectic analysis the role played by Hermite functions in usual analysis. But the collection (ϕ^j) of such functions is parametrized by $j \in \mathbb{Z}$, not $\frac{1}{2} + \mathbb{N}$: the spectrum of the "harmonic oscillator" is now $\mathbb{Z}$. The more general operators

$$A_z = \pi^{\frac{1}{2}}\left(Q - \bar{z}\,P\right), \quad A_z^* = A_{\bar{z}} = \pi^{\frac{1}{2}}\left(Q - z\,P\right), \quad z \in \Pi, \tag{7.2.16}$$

are invertible as well, and linked to one another by the relation

$$\text{Ana}\left(\left(\begin{smallmatrix} a & b \\ c & d \end{smallmatrix}\right)\right) A_z\, \text{Ana}\left(\left(\begin{smallmatrix} d & -b \\ -c & a \end{smallmatrix}\right)\right) = (c\bar{z} + d)\, A_{\frac{az+b}{cz+d}}. \tag{7.2.17}$$

We have now all the elements needed to define the alternative pseudodifferential analysis. Note that, unlike the p-Weyl calculus, it is not a generalization of the Weyl calculus: despite the formula (7.2.14), it does not define general operators as integral superpositions of the Heisenberg operators $e^{2i\pi(\eta Q - yP)}$, for if such were the case, we would only be led to the formally usual action of $SL(2,\mathbb{R})$ on symbols, not the action (7.2.2). On the other hand, the quantization rule to be introduced now was not at all the result of a guess: as already mentioned, the alternative pseudodifferential analysis is uniquely characterized, up to a normalization constant, by its pair of covariance formulas, and its definition was obtained at the end of a rather lengthy process, summed up in [38, p.35-37].

First, observe from the paragraph around (7.2.4) that each representation $\mathcal{D}_{m+1}$ (with $m = 0, 1, \dots$) occurs twice in the decomposition of $L^2(\mathbb{R}^2)$. Forgetting the term $L_0^2(\mathbb{R}^2)$, one can define two distinct (easily related) pseudodifferential analyses, an ascending (resp. descending) one, keeping all isotypic components $L_m^2(\mathbb{R}^2)$ with $m \geq 1$ (resp. $m \leq -1$). The ascending alternative pseudodifferential analysis is defined as follows. If $h = \sum_{m \geq 1} h_m \in \oplus_{m \geq 1} L_m^2(\mathbb{R}^2)$ lies in $\mathcal{S}^A(\mathbb{R}^2)$ (cf. (7.2.3)), one sets

$$\text{Op}^{\text{asc}}(h) = \sum_{m \geq 1} \text{Op}_m^{\text{asc}}(h_m) \tag{7.2.18}$$

with

$$\mathrm{Op}_m^{\mathrm{asc}}(h_m) = \frac{m}{4\pi}\,\pi^{-\frac{m+1}{2}} \int_\Pi (\Theta_m\, h_m)(z)\, A_z^{-m-1}\, (\mathrm{Im}\, z)^{m+1}\, dm(z), \qquad (7.2.19)$$

where dm is the invariant measure in Π. The alternative pseudodifferential analysis satisfies the two desired covariance identities

$$\mathrm{Ana}(g)\, \mathrm{Op}^{\mathrm{asc}}(h)\, \mathrm{Ana}(g^{-1}) = \mathrm{Op}^{\mathrm{asc}}(\widetilde{\pi}_1(g)\, h)\,, \quad g \in SL(2,\mathbb{R}),$$

$$e^{2i\pi(\eta Q - yP)}\, \mathrm{Op}^{\mathrm{asc}}(h)\, e^{-2i\pi(\eta Q - yP)} = \mathrm{Op}^{\mathrm{asc}}(\widetilde{\pi}_2(y,\eta)\, h)\,. \qquad (7.2.20)$$

The alternative pseudodifferential analysis has been developed in [38, chapter 3], but not for automorphic symbols (i.e., symbols invariant under the transformations $\widetilde{\pi}_1(g)$ with $g \in SL(2,\mathbb{Z})$). However, we have indicated in [38, chapter 5] what are the anaplectic substitutes for the Dirac comb and Eisenstein distributions, as well as the object replacing the Bezout distribution (5.3.9). The L-functions can again be introduced in a spectral-theoretic role, to wit as coefficients of decompositions into functions $h_{m,n}$ playing the same role as the functions $\hom_{\rho,\nu}^{(\varepsilon)}$ in the non-holomorphic theory (1.1.22). One must replace the pair $(2i\pi\mathcal{E}, 2i\pi\mathcal{E}^\natural)$ by $(2i\pi\mathcal{R}, 2i\pi\mathcal{E})$ with $2i\pi\mathcal{R} = \xi\frac{\partial}{\partial x} - x\frac{\partial}{\partial \xi}$ and define, with $z = x + i\xi$, $h_{m,n}(z) = z^m \bar{z}^n$, a joint eigenfunction of the new pair for the pair of eigenvalues $(i(m-n), m+n+1)$. In alternative pseudodifferential analysis, the operator $\mathcal{R}$, which preserves decompositions into isotypic components, is the answer to the operator $\mathcal{E}$ from the usual pseudodifferential analysis, which preserves decompositions into homogeneous components: again, one has the general identity

$$\mathrm{mad}(P \wedge Q)\, \mathrm{Op}^{\mathrm{asc}}(h) = 2i\, \mathrm{Op}^{\mathrm{asc}}(\mathcal{R}h). \qquad (7.2.21)$$

Let us conclude this summary with the sharp composition formula. Given two functions $f_1 \in \mathcal{H}_{m_1+1}$ and $f_2 \in \mathcal{H}_{m_2+1}$ (cf. (7.2.4)), and setting $m = m_1 + m_2 + 1 + 2p$, define the "Rankin-Cohen bracket" of the pair f_1, f_2, depending on m as well, by the equation

$$\mathcal{K}_{m+1}^{m_1+1,\, m_2+1}(f_1, f_2) = \sum_{q=0}^{p} (-1)^q \binom{m_1+p}{q} \binom{m_2+p}{p-q} f_1^{\,(p-q)}\, f_2^{\,(q)}. \qquad (7.2.22)$$

Consider now two symbols h_1 and h_2, both in the space $\mathcal{S}^{\mathrm{A}}(\mathbb{R}^2)$, lying in the isotypic spaces $L_{m_1}^2(\mathbb{R}^2)$ and $L_{m_2}^2(\mathbb{R}^2)$ respectively, satisfying some mild technical extra assumptions. The composition $\mathrm{Op}^{\mathrm{asc}}(h_1)\, \mathrm{Op}^{\mathrm{asc}}(h_2)$ has a symbol h, the isotypic components of which are characterized by the equation

$$\Theta_{m_1+m_2+1+2p}\, h_{m_1+m_2+1+2p} = \left(\frac{i}{p}\right)^p \mathcal{K}_{m+1}^{m_1+1,\, m_2+1}(\Theta_{m_1} h_1, \Theta_{m_2} h_2). \qquad (7.2.23)$$

Only isotypic components of the orders just indicated can occur in h, and the series $h = \sum_p h_{m_1+m_2+1+2p}$ is a convergent, not only an asymptotic one.

A few words about the automorphic situation must be added. In anaplectic analysis, define a modular symbol of weight $m + 1$, with $m = 1, 2, \ldots$, to be any distribution h_m transforming under rotations in the way indicated by (7.2.1), such that the function $\Theta_m h_m$ (note that the transformation Θ_m can be applied to any tempered distribution) is a holomorphic modular form of weight $m + 1$. Given now two modular symbols h_1 and h_2, of weights $m_1 + 1$ and $m_2 + 1$, the functions $\mathcal{K}_{m+1}^{m_1+1, \, m_2+1} (\Theta_{m_1} h_1, \, \Theta_{m_2} h_2)$, with $m = m_1 + m_2 + 1 + 2p$ for some $p = 0, 1, \ldots$, will of necessity be holomorphic modular forms. This is so because the Rankin-Cohen machinery is covariant: but, before taking benefit of this, we must take advantage of the fact that (in contrast with the operator the integral kernel of which was introduced in (3.3.3)), it consists of bidifferential operators, which can be applied without difficulty to automorphic objects. However, this is far from giving an answer to the question of defining the sharp product, in alternative pseudodifferential analysis, of two modular symbols, and decomposing the result into modular terms, since the (hard) convergence problems have not yet been taken care of.

Bibliography

[1] F.A. Berezin, *Quantization in Complex Symmetric Spaces*, Math. U.S.S.R. Izvestija **9**(2) (1975), 341–379.

[2] F.A. Berezin, *A connection between co- and contravariant symbols of operators on classical complex symmetric spaces*, Soviet Math. Dokl. **19** (1978), 786–789.

[3] J. Bernstein, A. Reznikov, *Estimates of automorphic forms*, Mosc. Math. J. **4**, 1 (2004), 19–27.

[4] D. Bump, *Automorphic Forms and Representations*, Cambridge Series in Adv.Math. **55**, Cambridge, 1996.

[5] J.L. Clerc, T. Kobayashi, B. Ørsted, M. Pevzner, *Generalized Bernstein-Reznikov integrals*, Math. Ann. **349**, 2 (2011), 395–431.

[6] H. Cohen, *Sums involving the values at negative integers of L–functions of quadratic characters*, Math. Ann. **217** (1975), 271–295.

[7] P.B. Cohen, Y. Manin, D. Zagier, *Automorphic Pseudodifferential Operators*, in *Algebraic Aspects of Integable Systems*, Progress in Nonlinear Diff. Equ. and Appl. **26**, Birkhäuser, Boston, 1996, 17–47.

[8] C.F. Dunkl, *Reflexion groups in analysis and applications*, Jap. J of Math. **3**, 2 (2008), 215–246.

[9] P.B. Garrett, *Decompositions of Eisenstein series, Rankin triple products*, Ann of Math. **125**, 2 (1987), 209–235.

[10] S. Gelbart, S.D. Miller, *Riemann's zeta function and beyond*, Bull. Amer. Math. Soc. (N.S.) **41**, 1 (2004), 59–112.

[11] S. Helgason, *Groups and Geometric Analysis*, Academic Press, New York, 1984.

[12] L. Hörmander, *The analysis of partial differential operators, III, Pseudodifferential operators*, Springer-Verlag, Berlin, 1985.

[13] A. Ichino, *Trilinear forms and the central values of triple product L-functions*, Duke Math. J. **145**, 2 (2008), 281–307.

[14] H. Iwaniec, *Introduction to the spectral theory of automorphic forms*, Revista Matemática Iberoamericana, Madrid, 1995.

[15] H. Iwaniec, *Topics in Classical Automorphic Forms*, Graduate Studies in Math. **17**, A.M.S., Providence, 1997.

[16] H. Iwaniec, E. Kowalski, *Analytic Number Theory*, A.M.S. Colloquium Pub. **53**, Providence, 2004.

[17] H. Iwaniec, P. Sarnak, *Perspectives on the analytic theory of L-functions*, in *GAFA 2000* (Tel-Aviv, 1999), Geom. Funct. Anal. 2000 special volume, II, 705–741.

[18] A.W. Knapp, *Representation Theory of Semi–Simple Groups*, Princeton Univ. Press, Princeton, 1986.

[19] T. Kobayashi, B. Ørsted, M. Pevzner, A. Unterberger, *Composition formulas in Weyl calculus*, J. Funct. Anal. **257** (2009), 948–991.

[20] S. Lang, *Algebraic Number Theory*, Addison-Wesley, Reading (Mass.), 1970.

[21] N. Lerner, *Metrics on the phase space and non self-adjoint pseudodifferential operators*, Birkhäuser, Basel–Boston–Berlin, 2010.

[22] W. Magnus, F. Oberhettinger, R.P. Soni, *Formulas and theorems for the special functions of mathematical physics*, 3rd edition, Springer-Verlag, Berlin, 1966.

[23] M. Mizony, *Algèbres de noyaux sur des espaces symétriques de $SL(2, \mathbb{R})$ et de $SL(3, \mathbb{R})$ et fonctions de Jacobi de première et deuxième espèce*, Publ. Dépt. Math. Univ. Lyon **3A** (1987), 1–49.

[24] C.J. Moreno, *The spectral decomposition of a product of automorphic forms*, Proc. A.M.S. **88**, 3 (1983), 399–403.

[25] A. Selberg, *On the Estimation of Fourier Coefficients of Modular Forms*, Proc. Symp. Pure Math. **8** (1963), 1–15.

[26] R.A. Smith, *The L^2-norm of Maass wave functions*, Proc. A.M.S. **82**, 2 (1981), 179–182.

[27] G. Tenenbaum, *Introduction à la théorie analytique et probabiliste des nombres*, Belin, Paris, 2008.

[28] A. Terras, *Harmonic analysis on symmetric spaces and applications* I. Springer-Verlag, New York–Berlin, 1985.

[29] F. Treves, *Introduction to pseudodifferential and Fourier integral operators: I, Pseudodifferential operators*, Plenum Press, New York–London, 1980.

[30] A. Unterberger, *Oscillateur harmonique et opérateurs pseudodifférentiels*, Ann. Inst. Fourier Grenoble **29**, 3 (1979), 201–221.

[31] A. Unterberger, J. Unterberger, *La série discrète de SL(2, ℝ) et les opérateurs pseudo-différentiels sur une demi-droite*, Ann. Sci. Ecole Norm. Sup. **17** (1984), 83–116.

[32] A. Unterberger, J. Unterberger, *Quantification et analyse pseudodifférentielle*, Ann. Sci. Ecole Norm. Sup. **21** (1988), 133–158.

[33] A. Unterberger, J. Unterberger, *Algebras of symbols and modular forms*, J. Anal. Math. **68** (1996), 121–143.

[34] A. Unterberger, *Quantization and non-holomorphic modular forms*, Lecture Notes in Math. **1742**, Springer-Verlag, Berlin–Heidelberg, 2000.

[35] A. Unterberger, *Automorphic pseudodifferential analysis and higher-level Weyl calculi*, Progress in Math. **209**, Birkhäuser, Basel–Boston–Berlin, 2002.

[36] A. Unterberger, *The fourfold way in real analysis: an alternative to the meta-plectic representation*, Progress in Math., Birkhäuser, 2006.

[37] A. Unterberger, *Quantization and Arithmetic*, Pseudodifferential Operators 1, Birkhäuser, Basel–Boston–Berlin, 2008.

[38] A. Unterberger, *Alternative pseudodifferential analysis: with an application to modular forms*, Lecture Notes in Math. **1935**, Springer-Verlag, Berlin, 2008.

[39] A. Unterberger, *Pseudodifferential analysis, automorphic distributions in the plane and modular forms*, Birkhäuser, Basel-Boston-Berlin, 2011.

[40] A. Unterberger, *Aspects of the zeta function originating from pseudodifferential analysis*, J. Pseudodifferential Operators and Appl., **5** (2014), 157–214.

[41] T. Watson, *Rankin triple products and quantum chaos*, Thesis (ph.D.), Princeton Univ., 2002.

[42] A. Weil, *Sur certains groupes d'opérateurs unitaires*, Acta Math. **111** (1964), 143–211.

Index

Index of notation

Subject Index

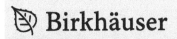 **Birkhäuser** | **www.birkhauser-science.com**

Pseudo-Differential Operators (PDO)
Theory and Applications

This is a series of moderately priced graduate-level textbooks and monographs appealing to students and experts alike. Pseudo-differential operators are understood in a very broad sense and include such topics as harmonic analysis, PDE, geometry, mathematical physics, microlocal analysis, time-frequency analysis, imaging and computations. Modern trends and novel applications in mathematics, natural sciences, medicine, scientific computing, and engineering are highlighted.

Edited by
M. W. Wong, York University, Canada
In cooperation with an international editorial board

■ **PDO 10: Nazaikinskii, V., Schulze, B.-W., Sternin, B.**
The Localization Problem in Index Theory of Elliptic Operators (2014). ISBN 978-3-0348-0509-4
The book deals with the localization approach to the index problem for elliptic operators. Localization ideas have been widely used for solving various specific index problems for a long time, but the fact that there is actually a fundamental localization principle underlying all these solutions has mostly passed unnoticed. The ignorance of this general principle has often necessitated using various artificial tricks and hindered the solution of new important problems in index theory. So far, the localization principle has been only scarcely covered in journal papers and not covered at all in monographs. The suggested book is intended to fill the gap. So far, it is the first and only monograph dealing with the topic. Both the general localization principle and its applications to specific problems, existing and new, are covered. The book will be of interest to working mathematicians as well as graduate and postgraduate university students specializing in differential equations and related topics.

■ **PDO 9: Cohen, L.,** The Weyl Operator and its Generalization (2013). ISBN 978-3-0348-0293-2
The discovery of quantum mechanics in the years 1925-1930 necessitated the consideration of associating ordinary functions with non-commuting operators. Methods were proposed by Born/Jordan, Kirkwood, and Weyl. Sometime later, Moyal saw the connection between the Weyl rule and the Wigner distribution, which had been proposed by Wigner in 1932 as a way of doing quantum statistical mechanics. The basic idea of associating functions with operators has since been generalized and developed to a high degree. It has found several application fields, including quantum mechanics, pseudo-differential operators, time-frequency analysis, quantum optics, wave propagation, differential equations, image processing, radar, and sonar. This book aims at bringing together the results from the above mentioned fields in a unified manner and showing the reader how the methods have been applied.

A wide audience is addressed, particularly students and researchers who want to obtain an up-to-date working knowledge of the field. The mathematics is accessible to the uninitiated reader and is presented in a straightforward manner.

■ **PDO 8: Unterberger, A.,** Pseudodifferential Analysis, Automorphic Distributions in the Plane and Modular Forms (2011). ISBN 978-3-0348-0165-2
Pseudodifferential analysis, introduced in this book in a way adapted to the needs of number theorists, relates automorphic function theory in the hyperbolic half-plane Π to automorphic distribution theory in the plane. Spectral-theoretic questions are discussed in one or the other environment: in the latter one, the problem of decomposing automorphic functions in Π according to the spectral decomposition of the modular Laplacian gives way to the simpler one of decomposing automorphic distributions in $\mathbf{R}^2$ into homogeneous components. The Poincaré summation process, which consists in building automorphic distributions as series of g-transforms, for $g \in SL(2;\mathbf{Z})$, of some initial function, say in $S(\mathbf{R}^2)$, is analyzed in detail. On Π, a large class of new automorphic functions or measures is built in the same way: one of its features lies in an interpretation, as a spectral density, of the restriction of the zeta function to any line within the critical strip.

■ **PDO 7: de Gosson, M.,** Symplectic Methods in Harmonic Analysis and in Mathematical Physics (2011). ISBN 978-3-7643-9991-7
The aim of this book is to give a rigorous and complete treatment of various topics from harmonic analysis with a strong emphasis on symplectic invariance properties, which are often ignored or underestimated in the time-frequency literature. The topics that are addressed include the theory of the Wigner transform, the uncertainty principle, Weyl calculus and its symplectic covariance, Shubin's global theory of pseudo-differential operators, and Feichtinger's theory of modulation spaces. Several applications to time-frequency analysis and quantum mechanics are given.

Printed in the United States
By Bookmasters